WAKEFIELD PRESS

Nature's Line
George Goyder

Janis Sheldrick is a lifelong resident of Melbourne who has always been strongly attracted to the landscapes of South Australia. She studied philosophy at the University of Melbourne, has a Graduate Diploma in Librarianship, and was awarded a PhD by Deakin University in 2000 for work on George Goyder and Goyder's Line. Working as an independent scholar, she completed the rest of the biography in the years that followed.

George Woodroffe Goyder,
as leader of the expedition to survey the Northern Territory

Nature's Line

GEORGE GOYDER

Surveyor, environmentalist, visionary

JANIS SHELDRICK

Wakefield
Press

Wakefield Press
16 Rose Street
Mile End
South Australia 5031
www.wakefieldpress.com.au

First published 2013
Reprinted 2014
This edition published 2016

Cover designed by Liz Nicholson, designBITE
Edited by Penelope Curtin
Typeset by Wakefield Press

National Library of Australia Cataloguing-in-Publication entry

Creator:	Sheldrick, Janis M., author.
Title:	Nature's line: George Goyder: surveyor, environmentalist, visionary / Janis Sheldrick.
ISBN:	978 1 74305 466 6 (paperback).
Notes:	Includes index.
Subjects:	Goyder, G.W. (George Woodroffe), 1826–1898.
	Environmentalists – South Australia.
	Surveyors – South Australia.
	Rain and rainfall – South Australia.
	Architecture and climate – South Australia.
	Goyder's Line (S. Aust).
Dewey Number:	994.232

C
CORIOLE
McLAREN VALE

Australian Government

Australia Council for the Arts

Publication of this book was assisted by
the Commonwealth Government through the
Australia Council, its arts funding and advisory body.

Contents

Timeline   vii

Acknowledgements   xiii

Preface   Never to be forgotten   1

PART ONE: In search of the rainfall

Chapter 1   Receiving the life of heaven   19
Chapter 2   The climate of paradise   33
Chapter 3   As far as the eye could reach   51
Chapter 4   Systematic observation   73
Chapter 5   Taking charge   95
Chapter 6   Bird's-eye view   111
Chapter 7   Magnum opus: the people's grass   119
Chapter 8   In search of the rainfall   141

PART TWO: The dark divide

Chapter 9   Colonial morality: 1861–63   165
Chapter 10   Transition: 1866–68   189
Chapter 11   Larrakia country: the founding of Darwin   209
Chapter 12   Going home   232

PART THREE: Universal genius and Clerk of the Weather

Chapter 13   Nature's Line   249
Chapter 14   Following the plough   267
Chapter 15   Fresh water and peculiar country   279
Chapter 16   Tree theories   292
Chapter 17   The universal genius   309

PART FOUR: Enduring marks

Chapter 18   A house in the hills   323

Chapter 19   Steward of all the Crown Lands   334

Chapter 20   Final years   347

Chapter 21   A gentleman of the Civil Service   359

Chapter 22   Remembering   379

Notes and abbreviations   395

Source list of South Australian Parliamentary Papers   396

Notes   400

List of illustrations   444

Index   448

Timeline

Event	Year	Event
George Woodroffe Goyder born in Liverpool.	1826	
Goyder family lives in Preston and Accrington (Lancashire), Hull and Newcastle.	1828–34	
Family settles in Glasgow.	1834	
	1836	Colony of South Australia founded with Colonel William Light, RE, first surveyor general.
	1839	Light dies, explorer Charles Sturt is invited to become surveyor general, but the position is taken by Lieutenant Frome, RE, who has been despatched from England.
Goyder trains as a railway engineer, works in Liverpool and London.	1840–47	
Migrates to Melbourne.	1848	
Smith family arrive in Adelaide.	1849	Captain Arthur Henry Freeling, RE, becomes surveyor general of South Australia.
	1850s	
Goyder travels overland to Adelaide, marries Frances Mary Smith.	1851	Goyder enters SA public service for the first time.
Frances gives birth to first child.	1852	Takes position as secretary of the Adelaide Exchange.
	1853	Rejoins public service as chief clerk of Land Office.
	1854	Appointed second assistant surveyor general.
	1855	Appointed first assistant surveyor general.
	1857	Exploring north of the Flinders Ranges, Goyder encounters Lake Blanche in flood. His reports of permanent water lead to a pastoral land rush.
		First fully elected government in South Australia.
	1858	Goyder officially appointed deputy surveyor general.

	1859	Goyder as guide for A.R.C. Selwyn on brief survey of mineral resources, and later leads party to triangulate country south of Lake Eyre South.

Frances gives birth to her fifth surviving child (of six).

1860s

	1860	Goyder continues triangulating south of Lake Eyre South. Applies for position of surveyor general in August, is accepted, and returns to Adelaide.
	1861	Officially appointed surveyor general, chief inspector of mines, and valuator of runs (and JP).
		Copper discovered at Tipara (Moonta) and dispute over original claims begins.
	1862	Goyder investigates country behind Fowlers Bay.
		Mitford makes implied allegations of corruption about Goyder and Strangways in relation to the Moonta claims, and distributes pamphlet making false allegations about Goyder.
		First letter of resignation (essentially over salary).
Frances critically ill after giving birth to twins who do not survive.	1863	Goyder visits South-East with commissioner of public works Milne and proposes plan for draining the region.
		Tipara Inquiry exonerates Goyder in relation to disputed Moonta mining claims.
	1864	Goyder begins revaluation of over 80 runs. The first valuations, of runs in the Mid-North, provoke political uproar, which continues to the end of the year.
		Dissolution of parliament over valuations.
	1865	Drought reported in the north at the beginning of the year.
		Squatters demand line defining drought-affected area for drought relief in late June.
		Six days later, Goyder describes a line of reliable rainfall which could be used to define the limits of agricultural land.

		Goyder leaves to inspect the drought in November and presents map showing his line to the government on 6 December. The government only want a tool to administer drought relief but Goyder has taken the opportunity to modify the line of reliable rainfall so that, in coming years, he can determine where to end the survey of agricultural land.
	1866	The Line makes its first public appearance in maps presented to parliament.
Goyder's cottage in Medindie extensively renovated and extended.		Goyder reports on making scrublands available for settlement.
Frances gives birth to her ninth surviving child (of 12 births).		Having ridden an estimated 30,000 miles, Goyder warns of the impact of the valuations on his health.
	1867	Goyder put in charge of drainage in the South-East.
		Mitford begins to publish *Pasquin*, in which Goyder is regularly libelled (until Mitford's death in late 1869).
		Scrublands Act passed.
Because of illness, Frances returns to England to recuperate accompanied by her sister Ellen and all the children.		
	1868	Goyder presents confidential memo recommending the creation of agricultural areas and new Regulations, before leaving with a party to select and survey the site of a town and surrounding regions in the NT.
	1869	New land Regulations (the Strangways Act) allowing for purchase of land on credit.
		Goyder arrives in the Northern Territory
		The site of Darwin (then Palmerston) selected and surveyed, along with 650,000 acres of surrounding land.
		Goyder returns to Adelaide in November and soon returns to work in the South-East.

1870s

		1870	Visits Victoria to study and report on impact of free selection of land.
	Frances dies in Bristol on 8 April.		Report on Victoria Land Regulations refers to a natural demarcation dividing agricultural from pastoral land.
			Goyder expresses concerns over his health to the government.
	News of Frances's death is published in Adelaide on 4 June, but the mailbag containing the letter informing Goyder has been misdirected to Melbourne.		Goyder suggests forest reserves.
	Ellen and the children return to Adelaide.		
			Goyder relieved of responsibility for drainage of the South-East at his request.
			Krichauff raises the issue of trees and timber in parliament and Goyder reports on locations of forest reserves.
		1871	Goyder requests leave of absence on grounds of ill health and sails for England on the *Queen of the Thames* in February.
			Queen of the Thames stranded off the coast of South Africa.
	Goyder returns to Adelaide and marries Ellen Smith in November.		
		1872	15 August: the Line becomes the limit of the agricultural land as the First Schedule of the *Waste Lands Alienation Act*.
			Overland Telegraph completed.
		1873	Writes letter of resignation over inadequate salary, effectively raising issue of salaries throughout the public service.
	After giving birth to a son in September 1972, Ellen gives birth to twins. They are Goyder's last children, making 12 surviving children of 15 births.		Goyder reports on forest trees and reserves and plans for forestry in South-east.

	1874	From 6 November, the *Waste Lands Amendment Act* removes any limit to agriculture. Agricultural expansion into the north begins.
	1875	Goyder chairs Railways Commission.
		Requests leave because of ill health.
		Forest Board Act passed. Goyder becomes chairman.
	1876	Requests leave again to travel because of ill health.
		Visits New Zealand on full pay.
	1878	Resigns due to collapse of life insurance.
		J.E. Brown arrives to become conservator of forests.
Goyder purchases Wheatsheaf Inn.	1879	In an appendix to Brown's annual report, Goyder attacks belief that planting trees will modify the inland climate.

1880s

	1880	Dry period begins and agriculture in the north begins to falter.
Wheatsheaf Inn modified to become Warrakilla.		
		Responsibility for drainage of South-East returned to Goyder. He remains in charge to the end of his career.
	1881	Continues to oppose Brown over tree planting in the north.
	1882	Forest Board dissolved.
		Goyder travels to England and America to purchase boring equipment.
		The Line is remapped, with some alterations.
Goyder takes up residence at Warrakilla.	1883	Report 'Water Conservation and Development' written at Warrakilla.
Visits New Zealand, May–July.		
	1885	First meeting of Royal Geographical Society, SA Branch; Goyder part of first provisional council.

1887	Goyder elected chairman at first meeting of Board of Examiners under *Licenced Surveyors Act 1886.*
	Floods in the north-east recreate the conditions seen by Goyder in 1857.
	Goyder's health breaks down again and he resigns. He is granted eight months leave of absence, but, because of important business, is not able to take this immediately.
1888	Commission on the Land Laws of South Australia addresses the aftermath of the failed expansion of wheat farming into the north.
	Goyder takes eight months leave of absence and returns to work on 1 September.
1889	Made Companion of St Michael and St George.
1890s	
1892	Suffers influenza followed by acute bronchial attack.
1893	Goyder's last journey. Travels to South-East and Pinnaroo.
	In late November Goyder is attacked by premier C.C. Kingston for allegedly aiding pastoralists whose leases Kingston was disputing.
	Treasurer (and former commissioner of crown lands) Playford joins Kingston in attacking Goyder behind closed doors.
	On 13 December Goyder resigns on grounds of ill health and is granted the six months paid leave he requests, so that he can set himself up in private practice as a surveyor.
1894	Goyder begins six months leave. Strawbridge takes over as surveyor general.
	Goyder's resignation takes effect on 30 June.
	Becomes second president of Royal Geographical Society SA Branch.
1898	Goyder dies at Warrakilla on 2 November.

Acknowledgements

My first thanks are to Hank Koderitz, who brought about my introduction to Goyder and without whose support the work could not have been completed. Thanks also to Enid and Pat Rehn, who furthered that introduction.

In converting a personal interest into the written outcome of research, I am indebted to the encouragement provided, first of all, by Judith Rodriguez, of Deakin University, and later by Tom Griffiths and Libby Robin of the Australian National University, and finally by Michael Bollen of Wakefield Press. I would also particularly like to thank the editor, Penelope Curtin.

Although I initially began working purely from the public record, I am very grateful to Ian Woodroffe Goyder, a great-grandson of George Goyder, and to Dan Farmer, a descendant of Edwin Smith, for assistance they gave in the later stages of the research, and I wish to express my thanks to Mr and Mrs Schofer, who kindly introduced me to Warrakilla. I would also like to thank Chris Goddard and I am especially indebted to the generosity of Vaughn Smith, a descendant of A.H. Smith, who not only made documents and photographs in his possession available, but delivered them to my door. I acknowledge the research of Smith family historian, Vladimir Derewianka.

Special thanks are due to the Peter Kentish, who, as Surveyor General of South Australia, arranged to have the surviving image of Goyder's original map of the Line, an old microform, converted into a digital image, so that it would be available for reproduction. I would also like to thank Merridy Lawlor of the State Library of South Australia for her assistance, especially for locating maps of the horseshoe lake in the Library's collection.

Many people at various institutions have provided invaluable assistance in various ways, in particular (in chronological order) Richard Gillespie of Scienceworks Museum, Melbourne, Matthew Gordon-Clark and Kathy Gargett of State Records South Australia, Mick Sincock (now retired) of the Land Services Group, Department of Transport, Energy and Infrastructure, Dale Turner of the Department for Environment and Heritage, Georgia Hale and Barbara Fargher of the Art Gallery of South Australia, and Lyn O'Grady of the Walkerville Council.

I would also like to acknowledge the helpfulness and companionship of Frank Williams (by coincidence, a former pupil of Brother Romuald) and Anne McArthur.

Many other people helped in various ways – making suggestions, indicating sources, taking photographs, and even having me stay in their home.

For their interest and help I wish to thank Stan Cornish, Elaine and the late Ron Ellis, Alan Lovejoy, Kelley Henderson and Richard Shapcott, and also Elsie Anderson and James Beattie (of New Zealand), and Monika Koderitz and Kelvin Callaghan.

It would not have been possible to carry out the research while based in Melbourne without access to the collections and services of the Latrobe Library of the State Library of Victoria, the Baillieu Library of the University of Melbourne, the Boroondara Library Service and stack collection of the former Melbourne City Libraries (now the Melbourne Library Service).

The first part of *Nature's Line* and sections of the following parts were completed as part of a PhD (by creative thesis) awarded by Deakin University in 2000. Working as an independent scholar, I completed the rest of the biography in the years that followed.

The story of Goyder's Line, as told in *Nature's Line*, was first presented at the Climate and Culture Conference, Canberra, 2002 and was presented again at the ANZMaps Conference, Adelaide, 2010. The papers were published as 'Goyder's Line: the unreliable history of the line of reliable rainfall' in *A Change in the Weather: climate and culture in Australia*, edited by Tim Sherrat, Libby Robin and Tom Griffiths (National Museum of Australia, Canberra, 2005), and '1855–56: George Goyder's long ride to mapping reliable rainfall', *The Globe: journal of the Australian and New Zealand Map Society*, no. 65, 2010.

PREFACE

Never to be forgotten

By opening our eyes, we do not necessarily see what confronts us.

– Iris Murdoch, *The Sovereignty of Good*

Scattered across the landscape of central South Australia, at the edge of the agricultural country, are the crumbling shells of long-abandoned dwellings. Simple nineteenth-century structures, they are all much alike: rectangular, with long windows flanking a central doorway and, usually, a chimney to one side. Any distinguishing character lent by their inhabitants is long gone. Now they are differentiated only by size and the degree of their disintegration.

My first encounter with such a ruin was in the mid-1960s. A visitor to South Australia and not many years beyond childhood, I was on my way up to the Murray with a friend, when we stopped to stretch our legs. The country we stepped out into was flat and empty, marked only by shallow, sandy rises. This topography was mirrored overhead by rippled bands of low grey cloud. The only thing to be seen was the roofless remains of a small stone cottage, set back from the road, not far from where we had stopped. Apart from a collar of tiny bleached bones spilling down the slope below a deserted foxhole, there was no other sign of habitation. Even the ground was bare except for a litter of small stones and dead grass stalks. It was a world in which desolation had reached a disturbing pitch of intensity, and we did not stay long.

Despite our hurrying away, the lonely ruin became fixed, not only in my memory, but in my imagination, where it floated disconnected from any context, seeming to demand a response. My sense of puzzlement about the image increased some years later, when, approaching the Flinders Ranges, I travelled along a road that was lined, at almost regular intervals, with similar ruins, all set in the whitened remains of a crop. They seemed like the way stations of a lost civilisation, though no Ozymandian head protruded from the earth to explain.

About a quarter of a century after encountering the first ruined house, its image, which had remained with me, made an unexpected and forceful return. This took place in 1990, on a hot summer evening in Coffin Bay, at the western tip of the Eyre Peninsula, where a conversation about farming in marginal country was taking place. Eventually, out of consideration for me, a visitor, a young teacher who had studied agriculture was delegated the task of elucidating a key point of reference that kept recurring in the conversation: *Goyder's Line*. He explained that in the nineteenth century a man named Goyder had attempted to establish a northern limit for wheat farming, based on rainfall. But he had failed, and despite his warnings, farming had surged beyond that limit, only to fall back after a run of bad years. Images of the ruins came immediately to mind, their meaning at last disclosed. But who was Goyder? How had he drawn his line? And why had he been the only one to recognise this limit? Unfortunately, the young teacher could only add that the line had been drawn in 1865, although he had no idea how or on what basis. He also explained that Goyder's life had a tragic aspect: his line had been derided, he had lost his health through overwork, and his wife had died in unfortunate circumstances.

The curious persistence of the image of the ruined house was transformed into a determination to find out all I could about Goyder and his line. Not long after, while I was staying as a guest in an old stone house on the Eyre Peninsula, north of the Line and surrounded by both wheat and sheep, I was shown an old cloth wall map, hanging between rods, of the pastoral leases of South Australia. Extending right across it was a thick and uncompromising red line which rose precipitously out of two shallow curves to an indented peak in the centre. The stations were piled above it. Goyder's Line, a 'special note' explained, was 'not in the ordinary sense a rainfall line': it had been mapped in the drought of 1865, and although its location could be disputed around Franklin Harbour (just below where we were) and around Pinnaroo (in the east), 'except for slight modifications, the line holds good to this day'. A speculative afterthought wondered if it was perhaps 'more than a coincidence' that the Line 'practically followed the … edge of the saltbush country'.

One of our hosts was also able to identify a place, in country near where she had grown up, where the Line was clearly visible on the ground as a road. Outside the most easterly line of the spine of hills that travels up from behind Adelaide to meet the Flinders Ranges, a road runs north–south between Burra and Robertstown. To the west, uninterrupted wheat fields spread from the road up to the low, rounded hills. To the east, a flat plain covered in stubby saltbush extends to the horizon. (An early pastoral run located here was named 'World's End'.) It was not until March 2005, however, when the whole of southern Australia was affected by drought, that I was able to experience, while travelling

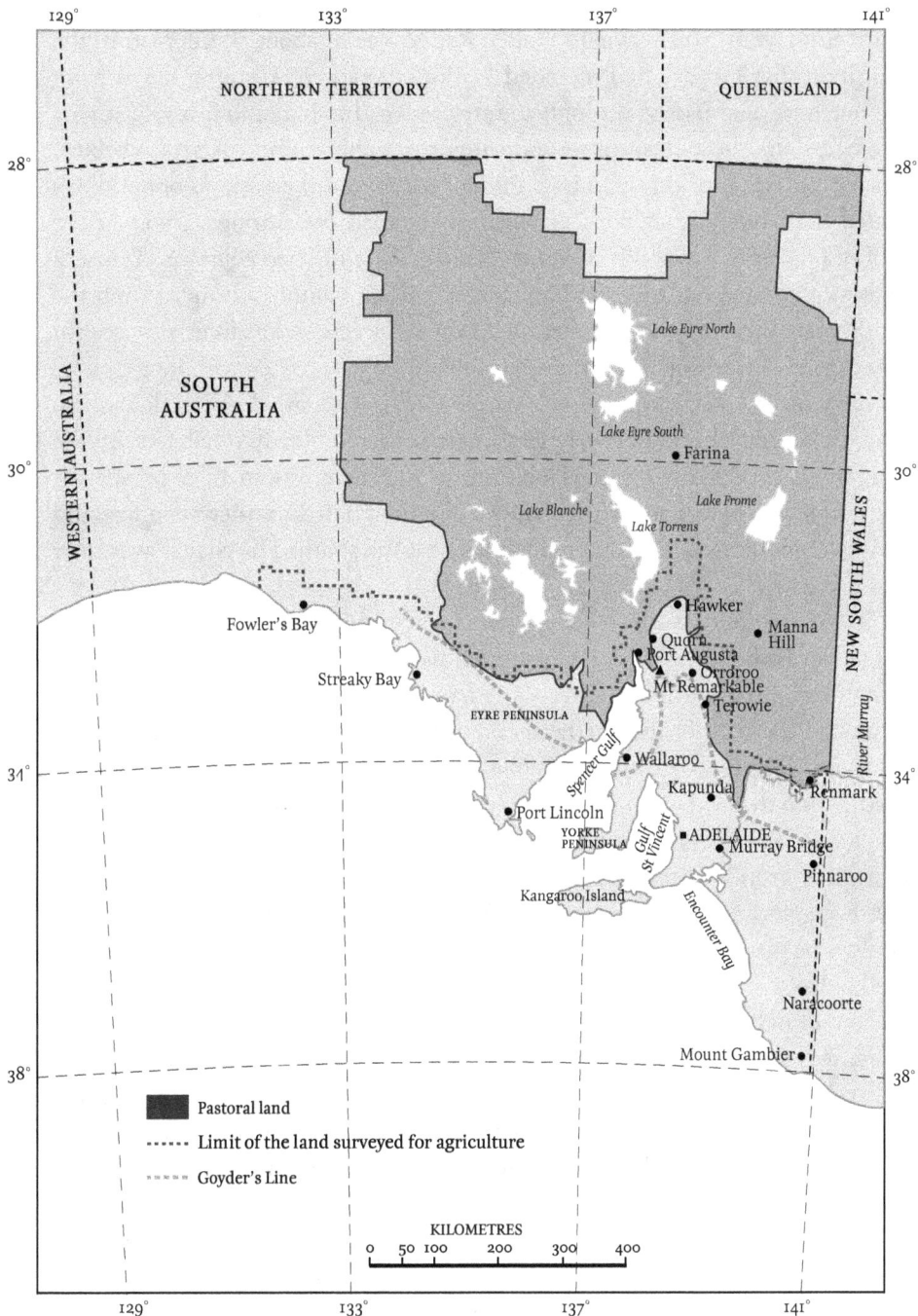

Goyder's Line, the limit of land surveyed for agricultural purposes, and pastoral land.

south from Whyte Yarcowie to Hallett, a distance of about 20 kilometres, the transition that Goyder had intended his line to express. Coming down from the north we had passed through country reduced to bare reddish soil, stirred into occasional little dust devils and supporting only a thin cover of bluebush or saltbush and the odd small tree. Piles of stones and the outlines of buildings dotted the landscape. A flock of pigeons streamed out through a hole in the roof of one ruin, just south of Whyte Yarcowie, while two columns of dust, a majestically towering adult with an infant trailing behind, advanced from the south-west. But not much further along the road was an apparently successful wheat farm. Pines and taller trees began to appear, together with long grasses on the roadside. Soon the soil was covered again and small plants flourished. The remaining old houses were freshly painted and clearly inhabited. By the time we reached Hallett, things looked tidy and there was an air of prosperity. We had left Whyte Yarcowie in badly drought-affected northern country and arrived at Hallett, on the border of the agricultural south. The change was clear and dramatic.

It was another journey that signalled something bizarre and compelling, not so much about the Line, but about the world in which it had emerged. The remains of a town called Farina – the name is Latin for flour, or meal – are located to the west of the northern end of the Flinders Ranges, 53 kilometres south of Marree, where the Birdsville Track begins, and over 300 kilometres further into the arid inland than the ruins clustered just beyond the Line. First known as 'Government Gums', or sometimes just 'The Gums', from the trees that grow around the water tank established there, it was renamed 'Farina', at the suggestion of the governor, at the height of the expansion of the wheat lands. The place is an oasis for birds, and when I camped there, budgerigars (or shell parrots, as they were known in the nineteenth century) peered proprietorially out of hollows in the trees. Nearby, a rainbow bee-eater, cheered on by its mate, was vigorously excavating in the side of what appeared to be a railway embankment. On investigation, this 'embankment' proved to be a tiny escarpment at the edge of a plateau, absolutely flat and covered in small, sharp lumps of rock, which extended to the horizon. An old plough had been set up at the edge of the camping ground and a sign among the ruins explained that the original plan had been to grow wheat, 'but the climate' – not to mention the stones, presumably –'proved unsuitable'. The presence of the old plough only ensured that seeing was disbelieving. As a railhead and later a stop on the line from Port Augusta, the town did experience a modest boom, but as a centre for the transportation of cattle, not wheat.

Goyder and the Line

Goyder, it turned out, had been the surveyor general of South Australia from 1861 until 1894 – about half of the colonial history of the state – and his attempt to confine the cultivation of wheat to the southern areas had met some success at first. His line had been enshrined in law in August 1872, but this had only lasted for a period of marginally over two years. In November 1874 the government dispensed with it in response to the demands of land-hungry settlers, who, as Goyder himself put it, raised 'such a clamour ... that no Ministry could remain in office that declined to bring the land into the market'.[1] Goyder continued to issue his warnings, but no one was paying attention; as he observed, the land beyond his line was taken up 'even more readily than before'.[2]

The advance of the plough was halted by a protracted drought, which began with the 1880s, taking South Australia into depression years ahead of the slump that affected the whole of Australia in the 1890s. The Line – 'that buried demon of agricultural settlement', as it was described at the time – emerged from the grave in which it had been hastily dumped.[3] Its author was seen to accompany it. 'Goyder's ghost seems to hover about', one unhappy farmer lamented in 1882.[4] Goyder, however, was still very much alive, even if exhausted by years of overwork. In the mid-1890s, when drought returned again, one parliamentarian reminded his fellows that: 'Nature had clearly indicated how far we should go for wheat-growing. Goyder's Line was really Nature's line of rainfall.'[5] It was an idea that Goyder had suggested himself well before the Line had been given official status, when he had observed that 'nature has clearly established a line of demarcation' to divide the agricultural from the pastoral land.[6]

And so Goyder's Line, or 'Goyder's line of rainfall', or even 'Goyder's rainfall', as it was sometimes known, became a part of South Australia. For many years it was routinely included in maps of the state (even on some road maps, to the mystification of post-Second World War immigrants) and it continues to be an occasional point of reference in weather forecasts. Before the sesquicentenary of the founding of South Australia was celebrated in 1986, the South Australian branch of the Royal Geographical Society had a series of roadside cairns constructed along the Line, each bearing a commemorative plaque. More recently, in 2003, Goyder's Line was declared a 'heritage icon' by the National Trust of South Australia.

Goyder has been celebrated as a man of wisdom and foresight, but neither he nor his line is well known outside their home state. Long-running drought in southern Australia has meant that his name is once again being heard, and the Line has begun to impinge on the national consciousness. As an attempt to recognise and define a natural limit, the story of Goyder's Line resonates remarkably with contemporary concerns, while the threat climate change poses to agriculture, especially in already marginal country, has also brought it to public attention, with newspapers carrying reports that the Line is 'moving south'.

As the man central to an early and major environmental debate – perhaps Australia's first – Goyder seemed to me, from the moment I encountered him, to be a figure of enduring significance and one who was more intriguing for having no obvious place in either of the dominant versions of the story of the settlement of the land.[7] He did not belong in the myth of the brave pioneer determinedly battling the land, nor did he quite fit in the alternative account of wanton environmental destruction. But in 1990, when I first learnt of his existence, he was a figure of essentially local interest, and little had been written about him. There was, of course, an entry in the *Australian Dictionary of Biography*, but the only book-length study devoted to his work had been written by a grand-daughter, Margaret Goyder Kerr, around the time of the centenary of the founding of Darwin, and that was concerned purely with his expedition to the Northern Territory. An investigation of the volumes of Australiana published to celebrate the nation's bicentenary in 1988 suggested his status was in flux. While one ten-volume reference work accorded the Line a brief entry but had no place for the man, his biography was one of the 200 which make up *The People Who made Australia Great*, a work published in association with the *Australian Dictionary of Biography* and aimed (apparently) at school students. In this volume the situation was reversed. Goyder was categorised as one of the 'Builders of the Nation', and possibly Australia's first 'greenie', while the Line was treated as an afterthought, of little more than historical interest.

The idea that Goyder was a figure of enduring importance and national relevance had been voiced in the early 1970s by Geoffrey Dutton, a literary figure and member of an old South Australian pastoral family. In a brief popular history of the state, Dutton had unhesitatingly identified Goyder as one of South Australia's 'few major figures', citing 'his unshakeable understanding of the nature and perils of the environment' as the core of his contribution.[8] Later, Dutton suggested that the Line could usefully be extended to the entire continent. Inside the Line, he explained, 'the rainfall will support agriculture, outside the line it is too dry for anything but grazing. So in South Australia people talk about "inside country" and "outside country", and they are handy terms for the rest of Australia.'[9]

The strongest claims for Goyder's relevance and importance were made by the historical geographers Donald Meinig, Michael Williams and Joseph Powell. Meinig, an American scholar with a particular interest in wheat colonisations, had visited South Australia in the late 1950s and produced a work on the events of the 1870s and 1880s that has since come to be regarded as a classic of historical geography, *On the Margins of the Good Earth*. The book ends where I began, with the ruins and their landscapes, which seemed to Meinig, too, to possess 'a curious archaeological appearance of an older and distant landscape'.[10] Michael Williams, author of *The Making of the South Australian Landscape* (and later to become Professor of Geography at Oxford University), and Joseph M. Powell, whose writings include *Environmental Management in Australia, 1788–1914* and *An Historical Geography of Modern Australia*, both presented Goyder as a major figure. Williams estimated that, while undertaking research for *The Making of the South Australian Landscape*, he had read about three-quarters of a million pieces of official correspondence, making use of much of this when he made Goyder the subject of his presidential address to the South Australian Branch of the Royal Geographical Society (of which Goyder was himself a founding member and past president) in 1978. Contrasting him with the well-known figures of explorers, whose behaviour suggested they were driven by a death wish – 'impractical geographers' in his view – Williams characterised the little-celebrated Goyder as a 'practical geographer', and observed that he was: 'literally writ large over the correspondence and decisions of nearly four decades, so that land settlement in South Australia during these years is really the story of Goyder'.[11] Williams saw the survey, which Goyder had largely overseen, as leaving 'the greatest and probably the most enduring imprint of man on the land in South Australia'.[12] He also acknowledged Goyder as a public servant of unusual scope, whose accession to the position of surveyor general was 'the beginning of an era of public service supervision and control by one person that probably had no counterpart in any other part of colonial Australia'. (The position of surveyor general as Goyder held it certainly has no modern equivalent.) It seemed likely that 'no one bequeathed more to the living present of South Australia', Williams declared, adding:

> One cannot but admire his professional and personal qualities. Here was a man in the midst of the hurly burly of early colonial society, whose dealings were fair, whose probity was beyond question, and who never stooped to the unkind word or the underhand action which were, from time to time, the tactics of his adversaries. In short, he was a man to be admired.[13]

Joseph Powell echoed William's words, describing Goyder as:

> a man of considerable historical and geographical vision. The South Australian landscape bore the massive imprint of his major management decisions during his energetic tenure and for long after his departure.[14]

In 1991 Powell unequivocally projected Goyder into a larger, national context when he stated that:

> the contributions of a few powerful bureaucrats in key resource management roles – Surveyors-General Thomas Mitchell and George Goyder, for example, and Government Botanist Baron von Mueller – are properly described as instances of 'landscape authorship'.

Drawing on Williams's earlier comments, he went on:

> these talented individuals and their senior colleagues must be seen as archetypal public servants *and* practical geographers. Their bold signatures across the land proclaim an intrusive human agency in the modification of enormous tracts which, taken together, easily exceed the entire land area of Western Europe.[15]

A year later Powell again invoked Mitchell and Goyder and declared:

> Key bureaucrats (I mean leading public servants) were intimately and continuously involved in the interpretation and management of Australia's natural resources, and in the development of urban and regional planning ... these under-researched individuals played leading roles.[16]

With access to relatively high-quality data, Powell saw these bureaucrats as enjoying 'enough seniority to challenge contemporary goals and assumptions'.[17] Mainstream history, he believed, gave too much status to prominent politicians.

Goyder's contemporaries had no doubt that the Line was a central achievement of his career and an ongoing reason to remember his achievements. 'His deeds are written in the records of the past', the *Observer* had declared after his death:

> but who, in these times of drought and agitation for bringing the farmers of the parched northern country into areas where climatic conditions are more certain, has not heard of Goyder's line of rainfall? If for no other reason than that he drew that famous line of demarcation his name is one never to be forgotten.[18]

Sixty-three years later, the visiting scholar, Meinig, would more or less agree, commenting that:

> Many a man has left his name upon our maps, but while it is commonplace that a myriad of visible discrete features are specified by personal names, it is rare to find a qualitative, geographic concept so identified and displayed.

That 'Goyder's Line' should be a so prominent and persistent cartographical feature suggests a singularly important local concept. That it should be so titled suggests a singularly important man.[19]

The problem with 'Nature's line' was that it was by no means clear exactly what it expressed. The Line's *function* was to serve as a land-use border – to divide agricultural from pastoral land – but what did it actually map? It was commonly said to approximate an isohyet of average annual rainfall, and generations of South Australian schoolchildren had been taught to equate it roughly with the ten-inch isohyet – although it was also seen to resemble the 12-inch and even 14-inch isohyets. With the introduction of metrics, the 250, 300, and 350-millimetre isohyets took their place. In 1918, the perceptive geographer T. Griffith Taylor described it as an 'ecological isopleth'.[20] A related view linked the line to changes in the vegetation. In the early twentieth century, the historian A. Grenfell Price was a strong proponent of this idea, even going as far as to write of 'Goyder's vegetation line'. According to Price, the Line divided the mallee, saltbush and bluebush from the southern flora; corresponded to the 12-inch isohyet (as Taylor had stated); and was regarded as an excellent indicator of the limits of safe wheat farming.[21]

Document-based history provided a quite different account. According to this approach, the Line represented the border of a particular drought, the great drought of 1865, and defining it had nothing to do with establishing an agricultural limit. Goyder had been sent out with instructions to identify the areas where the drought was at its worst and where pastoralists would not be able to carry on without changes to the conditions of their leases. According to Meinig, who accepted this version of events, it was not long 'before some began to wonder if the immediately renowned "Goyder's Line" might not have broader implications'. Goyder, Meinig claimed, agreed that it did, 'to a certain extent', although by 1870 he had become so committed to the idea that he was advocating that the government take responsibility for preventing attempts at settlement in unsuitable areas. 'Thus,' Meinig concluded, 'Goyder, himself, gave his "Line" a significance quite beyond its original intent, and a role which was to make of it a persistent issue in South Australian colonization history'.[22] But even while asserting Goyder's 'singular importance' as the author of the Line, Meinig had not written as if entirely convinced by this account, and chose to describe the Line not as a drought border but as a 'very personal concept' – Goyder's 'own specific qualitative assessment of land potential'.[23]

Undertaking the research for this biography revealed a quite different account of the Line's meaning and origin – and one which shows Goyder to have been even wiser and more perspicacious than previously acknowledged. His writings, and his answers to questions posed by parliamentary committees and commissions, make clear that he had arrived at a remarkable early insight: that the distinctive aspect of the inland Australian climate is its highly variable rainfall, and that accepting this was key to settler society's adjustment to the Australian inland environment. Goyder's understanding of the singular nature of the Australian climate was the outcome of both fortuitous experiences and his approach to satisfying the demands of his work. During exploring journeys early in his career he had seen the arid inland twice transformed by floods – apparently the first colonist to do so. He had also spent years on horseback, examining land prior to survey. To evaluate leased pastoral land for revaluation, he travelled thousands of miles across runs, systematically recording his observations of vegetation along transects. In 1978 Michael Williams described the resulting material as constituting the 'most important untapped documents of the historical geography' of South Australia, which needed to be interpreted 'in their entirety'. Today these records provide researchers and workers in the field of biodiversity conservation with coherent observations spanning large areas to which they can refer.[24] Although Goyder did not say so explicitly – he said little that he was not called upon to say – it is evident that he had not only come to understand that rainfall in the inland was highly variable, he had also learnt to recognise seasonally unreliable rainfall as it was expressed through changes in vegetation. By the time he had finished examining runs in the Mid-North and southern Flinders Ranges in 1864 he already had a definite idea of where the reliable rainfall ended.

The documents show that in 1865 Goyder requested that he be sent out, ostensibly to assist the drought-stricken northern pastoralists, but privately with the intention of seeing the extent and impact of the worst drought the settlers had so far experienced. He needed this new information to adjust and refine the line he had already drawn in his mind, to enable him to use it, eventually, to ensure that small farmers, who were entirely dependent on their crop for an annual income, were only offered land they could farm with confidence. Such a line would also provide security for pastoralists in the north, who could then develop their leases, safe from the threat of having the land taken back and divided into agricultural lots. As surveyor general in a colony where that position was unusually central to planning, Goyder knew he would be involved in any decision to define the extent of the agricultural land, and he intended to be well prepared.

Goyder's genius was to grasp that in the arid inland 'it is seasonal reliability or variability of rainfall, and not its average, which is the true gauge to the ecological situation', as the prehistorian John Mulvaney has succinctly put it.[25]

His tragedy was that he understood this phenomenon more than a hundred years in advance of the science of both the climate and the ecology which would explicate this insight. The way in which the vegetation and animal life of the arid regions of Australia are adapted, not just to aridity, but to the prolonged and unpredictable absence of rainfall, as well as to episodes of severe flooding, is now widely understood, but researchers into desert ecology had not begun to recognise this until the 1970s. In recent decades official attempts have been made to address climate variability and its effect on agriculture, and in explaining our highly variable climate it is now usual to invoke El Niño, or the El Niño-Southern Oscillation (ENSO), the global climatic event characterised by changes in the temperature of the Pacific Ocean, which has far-reaching and dramatic effects and causes much of the drought experienced in Australia. The Southern Oscillation Index is routinely reported for the benefit of farmers. But the first successful computer-model prediction of an El Niño event did not take place until 1986, and it was not until the beginning of the 1990s that the Australian Bureau of Meteorology began issuing predictions of such events. It was only then that scientists began considering the role of the Indian Ocean Dipole in drying southern Australia, and the importance of this, too, has been confirmed.

Throughout his life Goyder tried to convey his understanding of the country and the climate to his contemporaries, without success. The colonists had brought with them a culture in which it was taken for granted that four seasons followed each other in a continual and reliable round; departures could only be viewed as anomalies. In the absence of technical jargon to draw attention to the point being made, or any framework of understanding to which he could appeal, Goyder had to rely on ordinary language, and it didn't serve him well. He invariably referred to rainfall in the north as 'unreliable' or 'uncertain' and, as late as 1888, appearing before the Land Laws Commission convened especially to deal with the aftermath of the disastrous attempt to farm the north, Goyder was asked by the chairman if rainfall beyond his line was 'doubtful in quantity'. He replied that it was 'precarious'. Sometimes, he explained, the rainfall was 'very heavy, and sometimes light'.[26] But his language was not taken up by his contemporaries and ultimately his point was lost, obscured by the mass of assumptions about weather and climate on which everyday language was based. To make matters worse, the Line had lost its proper name. Although to Goyder it was the line of reliable rainfall, it was, as he complained, 'commonly, though erroneously, called Goyder's line of rainfall'.[27] As a result it was understood only in terms of its function – separating agricultural from pastoral land – and for most South Australians, its basis remained, as for Meinig, a mysteriously personal concept of the surveyor general's.

That the colonists could not grasp the idea of a land-use boundary based
on the seasonal reliability of rainfall – the reliable limit of a sufficient rainfall,
as Goyder defined it toward the end of his career – might seem strange in a
country in which 'droughts and flooding rains' have long been a climatic cliché
and where the overlander of the old folk song described himself as coming from
the northern plains: 'Where the creeks run dry or ten foot high / And it's either
drought or plenty'. One climate researcher has pointed out that an almanac
published in Sydney in 1859 stated that:

> the Australasian climate is one of irregular rains, and is thus distinguished
> from most climates where the rains are pretty equable and constant although
> less in total amount.[28]

Nevertheless, the work and writings of one of Goyder's contemporaries
display an irrefutable example (presented in Chapter 16) of this inability to
grasp and integrate an idea which had no theoretical framework and which
ran counter to the assumptions of culture. The forester J.E. Brown wanted to
plant trees in the inland in the belief that this would make the climate less arid,
but was persuaded by Goyder that the rainfall in these regions was irregular.
Brown's response was to plant seeds rather than seedlings, believing that seeds
could wait patiently in the soil until the rain finally fell. Brown seems to have
been unable to comprehend that the rainfall would continue to be irregular –
that even if enough rain fell for his seeds to sprout and the seedlings to reach a
modest height, they would still ultimately perish (as they did). He simply could
not think in terms of an irregular rainfall as an ongoing phenomenon.

Another apparent example, from the mid-twentieth century, appears in a
Master of Arts preliminary thesis in history on the subject of Goyder's Line
presented in 1952 to the University of Adelaide by F.J.R. O'Brien (or Brother
Romuald). O'Brien was obviously aware of the point at which Goyder, months
before going out to survey the extent of the drought, offered to define the
agricultural lands on the basis of the reliable rainfall. In his thesis O'Brien
even described the line that Goyder had verbally sketched, but he remained
convinced that the final line defined a single drought and that Goyder had had
'no intention of giving [it] any prophetic value for future conditions'.[29] He even
used a passage in which Goyder referred to the Line as showing the localities
where 'a tolerably reliable rainfall is generally secured' as evidence that he had
come to regard the Line as 'something in the nature of a rainfall isohyet and
agricultural boundary', and went on to criticise Goyder on the basis that a 'line
based on extreme drought conditions could not also represent average rainfall

conditions'![30] O'Brien's approach can be partly attributed to his having fallen under the spell of an earlier geographer, John Andrews, who claimed to have investigated the 'vanished frontier' of the Line (although he referred to only two of the documents relating to it) and demonstrated it to be a drought line.[31] If O'Brien had actually understood what Goyder was saying, he would not have been persuaded by Andrews, but in 1952, O'Brien, like Goyder's contemporaries, was construing Goyder's words in the absence of today's understandings of climate and ecology. Charles Darwin, in his autobiography, gives a compelling example of this inability to recognise what is not already named and explained. As a student, Darwin hunted for fossils with Adam Sedgwick, a great early geologist, in a valley in Wales at a time when it was not yet understood that these valleys had been formed by glaciation. The result, as Darwin explained, was that:

> neither of us saw a trace of the wonderful glacial phenomena all around us; we did not notice the plainly scored rocks, the perched boulders, the lateral and terminal moraines. Yet these phenomena are so conspicuous that ... a house burnt down by fire did not tell its story more plainly than did this valley.[32]

Although the nature of his line remained obscure for some time and its real significance not comprehended, Goyder has nevertheless been acknowledged for his early and astute contribution to the understanding of Australia's climate. In *The Weather and Climate of Australia and New Zealand*, published in 1996, he is credited with having made 'the first attempt at regional climate evaluation in Australia'.[33] The third edition of *The Australian Weather Book*, published in 2012, devotes almost a page to the 'surprising contribution' of Goyder's Line, which it describes as a 'bold prediction, as no detailed rainfall records were then available on which to base his belief'. Unfortunately, relying on established accounts of Goyder's Line, as the authors were inevitably forced to, the first book reports (without comment) that the Line was later equated with the 12-inch, or 300-millimetre, annual isohyet, while the second claims that the Line defines the point at which farming 'would not be sustainable because of *insufficient* rainfall' [my emphasis].[34] Incidentally, the authors of *The Australian Weather Book* also contribute the unexpected information that Goyder had red hair and a pallid complexion, a claim made recently by Michael Williams.[35] In terms of stereotypes, red hair fits nicely with Goyder's famous energy and drive – not to mention his quick temper – but there are no references to red hair, not even oblique ones, among the writings of his contemporaries, and a portrait in the possession of his descendants, painted late in his life, suggests that his hair was a very ordinary shade of brown. Perhaps in his earlier years it had a richer tone.

Goyder's friendship with Charles Todd also connects him to the history of weather and climate studies in Australia. Todd is best known for establishing the Overland Telegraph line through the centre of Australia, linking Adelaide with Darwin, and so, via underwater cable, with Indonesia and the rest of the world. Before this, he had already connected Adelaide with Melbourne and Sydney. As well as being superintendent of telegraphs in South Australia, Todd occupied the position of government astronomer – he had been an astronomer in England – and had responsibility for keeping weather statistics. He quickly put this combination to good use, organising the operators at telegraph stations to collect and send on meteorological observations and in the process becoming one of the important pioneers of Australian meteorology. Todd became interested in the nature and causes of droughts, and in 1876 he speculated that the droughts of central Australia would be found to coincide with the 'magnetic cycle of eleven and a quarter years, which is believed to determine the frequency of auroræ, magnetic storms, and solar spots'.[36] Later, in response to a request from India for information on atmospheric pressure in Australia at the time of the catastrophic drought that struck the subcontinent in 1877, Todd noted a similarity, and in 1888 he concluded that 'there could be little or no doubt that severe droughts occur as a rule simultaneously over the two countries'.[37] In doing so, Todd had noted the first of what would become many observed 'teleconnections', or links between climate anomalies in different parts of the world, that are essential features of the phenomenon of El Niño. In the context of understanding drought and the variability that distinguishes the Australian climate, Goyder and Todd made a singular pair of friends.

Surveyor general

The Line was only one of Goyder's many concerns. He had become surveyor general at a time when that position, far from being purely technical, was critically involved in planning the fledgling colony. Goyder's almost boundless self-confidence, determination, intelligence and energy meant that he could make a place for himself as head of an administrative empire covering land and natural resources, while as an individual he filled the role of the government's environmental information service, using the knowledge he had gained in his early years as a surveyor and explorer and when valuing pastoral runs. In his primary role he was responsible for laying out roads and establishing stock routes, but his involvement with transport extended to his being chairman of a railways commission that determined the shape of the colony's rail network. He established forest reserves and plantations, also founding the

administrative apparatus of forestry in the colony. He was preoccupied with
the issue of water, and promoted and developed artesian boring. In addition to
this he also pursued a parallel career as an engineer, initiating the drainage of
a vast seasonal wetland to create productive land in the South-East, although
this aspect of his work seems largely to be forgotten. He even founded South
Australia's own colony, the Northern Territory, when he selected and surveyed
the site of what is now Darwin. Goyder was a link between the city and the
bush, and as a result he 'was as well known in the desert wilds as in the thickly
populated centres'. His distinctive personality ensured that his name was not
only widely known, but a 'household word for energy and decision'.[38]

Goyder's expansive geographic interests have been viewed in ways that,
initially at least, do not seem to sit well together. J.M. Powell's 1991 characteri-
sation of Goyder as a 'landscape author' adopts a concept put forward in 1979
by the geographer Marwyn Samuels. The created landscapes of man, Samuels
argued:

> are much like any other product of human creativity. They have much in
> common with the manifold forms of human art and artifice. That is, they are
> constrained by need and context, but they are also expressive of authorship.[39]

Put simply, landscape authorship is about 'the who behind the facts of geography'.[40] Samuels identified two interconnected forms of landscape authorship:
the making of landscape *impressions* (by artists and surveyors, for instance)
and landscape *expressions*, or the making of actual landscapes. In the work of
colonial surveyors, landscape impressions generate landscape expressions, and
Goyder was clearly an author of both.

Another approach to his work is represented by a tiny brass plaque set into
the footpath of North Terrace, Adelaide – one of many that commemorate
figures in South Australia's history. Here Goyder is honoured as a 'surveyor and
conservationist', although 'conservationist' (like 'Australia's first greenie') seems
to be an unlikely way to describe a person bent on dividing the country for sale
and draining a vast and biologically rich seasonal wetland. Goyder was certainly
not a conservationist as we would now understand the term – although he did
preside over the declaration of a national park – but he foreshadowed contemporary environmental attitudes in his concern for the management and preservation of soil, water, forests, and the natural vegetation of the northern country,
which he perceived as resources. While settlement was his driving aim, his
enthusiasm for settlement was not unqualified. At a time when gross and transient exploitation of the land was widespread and taken for granted, Goyder
insisted that settlement should be what he termed 'permanent', a forerunner

of the contemporary term 'sustainable'. It is the line of reliable rainfall, then, that shows how these aspects of Goyder's work – landscape author and proto-conservationist or environmentalist – are related. Although created to divide the landscape into two distinct realms of primary production – agricultural and pastoral – the Line also implied that the pattern of settlement should conform from the outset to the realities governing life on the new continent.

In the biography which follows, Goyder is understood (in accordance with the observations of a Public Service Commission in 1890) as a land steward with responsibility for the government lands of South Australia, a role which incorporates without conflict his land authorship and land management concerns.[41] Because of the range of his activities, the story of his life and work is not presented in strict chronological order. While generally moving forward, this account follows themes, an approach which enables different areas of his work to be presented as coherent narratives. Part one gives an account of his activities as he developed the knowledge, skill and experience that resulted in the mapping of the line of reliable rainfall. Part two revisits the early years of his surveyor generalship to examine aspects not covered in part one and concludes with the Darwin survey and the years of personal tragedy which followed. The theme of the Line is taken up again in part three, which covers the agricultural expansion of the 1870s and its aftermath. This section also investigates Goyder's work and interest in forests and water, the drainage of the South-East, railways and other concerns. The final part, part four, reviews his career as a surveyor and as a public servant, and the ideas that underpinned his work, and gives an account of all his areas of activity in the last years of his career. It also presents the final years of his personal life at the house and estate he established in the Adelaide Hills.

Part one:
IN SEARCH OF
THE RAINFALL

The S.G. in pursuit of the Rainfall. This satirical sketch by Adam Gustavus Ball shows Goyder during the 1865 drought, on the journey that led to the appearance of the Line. The rest of the caption reads: 'Novr annus memorabilis 1865 small memento of senatorial wisdom scene Squatters Paradise N. E. Plains'.

[SLSA]

CHAPTER ONE

Receiving the life of heaven

The Kangaroo

Q. What is this?

A. A kangaroo.

Q. How large is this animal?

A. The size of a sheep.

Q. What place is it a native of?

A. New Holland.

Q. Is this animal anything like an opossum?

A. Yes; it has a false womb and a very long tail like the opossum.

Q. Is the tail useful?

A. Yes; it defends them against dogs and assists them to leap away from their enemies.

Q. What? does it beat dogs with its tail?

A. Yes; and they have been known to hit dogs so very hard as to make them give over hunting them.

Q. You said it has a false womb like an opossum, do the young ones take shelter in the false womb?

A. Yes.

Q. And how large are the young when born?

A. They are hardly an inch in length.

David Goyder, *A Manual of the System of Instruction pursued at the Infant School, Meadow Street, Bristol*, 1825

There is a doorway into the world that formed George Goyder. In 1857, the year that his explorations first brought his name to public attention in South Australia, his father's autobiography was published in London. David Goyder was best known as an exceptionally hard-working follower of the Swedish mystic, Emanuel Swedenborg. He was a minister of the New Jerusalem Church, founded on Swedenborg's visionary teachings, and one who was prepared to lead a peripatetic life and to travel long distances to preach and attract new congregations. He played a leading role in establishing the church in Britain and, even in a history of the New Church (as it is also called) in Australia, he is mentioned as 'that remarkable labourer in the vineyard'.[1] But David Goyder was a man of many enthusiasms, and what he intended to be his story, *My Battle for Life: The autobiography of a phrenologist*, is in fact a bulging portfolio of things that engaged him. Part narrative, part an assemblage of lectures and testaments to intellectual and religious fascinations and beliefs, not to mention a song composed by the author (with music) and comic poetry, it constitutes an immediate taste of the world from which George Goyder emerged. It is all that is available. Since he left no personal diaries, letters or reminiscences, the course and development of George Goyder's life must be gleaned from the public record.

The battle for life

The only lengthy personal narrative in David's autobiography describes his early years. Reading it is like encountering a character from Dickens who has strayed into the realm of reality and acquired an independent voice, but whose life remains dramatically full of pathos, curiosity, excess and coincidence.

David George Goyder was the twenty-first son and twenty-second, and last, child of his father, Edward Goyder, an official with the Exchequer in London who had been born in Glamorgan, Wales.[2] The Goyders believed that their name was a derivation of 'Gwydyr' or 'Gwydir' (the Welsh 'w' represents a vowel) and that they were connected with the Wynns of Gwydyr, descendants of the ancient lords of north Wales. However, the name 'Goyder' is now regarded as a later spelling of the Welsh 'coedwr' ('woodman'). The name first appears (as 'Cydowre') in the early 1500s in the records of the south-west of Carmarthenshire, the southern Welsh county from which George Goyder believed his family to have hailed.[3] Whatever his origins, Edward died in London when David was three, and the boy's earliest memory was of watching his father's funeral procession – he had been honoured with a public funeral by the Independent Order of Oddfellows – through a neighbour's window.[4]

David's Welsh-speaking mother, Margaret Lloyd, was his father's third wife and bore six of his crowd of sons.[5] After Edward's death, she continued to live in their large old house in Angel Court, near Westminster Abbey, but David seems to have had little contact with his brothers. Growing up in Westminster, without other children for company, David was drawn to the music and ceremony of the Abbey, which affected him so deeply that he set his heart on being ordained and took to play-acting at home with a prayer book and tablecloth. When the family was forced to move to more modest quarters, he could not bear to be parted from the Abbey sounds he had become so accustomed to – the magnificent choral singing and the solemn organ music – and often returned to linger in the cloisters and among the tombs.

When his mother died at the age of 54, on 1 March 1805 – his ninth birthday – the little boy suddenly found himself frighteningly vulnerable and alone. He was taken in, not by his mother's brother Lloyd Lloyd, a barrister, but by a relative he had not met until the day of his mother's funeral, Evan Evans. The family included children of around his own age, but David found his new situation both meaner and harsher. In particular, he was deeply disturbed by being punished severely for telling what was believed to be a lie, and then further punished for refusing to abandon it. The incident laid the foundations for a hatred of corporal punishment and a lifelong concern about the ways in which honesty could be cultivated in children.

To David's delight, he was able to return to Westminster when Lloyd Lloyd succeeded, by means of his influential position, in securing him a place in the Foundation of King Charles I at the Westminster School, one of the eight major public schools of the time.[6] These places were intended for the education of orphans, but David observed that he was the only orphan among the boys, many of whose parents were wealthy enough to keep carriages. Fortunately, once admitted, and dressed in the archaic green uniform of the school, the pupils were treated 'equally enough'.[7] But this still meant that like all juniors and newcomers he was first required to submit to the tyranny of the older boys in a process known as 'taming the fags'. The experience made him aware that his robust spirit was housed in an unusually small and 'fragile' (although healthy) physical frame.[8] The tiny boy was lucky to be otherwise ordinary and happily sociable, and he used his storytelling abilities to win himself a place among his fellows. Because he had learnt to write at home, David's handwriting was chronically unsatisfactory and as a result he was punished repeatedly, and violently. But even that eventually ceased and David at last found himself at ease in his world. He became head boy and was a good student, with every prospect of going on to fulfil his dream of becoming a clergyman.

These hopes were dashed when Lloyd Lloyd died without a will. David was 13 years old, alone and without resources. By the provisions of the Foundation, he was entitled to remain at the school until he was 13, and he was kept on an extra three months – by the same master who had beaten him – in the hope that some way would be found to keep him there. (It was suggested, presumably by one of the other boys, that he could make a fortune exhibiting himself as a dwarf.[9]) When it became clear nothing would eventuate, two of his brothers, one a printer, the other a bookbinder, assumed control of him and found a position for him as an apprentice to an ivory and bone brush-maker. It was a miserable disappointment that soon became a nightmare.

Despite having made a plausible initial impression, the master to whom

David was apprenticed, a man named Fleming, turned out to be drunken and vicious, given to the routine terrorising of his family and apprentices. To make matters worse, Fleming was illiterate and resented the education of his most recent indentured victim, often beating him for the most minor of misdemeanours. In the few hours when he was not required to work, David slept in the filth and stench of the workshop. On Sundays he had half a day free, and usually went to the house of one his brothers, where he had dinner and tea, the only two full meals of his week. He was little more than a skeleton, but although his sister-in-law plied him with food, his brothers apparently did nothing about his plight.

David survived this treatment for three years, until a chance encounter with his cousin, Evan Evans. Evans was appalled at his state and wanted to confront Fleming, but his wife suggested that David escape to their house – although this meant he would become a runaway apprentice and liable to punishment if caught. When he was restored to health, the Evans family helped him to find a position as an errand boy, taking advantage of his still very juvenile appearance. On one of his errands he was seen by Fleming's eldest son. Fleming dragged David to court, but the magistrate had the good sense to listen to the apprentice's complaints. David was returned to Fleming on the understanding that he must be treated properly, but within a couple of months his situation was as bad as before.

At his cousin's instigation, David escaped again, and again taking advantage of his still-childlike appearance, Evans had him registered as a lady's page who sought a position in the country. He was soon taken on by a countess of sorts, an English woman who had married an impoverished European aristocrat purely for his title. When the countess returned to Europe, David was passed on to her son, a captain in the army. Soon after, the captain was summoned to India. David returned to his cousins still 'dwarfish', but in good condition, with 10 pounds, plenty of good clothes and a well-stocked trunk of linen.[10] His happiness was complete when he learnt that Fleming had died.

David would have been content to continue as a page, but his brothers would not countenance his remaining in a menial occupation. He was invited to take up an apprenticeship as a printer with his brother Thomas, but once again his size presented difficulties. He was too small to operate all of the presses, but by choosing jobs carefully, he was able to support himself. It was during this long second apprenticeship that David was introduced to the New Church – properly titled the Church of the New Jerusalem, and commonly known as 'the Swedenborgians', while its members referred to its various groups as 'Societies'. Thomas was a minister in the church and David soon became involved.

The other-worldly scientist

The New Church was founded on the teachings of Emanuel Swedenborg, a brilliant Swedish scientist, whose first fascination had been with mathematics, but who had gone on to study mechanics. For 30 years he occupied a position, specially created for him by the Swedish King, with the Royal Board of Mines. As well as giving lifelong service to the nation's metal-mining industry, Swedenborg had founded the country's first scientific journal and published works on mathematics, philosophy, chemistry, physics and even human perception. Determined to demonstrate the immortality of the soul, he had expanded his studies to include human anatomy, physiology and psychology. In 1743–44 Swedenborg underwent an intense spiritual crisis, and in 1745, after a powerful vision, he gave up his scientific studies. From then on he turned his intelligence and energies to presenting his insights into the Bible and its proper interpretation, and to transmitting his experiences in the spiritual world. Swedenborg believed that these writings were a revelation from God that constituted *in themselves* the Second Coming of Christ. They would usher in a new age – the 'New Jerusalem' – of religious truth.

The best known of Swedenborg's teachings concern the heavens and hells of the spiritual world, which he taught, as the poet Borges expressed it in a poem about him, are 'in your soul, with all their myths'.[11] But this strong otherworldly belief did not mean that Swedenborg or his followers were detached from the concerns of daily life. In a chapter of his most famous work, generally known as *Heaven and Hell*, Swedenborg explained that spiritual development did not depend on renouncing the physical and social world. On the contrary:

> to receive the life of heaven a man must needs live in the world and engage in its business and employments, and by means of a moral and civil life there receive the spiritual life. In no other way can the spiritual life be formed in man, or his spirit prepared for heaven ...[12]

After Swedenborg's death in 1772, his teachings were introduced into England and became established among a small group of people, radical urban artisans who included, for a time, William Blake. One of Blake's biographers has described them as a group who stood 'quite outside the current dispensation' in various ways, but who all 'believed in the primacy of the spiritual world. They were often considered eccentric or mad, as remnants of a murky superstitious past or as small tradesman with absurd ideas quite above their station in life.'[13] By the Goyders' time, things had settled to a more everyday style, and the New Church had acquired ordained ministers and a liturgy borrowed from the Church of England. Nevertheless, its members remained disadvantaged in society.

Sarah Etherington

By the end of his seven-year apprenticeship David was comfortably off and preparing to get married to 'the most beautiful little woman in London'.[14] The bride-to-be was another New Church member, Sarah Etherington. Sarah's brother, John, was English secretary to the French General Bertrand, and in this role had accompanied Napoleon to St Helena. Unfortunately, Sarah, who was two years older than David, was considered to have an incurable case of tuberculosis, and her beau was warned that marriage was out of the question. David dismissed this objection – he did not even deign to mention it in his autobiography – and the resilient Sarah defied her prognosis and recovered.[15]

Sarah would live to display what can only have been a robust constitution. She had many children and raised them in the difficult circumstances of limited resources and constant relocation. Thirty-five years after their wedding David proclaimed: 'My admiration for her now is even greater than when she bestowed her hand upon me. It is impossible to speak in terms too high of her many admirable qualities.' In particular, he admired her 'tact, her prudence, her economy in the management of a very limited income, her truthfulness, her devoted attachment to the doctrines of the New Church ...' However, despite his admiration for her, David was not shy to nominate her faults. According to David: 'when she thinks herself right, she is very determined. She is firm, even to obstinacy.'[16]

Their grandson, the famous Victorian actor Sir John Martin-Harvey, has left a more engaging picture of Sarah, his 'charming grandmother', whom he remembered as a 'delightful little lady, full of humour and humanity'. He also records that, after a lifetime of devotion, in her mid-90s she instructed a young relative who had offered to read comforting hymns to her on her death bed to 'throw away that rubbish', and fought on to the end.[17]

The Pestalozzi schools

David and Sarah were married on 11 February 1821 at St John's Church, Westminster, and again before their New Church congregation in Waterloo Road, Lambeth, in London, where David's brother Thomas officiated.[18] (Only Church of England marriages and those of Jews and Quakers were legal at the time.) The year before, David had become acquainted with a man who had been engaged to establish a school for young children among the very poor (commonly known as 'ragged schools'), following the methods of the Swiss educational reformer, Johann Pestalozzi. Another such school was planned for Bristol, and a master and mistress were needed to run it. The man suggested that, once married, David and Sarah could take up these positions if they made themselves familiar with Pestalozzi's writings and methods.

They duly prepared themselves and were sent to Bristol to carry on the good work. Once there they became acquainted not only with members of the New Church in Bristol, which David preached to and presided over, but also with the well-to-do nonconformist families, mostly Unitarians and Quakers, who supported the school. David gave lectures on Pestalozzi's ideas and published a book about the school which contained materials used in the classes and illustrations showing the light-filled open space of the building and the play area.[19] A fourth edition, 'considerably enlarged', was published in 1825. After a year of leading the Bristol Society, David was ordained at the urging of his congregation, finally satisfying his desire to become a minister, although not quite as first envisaged.

It was while they were in Bristol that David was introduced to phrenology, the 'science', as it was thought to be, that related character and ability to the size and development of different areas of the brain, interpreted through the shape of the skull. Throughout his life, David's core intellectual interest was what would now be psychology and neuropsychology. He was curious about physiognomy and graphology and collected reports of the effects of brain injury, but phrenology was his major passion, and he studied and taught it for the rest of his life.

It was while they were in Bristol, too, that the Goyders' first children were born. Their first child, a son, died a year after his birth, in 1823, but that same year a daughter, Sarah Anna, was born. She was followed in the next year by another son, John Thomas.

The family moved to Liverpool in 1825 and on 26 June 1826 their second surviving son and third surviving child, George Woodroffe, was born. (The Woodroffes were friends from the Waterloo Road church where David and Sarah had met.[20]) He was 'officially' baptised into the established Church of England at the church of St Peter in Liverpool on 16 July. A portrait of David shows that George Goyder inherited his father's high forehead and strong gaze – along with, although to a lesser extent, his diminutive stature. (George Goyder was probably just over 160 centimetres – around five foot four inches. He is remembered, through a grandson, as having appeared to be of normal size until he stood up.[21]) David's legacy to his son also appears to have included gentleness, coupled with impulsiveness and a hot temper, a love of music and story-telling ability.[22] It was Sarah who seems to have been the source of the two poles of her son's temperament: sociability and charm at one extreme, and an intense, focused determination at the other, along with a natural skill for organising and managing – although both David and Sarah's lives are testimony to their enormous capacity to persevere. After his death, George Goyder was remembered as a man 'of such intense determination that if he had entered the army he would most probably have developed into a Napoleon'.[23]

Having abandoned printing, David had to support himself and a family on the income he received from his New Church congregations and from managing Pestalozzi schools, although his income from teaching was not adequate. In Liverpool he indentured himself again, this time to an apothecary. Adding the skills of a chemist and druggist to his repertoire would enable him to establish his own small business wherever his other concerns took him. In 1825–26 he had published two books on infant schools, and in 1827 he published *A Concise History of the New Jerusalem Church*. For the rest of his life he would continue to publish books on phrenology and psychology, and on Swedenborg and his teachings, as well as manuals of spiritual instruction, including a book of family worship, and, together with his brother Thomas, a two-volume set, *Spiritual Reflections for Every Day of the Year*.

In 1828 the family moved to Preston, Lancashire, where another son, Charles Stones, was born. This was followed by a move to Accrington, also in Lancashire, where David George was born in 1830. Although pastor of the Accrington New Church, David also joined the Freemasons, became secretary to the local Lodge, and soon published lectures on this subject as well.[24] The next move was across to Hull, on the east coast, and from Hull the family returned inland to Newcastle-upon-Tyne. Another child – a girl – was born in 1832 but she did not survive the year. The family was in Newcastle during an outbreak of cholera, and David treated victims brought to him with hot vapour baths and a mixture containing opium.

A quotation David selected as an epigraph for the title page of his life story illuminates his approach – evidently also a principle espoused by his son – towards self-imposed commitments. The words are attributed to a Doctor Hunter:

> My rule is deliberately to consider, before I commence, whether the thing is practicable. If it be not practicable I do not attempt it; if it be practicable, I can accomplish it, if I give sufficient pains to it; and having begun, I never stop till the thing is done.

He also explained his method:

> I get through my labor by system ... By varying my studies I give exercise to my various faculties, and while one class ... is in active labor, another is reposing ... I have lived to realise the truth long ago uttered by Dr. Sir Edward Clark: 'The true secret of human happiness lies in the cultivation

of all the faculties'. I therefore, never suffer any of my energies to stagnate. The old adage … of too many irons in the fire, is a libel and a falsehood. You cannot have too many; when one iron cools, or when one faculty is wearied, I give it rest, and take up another; thus, shovel, tongs, and poker, I keep them all a going.[25]

Early education

David had long experienced problems with his hearing, which he believed to be the result of the blows he had received at school, and by the time the family had reached Newcastle, probably in 1831, he was too deaf to teach. He established a Pestalozzian school there and mentions in his autobiography that three of his children (presumably the eldest three, Sarah, John and George) were pupils there, but that a teacher had to be engaged.[26] Everything went well until the first teacher left. The second, a budding De Quincey, was discovered to be composing poetry in a state of 'ecstatic delirium' achieved by chewing opium and downing draughts of laudanum, or tincture of opium (opium dissolved in pure alcohol).[27] The consumption of opium was not illegal at the time, and David seems to have regarded the incident with nothing more than sympathetic regret.

In the first five years of his life, then, George Goyder had lived in a constantly changing world, but with the unwavering influence of his parents', in particular his father's, beliefs about the proper development of children. David had taken Pestalozzi as a master, and according to his theories, learning in young children should proceed from the familiar. By direct experience, the whole range of a child's faculties should be engaged, including the emotions, while activities were paced to facilitate the unfolding and strengthening of the child's abilities. Pestalozzi emphasised group and participatory activity, but saw the capacity to think for oneself as one of the goals of education. He gave a pre-eminent place to moral education, which he viewed as proceeding from the ideal unit of the family circle. Accordingly, David attempted to make his schools function as much like a loving family as possible. As he saw it:

> It was the aim of Pestalozzi to combine the powers of the understanding with the will – of thought with affection – and to bring them both into actual existence in the life. Hence, his system is one of faith and love, or, in other words, he united the cultivation of heart and understanding with the labor of the hand. His motto in education was – *Heart, Head,* and *Hand*.[28]

As an educationist with a neuropsychological bent, David also diverged from what he perceived to be the growing general practice of 'forcing the infant brain to premature maturity'. Precocity impressed him not at all – he believed there was always a price to be paid – and 'book learning' was kept to a minimum of one hour a day at his establishments. His charges, mostly aged from three to seven years, were verbally introduced to 'the elements of history, natural history, geography, astronomy and music', as well as being given a practical introduction to 'every article' of domestic object and food. Their day included an hour of unrestrained play, joined by their teacher, and general vigour of demeanour was developed through a good deal of marching about and singing and shouting. One visitor observed ironically that, 'if the children learn nothing else, they learn to make a glorious noise', a remark that David recounted with pride.[29]

David's early experience of corporal punishment had also led to a broad rejection of violence in any form or for any purpose. In his schools he impressed on his pupils the importance of treating all of creation, and therefore all animals, with kindness. The children themselves were treated with care. David believed that inflicting physical pain and humiliation on children in an attempt to change their behaviour only inculcated a propensity for lying and that children would develop in the right way morally by being allowed to experience the 'natural consequences' of their errors. The poison of error would produce its own antidote: 'for no one indulges in evil, abstractly considered as such; every fault committed arises from the idea that there is something, at least for the time, pleasant in it'. Only if the connection between act and consequences were too remote for the child to grasp would punishment be necessary. Even then, David stressed, the 'only effectual prevention of error is to show WHY IT IS ERROR …'[30] It does not take any great leap of the imagination to see the unfailing curiosity that George Goyder displayed toward the environment of South Australia as the outcome of the method of his early education, and his foresight and the willingness he would show in admitting his own mistakes, without feeling belittled or being deterred by them, as having been fostered by these attitudes.

With their curiosity 'unrepressed by the chilling sensation of fear' the Goyder children were presented with the example of Swedenborg, a spiritual seer with an extraordinary inner life who was also a scientific genius, whose interests and investigations ranged from the life sciences to engineering, and from pure investigation and theorising to practical technical applications.[31] It was natural for them to see no conflict between the vigorous investigation of the physical world and the cultivation of a spiritual life.

Glasgow and the Industrial Revolution

In 1834 the family moved again, this time away from the industrial Midlands

and north of England to the industrial north of Scotland.[32] They settled in
Glasgow, where David studied medicine at the University of Glasgow. He went
on to practise as a physician, although eventually deafness forced him to return
to pharmacy.

The Goyders' last child, a daughter, Margaret Diana Mary, was born during
their first year in Glasgow. Through the kindness of friends of the family, she
was educated at a convent in Ghent – unlike the other children in the family –
and acquired the tastes of a cultured lady. George Goyder, who would have
been eight or nine when the family moved to Scotland, attended the Glasgow
High School. Originally the Glasgow Grammar School, it had been established
in the medieval period and is possibly the oldest school in Britain. Predictably,
he is reported to have passed through this institution 'with credit'.[33] But he did
not follow his father into the practice of pharmacy or medicine. Instead, he was
articled to a leading Glasgow engineering firm, Rundolph, Elliot and Company,
in early adolescence. David taught that it was the duty of parents to become
familiar with the varying dispositions of their children and claimed that the
choice of career of all his children had been made phrenologically – by the
shape of their heads – and had proved successful in every case.[34] Despite the no
doubt precise accord that David must have discovered between the shape of the
boy's skull and the interests he displayed, external influences are not difficult to
identify. The first railway boom in England had begun in 1835, when Goyder
was about nine and susceptible to being enthralled by the new machines and
the world of hitherto unseen landscapes, where vast bridges spanned valleys
and rivers and the experience of space and time was radically altered. In the
early nineteenth century, the great engineers, most famously Isambard Brunel,
were celebrated and inspiring figures, heroes of human creativity and power,
the progress of whose projects was attended with excitement throughout society.

Goyder later recalled that with his first employer he had been able to gain
a 'large experience and technical knowledge' of surveying and engineering
instruments, and of drawing and designing. Certainly, when the family
completed their census schedule in 1841, when Goyder was nearly 15, his occupation was confidently put down as 'engineer', whereas his brothers John (who
was older) and Charles were described only as an apprentice accountant and
an apprentice cabinet maker respectively. At the time the family were living in
the Gorbals, a village south of the Clyde that had been developed earlier in the
century.[35] (The Gorbals did not become notorious as an area of over-crowded,
unhealthy and crime-ridden tenements until later.)

Goyder was next employed as an assistant by a civil engineer in London, J.
Brown, but went north again to undertake 'the partial direction of the large

works of Messrs. Tayleur and Co., of Liverpool and Warrington', a firm of railway engineers.[36] Later in life Goyder claimed to have gained 'considerable practical experience [as an engineer] on … such contract works as the Britannia and Conway Bridges, the Liverpool landing stage, and works of that kind', although he added, 'for the most part my experience was confined to the office'.[37] The Britannia Bridge, over the Menai Strait, and the Conway Bridge, across the mouth of the Conway River, are two tubular railway bridges erected on the line between Chester and Holyhead, across the top of Wales. The engineer for the line was Robert Stephenson, who worked on the Britannia Bridge with other eminent engineers. The two bridges were among the great feats of nineteenth-century English civil engineering, and in mentioning them, Goyder was indicating the quality of his engineering background. Having worked on these projects also meant that he had been directly exposed at a formative period of his life to the atmosphere of restless inquiry and triumphant determination that characterised civil engineering during the period when great works captured the collective imagination and confounded the public understanding of what was humanly possible. That grand sense of scale and import is evident throughout Goyder's own work, and it undoubtedly contributed to his ability to rally and inspire those he was leading with a sense of the worth of what they were doing.

Apart from his involvement in these glamorous projects, Goyder testified to having 'superintended a large number of bridges. The firm I was with in England had various contracts along the Crewe line [to the south-west of Liverpool] and several bridges were erected there under my superintendence.'[38] For him, the second railway boom of 1844–47 had created an opportunity to take on responsibility and develop his capacity for organising, supervising and exercising authority.

The railways and surveying

Goyder's training as a railway engineer would have ensured that he had acquired high-level skills in surveying. The railway booms had led a rush of people, many ill-equipped and even downright fraudulently unprepared, to offer their services as 'railway surveyors'.[39] Others set themselves up to teach surveying in a short time, although, according to respectable sources, to be 'perfect in surveying fit for practising on a line of railway' required serving articles for three or five years with an established land surveyor.[40] As a result, the business of 'land measuring', as simple plane surveying was called, fell into disrepute. (Plane surveying treats the earth as flat.) The enduring effect of the rush of bogus 'railway surveyors' was that land measuring was relegated to a technical activity subsidiary to civil engineering, while civil surveyors tended to focus on the more financially rewarding tasks of valuation, land agency and associated activities such as auctioneering. Planners turned instead to the maps of the Ordnance Survey, the trigonometric survey of England undertaken by the Royal Engineers, which had been made available for sale. While the Ordinance Survey maps were used to select overall routes, the task of preparing competent surveys of sufficient detail for displaying the engineering work required was undertaken by the railway engineers themselves.[41] Goyder's training as an articled railway engineer was therefore a reputable route to acquiring higher-level surveying skills.

Given the conflicts that broke out over the routes of proposed new lines, he would also have become accustomed to working in an environment in which it was taken for granted that different forces in society were in fierce competition. Angry farmers had torn up surveyors' pegs and members of the aristocracy fought, sometimes literally, to protect their lands. In 1844, bloodshed resulted when Lord Harborough sent out a party of servants to prevent surveyors from entering his estate at Saxby. The next year he sent a party of gamekeepers prepared for battle and a surveyor was sent to prison for drawing a gun. The line was routed around Harborough's estate.

Emigration

Despite all this excitement, something other than a desire to construct railway bridges must have been fermenting in the young engineer, because in 1848 – the year when uprisings and revolutions broke out across Europe – he migrated to a country in which not a single foot of track had been laid to accommodate steam-powered vehicles, where there were no large-scale engineering projects, and where the demand for fine instruments was satisfied by import. For David and Sarah, who left Glasgow for Ipswich that year, their son's departure for Australia meant the effective loss of two sons, as their eldest, John, who was to have taken over his father's medical practice in Glasgow, became ill and died. The couple, without resources, had to begin again, establishing another pharmacy business.

As the son of a 'poor parson, with a huge family' (as Sir John Martin-Harvey remembered his grandfather), George Goyder had to make his life entirely through his own effort, and his knowledge and abilities stood the best chance of being rewarded where they had been nurtured – in the new industrial frontiers of his homeland.[42] Nevertheless, he decided to make the long voyage to a place from which even return would be difficult if his life there proved to be unsuccessful. He already possessed all the elements, except experience of the natural environment of Australia, from which he would fashion his career on arrival: the key skill of trigonometric surveying and experience superintending engineering projects; the curiosity, attentiveness and respect for consequences that his upbringing had fostered; and the energy, resilience and determination, which, given what is known of the lives of both his parents (and his sister Sarah), were his legacy as a Goyder.

In his autobiography David acknowledged with some pride that he had a son who had 'risen to eminence' and who was deputy surveyor general in the Province of South Australia.[43] Elsewhere in the work, he offered a phrenological profile of a successful colonist, couched in the jargon of the 'science', in which 'Destructiveness' refers not to an enthusiasm for mayhem, but to 'physical energy, determined to overcome in a righteous cause', or a willingness to have dominion over the earth, in accordance with the biblical direction.[44] As comical as the description now sounds, it was probably a proud father's portrait of a successful son. To make the most of the benefits of emigration, David declared:

> the following seems to be the best combination:- A bilious nervous temperament, moderate Inhabitiveness, full Combativeness and Destructiveness, rather large Secretiveness, Acquisitiveness, Firmness, and Self-Esteem, with large Locality, and well-developed powers of reflection and observation.[45]

These qualities George Goyder possessed in ample proportions.

CHAPTER TWO

The climate of paradise

When he sailed for Australia, the 22-year-old was not striking out on an entirely independent course. While the family was in Glasgow, his older sister Sarah had met, and at the age of 18, married, a young Scotsman, Hugh Galbraith MacLachlan. Their first child, George, was born two years later, and another child, who probably died in infancy, was most likely born in the next few years.[1] The couple eventually decided to migrate to Sydney, leaving David Goyder to grieve over the loss of his favourite. Hugh, who was a minister in the Church of Scotland, seems to have shared the Goyders' interest in education: in Sydney he taught at St Andrew's Model School. He also embraced the family's unorthodox beliefs, and together Sarah and Hugh held small gatherings of Swedenborgians in their home in Sydney, maintaining contact with others in Adelaide, where the New Church had put down its first roots in Australia in 1838.[2]

George Goyder left Plymouth on 24 November 1848 together with three hundred or so other emigrants on the 768-ton ship, *Osprey*, bound not for Sydney, but for Port Phillip.[3] Although carrying goods bound for various merchants, the *Osprey* was a 'bounty ship', as they were known, commissioned by the government to carry assisted emigrants, who were the chief cargo. There were no passengers who had paid their own fare on board, and Goyder was one of 66 single men who would have been crowded into a hold together. In apparent contradiction to his claim to have had practical experience supervising the construction of railway bridges, Goyder's occupation was entered in the ship's register as 'millwright and draftsman'. The explanation seems to lie in the surprising number of millwrights registered as having been on the *Osprey* – so many that the *Geelong Advertiser* described the men disembarking there as 'mill-wrights, cabinet-makers, and farm labourers'.[4] This unlikely preponderance suggests that 'millwright' was a piece of bureaucratic labelling applied to men with mechanical and technical know-how, or even, perhaps, a description that the immigrants themselves had adopted in the belief that it presented their abilities in a desirable light. There was certainly no demand for railway

engineering experience. At about this time the Sydney Railway Company was
only in the process of becoming established, generating demand for one railway
engineer.

After a voyage of four months, the *Osprey* reached Geelong on 22 March
1849. Part of the cargo was unloaded, some of the passengers disembarked,
and five of the crew went missing, but Goyder was one of those who remained
with the ship, not disembarking until they reached Melbourne on 6 April 1849.[5]
From there the *Osprey* went on to Peru. While the site of Melbourne had once
been regarded as a place of beauty, the process of settlement had changed things
drastically. Arriving in Melbourne the year before, the Hungarian, Baron
Mednyánszky, recorded that: 'The first sight of Australia at Melbourne is not
gay. The flat shore – the town marked out, and not yet built – the temporary
huts – the naked landscape, without vegetation or water – are unattractive …'[6]
About 18 months after his arrival, Goyder left Melbourne for Adelaide. His
grand-daughter, Margaret Goyder Kerr, has provided something of an account
of the motives for his choice, claiming that: 'it was the thought of finding
rich minerals that appealed to him at that stage, rather than that of lining his
pockets with the wealth they might give'.[7] Since she has also commented on the
paucity of personal records left by Goyder, it can only be guessed that she was
drawing on family memory, or a reference in a letter. In any case, 'the thought
of finding rich minerals' is certainly congruous with the interest in mining and
geology that Goyder displayed for the rest of his life, and the claim that he was
keen to make discoveries rather than to line his own pockets with wealth is
equally consistent with his father's attitudes, David having lamented that:

> Money seems to be the chief thing, at least in the estimation of many, for
> which we are called into existence. Boys are taught its value from the moment
> they enter the counting-room; nay, almost from the moment they begin to
> think at all.[8]

The Goyder children, by contrast, grew up being able to recite whole chapters
of the Bible without faltering, and George Goyder would later teach his own
children to do likewise.[9]

Overland to Adelaide

The young traveller left Melbourne probably at the beginning of 1851, and it is
from that point that the process of his getting to know the climate and country
of southern Australia becomes traceable, because he chose to travel overland.
This was a challenging course. Since the mails went by sea, there was no mail
coach connecting the two southern cities, and travellers who decided against

the sea voyage had to make their own way. The route seems to have gone from Melbourne through what is now the Western District, where the squatters were already established, and where, according to a glowing report of the time, 'in some places ten thousand sheep are fed on a single station without being out of sight from the homestead, or making any sensible diminution of the luxuriant herbage'.[10] But Goyder was travelling in high-to-late summer, after a year of drought, and would have been somewhere on the road on the day commemorated in William Strutt's huge painting, *Black Thursday, February 6, 1851*, which shows squatters, travellers, wild animals and birds in what the painter described as a 'stampede for life' ahead of a wall of flames.[11] Small fires had been burning in pockets throughout the summer, and in the intense heat of Black Thursday, a furious hot wind from the north fanned these fires into one of the massive conflagrations that periodically consume much of the southern forest. Fires extended from Barwon Heads across the South Australian border to Mount Gambier. If Goyder was lucky, he was already past Portland, from where the route followed the coast around into South Australia, continuing to cling to the sea's edge all through the Coorong, to avoid the waterless Monster Scrub, also known as the Hundred-Mile Scrub and the Ninety-Mile Desert. (The better known overland route, used by the later gold escort, passed further north.[12])

Goyder reached Adelaide on 15 March 1851.[13] A painting by J.A. Gilfillan, now in the Art Gallery of South Australia, is a view of the city from the west made in that year. It shows a flat city of mostly single-storey buildings, without verandahs, or even eaves, or any decoration, arranged along broad and largely empty streets, although a light traffic of pedestrians and horse-drawn vehicles makes its way about. S.T. Gill's lithograph of Hindley Street, made in the same year, shows a slightly more crowded and lively scene that includes straw-hatted street vendors in the foreground, a disconsolate looking Aboriginal man sitting on the ground wrapped in a blanket, and a two-horse coach (an American-style stagecoach) with the word 'Comet' emblazoned across the back. In Gilfillan's broader perspective only a single tall spire and two columns of dark, industrial-looking smoke reach above the roofline of the city, and none of these breaches the horizon line of the hills that frame the city to the east. There are only two buildings to be seen that would clearly qualify as grand. The overall impression is one of solidity, ordinariness and order. In the census taken in January of that year, 63,700 colonists were counted, the majority resident in the town.

The New Church in Adelaide was flourishing, and had even opened a little church on land leased at Sowter's Gardens, Carrington Street, near King William Street and not far from the government offices on Victoria Square. If the Adelaide members were not already familiar with Goyder through Sarah and Hugh, his father's reputation would have been enough to establish his credentials. The shipboard diary of one prominent member, E.G. Day, an expert on Swedenborg's writings who had emigrated the year before, recorded his use of 'Goyder's *Spiritual reflections*' as part of his devotions.[14] At the core of the little community was the Chauncy family, and a friendship evidently developed between Goyder and one of its members, William Snell Chauncy, a surveyor and civil engineer employed by the government; Chauncy would be the best man at Goyder's wedding. Chauncy had first arrived in the colony in 1840 but had gone back to England for a few years before returning in 1849. He had been involved in attempts to establish a railway line from the copper mines at Kooringa (Burra) to Adelaide, and had laid out the site of the important early southern port, Goolwa, on the Fleurieu Peninsula at the mouth of the Murray. In 1851 he was investigating and reporting on a new road from Adelaide to Wellington, where the Murray reaches the Lower Lakes.

The climate of paradise

While in England, Chauncy had published *A Guide to South Australia*, and since he had formed such decided opinions that he had chosen to publish them, the fruits of his experience were probably poured into the new chum's ear as well. Chauncy was most impressed by the weather and the seasons. The ever-green trees of Australia realised a 'perpetual spring or summer', and the trees and wildflowers, though 'all different' from those in England, were not 'the less numerous and beautiful'.[15] But it was the skies – of such 'amazing beauty' that no one confined to England could imagine them – which expressed the overall excellence of the climate.

> On the whole, perhaps, there is no climate in the world which surpasses that of South Australia. The greater part of the year it is delightful: the winter is pleasant, and the autumn and spring are ... the perfection of climate. And even in the summer ... it is not humid, and consequently not oppressive. The air is clear and elastic and causes a buoyancy of spirits which is scarcely ever experienced in England ...[16]

Chauncy was not alone in his response. Extolling the perfection of the climate was such a commonplace that John Stephens, the editor of the colony's major newspaper and a savage critic of colonial officialdom, could use it to remark dryly that in South Australia: 'the climate of paradise at least seems to have survived the fall'.[17] However, there were aspects of the new climate that the colonists found unfamiliar and unsatisfactory, and lack of rainfall was one. The vocabulary of weather and climate descriptions the colonists brought with them had not been developed to describe the new conditions, and the colonists used the term 'drought' to characterise everything from several months without rain, to the sight of the Adelaide plains in summer, or the thin trickle of season-ally dried-up creeks and rivers.[18] Despite his enthusiasm for the seasons and the skies, Chauncy linked the vegetation not with the climate but with the soil, and put forward a simple guide by which the fertility of soil could be deduced from the vegetation it supported.

Apart from Chauncy's assessment of the natural environment, Goyder would also have had the benefit of the older man's experience in what was known as 'colonial surveying', or the technique of surveying land which did not come conveniently prepared with church steeples to act as widely visible points of reference, roads and lanes to move about on, and inns and taverns providing food and shelter. Presumably, Chauncy would also have had a network of professional contacts into which Goyder could be introduced, while the Chauncy family would have set before his imagination the instructive example of another of its members, Philip La Mothe Snell-Chauncy. Philip Chauncy had been the assistant surveyor general in Western Australia since 1841. But when Goyder set about establishing himself, he was still following his first career. On 10 June he entered the Office of the Colonial Engineer as a draughtsman. When he was married, in December, he gave his occupation as 'engineer'.

Grafted together

Goyder's bride was Frances Mary Smith, a young woman nearly a year and a half older than him who had arrived in the colony the year before. Frances, too, had travelled to South Australia alone and had family in the New World and connections with the New Church. In 1848, the year George Goyder emigrated, his parents had moved to Ipswich, Suffolk, so that David could take up the ministry of the New Church there, which had been established about 10 years before. Among its first members had been John Smith, a gardener, and his eldest son, also John Smith, a coach painter, both of whom had accepted the doctrines of Swedenborg in 1839.[19] According to Margaret Goyder Kerr, John Smith and his family were a branch of the Smith family whose company was a renowned manufacturer of clocks and precision instruments.[20] The elder John Smith, born in Ipswich around 1793, had been employed as a gardener by a local landowner, Dykes Alexander, from around the age of 20. Smith considered himself to be a practical and scientific gardener and became deeply involved in the cultivation of edible exotic plants in greenhouses. He would eventually become the author of a treatise on the greenhouse cultivation of melons and cucumbers, as well as asparagus, mushrooms and rhubarb.[21] John Smith had married another resident of Ipswich, a young dressmaker about three years his senior named Susannah Underwood, and the marriage had been fruitful. Frances Mary was the fourth of the couple's nine children, and their first girl. The 1841 census shows that Frances, like her mother, was a dressmaker.

David Goyder had been in Ipswich only a year or so when he began to lose his congregation, possibly in part because of good reports he had brought them about prospects in the new colonies of Australia and the example of the emigration of two of his own children and other followers of Swedenborg. Among the emigrants from Ipswich were almost the entire Smith family, whose migration began in March 1849, when the third son, William, and his wife left for Port Adelaide as assisted immigrants on board the *Sultana*. The independent Frances set out in July on the *Duke of Wellington*, arriving on 7 November. While she was still at sea, almost all of the rest of the family left England in two ships which departed within days of each other. In the first part of October, the second oldest son, Edwin Smith, also a gardener, left on the *Bolivar* with his wife Hannah, their two-year-old son Edwin Mitchell, and a baby daughter, Ellen Hannah. John and Susannah, now in their late fifties, together with 12-year-old Arthur Henry, 15-year-old Ellen Priscilla, 19-year-old Alfred, and adult daughter Mahala, followed only days later as assisted immigrants on the *Agincourt*. In the passenger manifest, Ellen and Mahala were described as domestic servants, while Alfred was listed as an agricultural labourer.[22] Only two of John and Susannah's children remained in England.

On 29 January 1850, the *Bolivar* reached Port Adelaide. When the *Agincourt* arrived three days later, most of the Smith family of Ipswich were reunited. It must have taken more than a good report from fellow Swedenborgians to prompt the family to emigrate, but it is only possible to guess at what attracted them. The Smiths, on the whole, were not so passionately nonconformist in their religious beliefs that the prospect of living in a 'paradise of dissent' (as colonial South Australia has been characterised) is likely alone to have lured them such a long distance – in fact, they retained their connection with the Church of England in South Australia.[23] As David Goyder himself observed, not all who accepted the teaching of Swedenborg felt compelled to leave the established church.[24] The opportunity to obtain assisted passage to a safe and ordered colony must have been attractive; but what may have clinched the decision for the two gardeners and their families could well have been the colony's much-promoted 'Mediterranean' climate, in which they would need, not hothouses, but shadehouses. The South Australian climate was considered beneficial to sufferers of tuberculosis of the lungs, or 'consumption' as it was known, and it is possible that this, too, may have played a role.

While it is unlikely that George Goyder had ever met any members of the Smith family in England, when he reached Adelaide and introduced himself into the small world of the followers of Swedenborg established there, the Smiths would have known who he was. The serene and reserved Frances – or so she appears in her portrait – evidently had no difficulty in immediately accepting this diminutive young engineer with his father's high forehead and dark, intense eyes. Through her he would come to occupy an important role in the Smith family, assuming responsibility for the care of John and Susannah as they aged. His presence would influence the career choice of the two young Smith boys in the parties of emigrants: both Arthur Henry and Edwin Mitchell would follow his model by becoming surveyors. Both would come to be employed by him as senior, first-class surveyors, and Edwin Mitchell would go on to become surveyor general himself. Eventually, Goyder would even claim not one, but two of the sisters. At the same time, he would absorb the botanical influence of the Smiths in a way that would markedly affect his contribution and career. The contrast in realms of interest of the young engineer and his wife's horticultural family was mirrored in physical appearance. While David Goyder was tiny, and George Goyder very short, the Smiths manifested the opposite tendency. At six foot six inches – over 198 centimetres – Frances's brother Edwin was reputed to have been the tallest man in the colony.[25]

In Adelaide, both John and Edwin set about establishing themselves as gardeners. Within a few years John had established a nursery and vines in Fuller Street in Medindie, a ward of the then rural suburb of Walkerville, which abuts the parkland to the north-east of North Adelaide.[26] Edwin Smith at first took a position in charge of the garden of The Grange, Captain Charles Sturt's house. Sturt was an explorer whose travels had played an important role in the founding of South Australia. Edwin worked at The Grange for 15 months, until he was able to move to his own property at nearby Findon and start his own nursery. He grew fruit trees and his specialty roses, in what he believed to be the first rose collection in the colony. Edwin Mitchell could not remember much about the place, but he could remember the Kaurna people, on whose lands Adelaide had been established:

> my recollection of a lot of blacks being there is very vivid. These people caught a lot of tadpoles in the creek, made a fire, placed stones in it until they were quite hot and then laid rushes on these and placed the tadpoles thereon until they were cooked. These embryonic frogs were evidently a great dainty, as was also snake, which they cooked. I remember stepping over a snake about thirty yards from the house.[27]

In June 1851, as Goyder was settling into draughting, Edwin Smith broke five acres of ground to begin farming. He tried barley first, which was sold for seed, and followed this with two seasons of potatoes and then wheat, from which he made a profit of £307 10s after threshing.[28] In November 1851, Goyder and Frances joined the New Church, presumably so that a ceremony could be held in the tiny church in Carrington Street.[29] Their official wedding took place at Christchurch, North Adelaide, then the Anglican pro-cathedral, on 10 December 1851. Ellen Smith, Frances's youngest sister, was one of the witnesses, and Goyder's best man, Chauncy, was the other.

The newly-weds settled in North Adelaide, an early residential section in which handsome houses were spaciously arranged along wide metalled roads. This enclave of wealth was separated from the rectangular grid of the main part of the city of Adelaide by the River Torrens and the ring of parkland surrounding the city that is a prominent feature of its design. However, the Goyders had settled on the east side, in Margaret Street, which is so narrow it is hardly more than a lane. In the early 1850s it was the most densely inhabited street in North Adelaide, lined with plain cottages only barely set back from the footpath, occupied by working people and members of the lower middle class. The families of professionals and senior administrators were more generously situated on the other side of North Adelaide, in the west. What may have influenced this choice of location, apart from considerations of economy, was the not-too-distant presence of Frances's parents, just a little further to the east. It is also possible that the couple had chosen to live with, or near, Mahala. In May 1851, Mahala had married John Isaac, a draper and tailor, but three days after Frances's wedding to Goyder, her husband had died. The young widow gave birth to a son in March 1852, but the child survived for only five weeks. Mahala was very ill by this time herself, and she died in August of consumption.[30] In the public notice, Mahala's death was recorded as having taken place in Margaret Street, North Adelaide. From the beginning of his marriage into the Smith family, it seems, Goyder had faced substantial family responsibilities.

In the six months between his taking up employment and getting married, partial self-government had been established in the colony, and, ironically, gold had been discovered in Victoria. Able-bodied men were setting off in the direction from which Goyder had recently come. As it happened, his employment with the public service proved as chancy as their ventures. Since enough land had been surveyed to satisfy needs in the immediate future and essential public works had been completed, most of the staff of the Survey Department were retrenched, and the entire Colonial Engineer's Office was abolished at the end of the year. But Goyder had readied himself for this, and on the first day of 1852 he took up a new position as secretary of the Adelaide Exchange.[31]

At the Exchange

Neale's Commercial Exchange, in King William Street, was the business of John Bentham Neales, an early, and active, colonist. Neales opened the Exchange, which became a central meeting place, in 1849, but in 1851 he became one of Adelaide's city commissioners and also entered parliament, making it necessary to employ a secretary to look after things in King William Street. The position required Goyder to become acquainted with land matters and with the commercial aspects of the agricultural and pastoral affairs of the colony.

In early September, the Goyder's first child, named Frances Ellen after both her mother and her maternal aunt, was born in Margaret Street. It may have been about this time that his sister Sarah and her children arrived to live in Adelaide. Her husband, Hugh MacLachlan, had died in Sydney on 30 May 1852, leaving the heavily pregnant Sarah alone. She had given birth to the couple's second daughter only 11 days later, on 10 June 1852. Once she had moved to Adelaide, Sarah became the superintendent of a school for young ladies.[32]

It is unlikely that Goyder was content with remaining in an office in the city. By 1852, he must have understood that he had the potential – once he had gained some familiarity with the affairs of the colony and with the country – to become a surveyor, and either set up in business or aim for a senior position with the government. It did not really matter that the work of the Survey Department had been brought to a virtual standstill at the end of 1851, because the continued growth of the colony ensured that it would inevitably start again. Moreover, in South Australia, the work of the top government surveyors involved a degree of responsibility for shaping the colony that he undoubtedly found appealing.

The surveyor general and the government survey

In South Australia the position of surveyor general was different from that in the other colonies. Unlike the colonies to the east and in Tasmania, South Australia had been planned according to the theory of 'systematic colonization' proposed by Edward Wakefield, even before a suitable site had been located. Once set in motion by initial investments, the engine of systematic colonisation was intended to run on independently, allowing the colony to expand without requiring financial support from home. Since land would simply be appropriated without cost, proceeds from its sale could be devoted to paying the fares of emigrants, ensuring a supply of labour. The price of land was to be adjusted to ensure that newcomers had to work for several years before acquiring enough money to purchase land of their own and their purchase would finance the emigration of more settlers. Control of the sale of land was therefore central to the success of the whole project and was also a means of containing the settlement: the goal was to create a community of prosperous, independent small farmers, living in close proximity and connected to a village structure. It was a democratic, nonconformist vision of a rural English paradise purged of those features perceived to disfigure the original – the monopolisation of land by a wealthy class leaving an attendant class of dependent workers and landless poor, and the claims of a national church established in law.

Not only did systematic colonisation require tight control over the disposition of land, it required a firm commitment to the policy of survey before selection, so that settlers would be able to purchase land only as it was offered to them in surveyed lots. This meant that the survey had to be running well in advance of land sales and that settlement had to be located and imagined before the plan could be marked out and the land divided and sold. The survey, therefore, was critical to the success of the whole project, and the role of the surveyor general, central. When another scheme of systematic colonisation was planned for New Zealand, Wakefield himself was reported to have exclaimed: 'The survey! The survey! The survey!', perhaps because the application of his theories in South Australia had not gone to plan.[33] The colony had not proved self-supporting and lost the special status bestowed at its creation, becoming an ordinary crown colony. The original vision continued to influence South Australia's development for decades, all the same. In 1894, writing on the lands systems of the various colonies, William Epps observed that the 'history of South Australia is essentially bound up with that of its land system'.[34] Apart from the hiccup of a number of special surveys for large wealthy purchasers that were permitted at the foundation of the settlement, the colonists remained faithful to the original vision and the policy of survey in advance of settlement it entailed. In central South Australia, settlement radiated from Adelaide, the surveyed units spreading contiguously. Rural land was divided into counties, hundreds (with an area of about one hundred square miles), and finally into individual sections. The plans were made available to the public and the land was put up for auction with a minimum price of one pound per acre. Beyond the hundreds – the term became a synonym for settlement – crown land could be leased for grazing, on terms that allowed for its resumption if a new hundred was required.

The harassed Colonel Light

The colony's first surveyor general was Colonel William Light, an officer of the Royal Engineers. Light was intelligent and artistic, a watercolourist of some accomplishment and a private poet, linguist and musician. A sailor as well as a soldier, he was a distinguished veteran of the Peninsular War, noted for his courage and daring and his unselfish nature. On his deathbed he is reported to have said that he had joined the South Australian venture because he had wanted to follow his father, an officer of the East India Company and a founder and governor of Penang, as a builder of empire. (Light had been born in Penang of a supposedly Malayan mother.) Light now possesses a unique status as the founder of Adelaide, the creative figure who brought the colony into physical existence and gave the capital its distinctive form, and, for his role in establishing the physical layout of the rest of the colony, Goyder has been likened to Light.[35] Apart from their roles as shapers of the environment, there were similarities, too, in their careers as surveyors general – similarities that, if they had been foreseen, would have put the young Goyder off his confident stride.

When Light was appointed, it was understood that the surveyor general would be required to make decisions of fundamental importance, but his position had been compromised by having him responsible to a resident commissioner, who represented the Colonization Commissioners in England, while the actual head of the colony was the governor. Worse still, his task was made next to impossible by commands that he inspect a vast area of coast and land and have a large amount of land actually surveyed before the governor and the first settlers arrived. To satisfy these requirements Light estimated he would have needed to set out two years in advance of the rest of the colonists, rather than several months. By the exercise of what has been described as his 'innate topographical instinct' and 'geographical genius' and by tailoring his interpretation of the instructions to suit the situation, Light succeeded in selecting a site for the capital and the port.[36] A struggle involving all the influential colonists broke out over his choice, but Light resisted every attempt to have it changed. With remarkable speed, Light designed and surveyed the city of Adelaide. The grid design, with its central and surrounding squares, one in each quarter, was placed south of the River Torrens and an additional area was added, overlooking it from an elevated area just beyond the north bank. A wide ring of parkland surrounded the whole. The result was to become renowned in the history of urban design, but at the time the wide streets and spacious squares prompted derision among the residents of huts and tents.

Light faced more criticism when inadequate staffing and difficult conditions delayed the surveying of the rural sections. Because of the slow process of the country survey, George Strickland Kingston, a young Irishman who had been appointed Light's deputy, despite an evident lack of training and ability as a surveyor, was sent back to England to consult with the commissioners at the request of others in the colony. Light had agreed to this absence because, in his opinion, Kingston was 'totally incapable of surveying – of triangulating a country he *knows nothing*'. He had instructed him not to return as his deputy but as a civil engineer. Light had also found Kingston to be 'more willing to direct others than to work himself', and to possess a 'haughty, imperious manner', a view echoed by other colonists.[37] But Kingston did not return as the civil engineer. He brought back instructions that Light was to abandon the trigonometrical survey in favour of a running survey, or to hand over the direction of the running survey to Kingston, if he would not. The new plan was to divide the country to be surveyed into parallelograms, measuring from point to point with chains, without constant reference to established trigonometric points. Even if executed with care and competence, a running survey was susceptible to error, especially those introduced by the distortion and wear of the chain. Since the survey was now almost keeping pace with the demand for land, Light refused the instructions to proceed in this slapdash manner, as the Colonization Commissioners had guessed he might. Most of the surveyors supported him, leaving Kingston to attempt to carry on virtually alone. Light was directed to continue examining the coastline, and his salary of £400 – the surveyor general of New South Wales received £1000 – was cut to £120. Finally, Light resigned and made it clear he would not return. He poured out his feelings to Wakefield:

> I am now completely tired of serving the Commissioners and, after founding their colony for them in spite of every abuse, I may now retire to seek a livelihood by my own industry when strength is gone. I will only make one remark ... in the shape of a question. Is it likely that the Commissioners could have found many surveyors to stand against powerful attacks from the Governor, the Press, and many others as firmly as I have done for *their* good?[38]

Together with Boyle Travers Finniss, his former assistant and deputy surveyor general who had also resigned in protest at the instructions to undertake a running survey, Light established a private surveying practice, which was contracted to carry out some of the special survey work, but his health was ruined. He died of tuberculosis in October 1839. Like Light, Goyder would have work piled on him and have to hold his ground in the face of criticism and conflict. He, too, would be worn down physically and retire only to have to set up in private practice. More curiously, the end of his career would also be brought about by a clash with the overbearing personality of a Kingston – in Goyder's case, the son.

The demise of the Survey Department led to near economic collapse. Gawler, the second governor, was dispatched to end the strife, and he immediately boosted the survey staff dramatically. Since the surveyors had remained loyal to Light, Kingston could not possibly be appointed surveyor general, and to fill the vacant role Gawler invited Charles Sturt to come from New South Wales, while the Colonization Commissioners in England, knowing only of Light's resignation, sent another officer of the Royal Engineers. To resolve the situation, another senior administrative post was created for Sturt, while Kingston abandoned surveying altogether, turning instead to private practice as a civil engineer and architect. He went on to became a prominent figure in parliament, a strong democrat who would serve as speaker in the House of Assembly for over 20 years and be knighted for his service. Kingston also took responsibility for maintaining meteorological readings until the arrival of Charles Todd to take up the posts of superintendent of telegraphs and government astronomer, and to take over the keeping of weather statistics. Todd considered him 'a careful observer', and evidently regarded his conclusions about weather patterns highly.[39]

The exhaustion of Edward Frome

Lieutenant Charles Edward Frome arrived in September 1839, together with a party of 16 sappers, and took up the office of surveyor general in early October. His succession was overshadowed by the death of Light, whose funeral was conducted with considerable pomp. Frome, who was quickly promoted to captain, was required first of all to complete the special surveys, which he considered 'a vicious system of cutting up the country', while half of his men had to redo Kingston's inaccurate work.[40] Even so, the main survey of the country was soon underway. Triangulation began from a base on the Adelaide Plains, and by 1845 extended about 150 miles north of Adelaide, 50 miles to the south and 40 miles east.[41] Enough country was surveyed to satisfy demand for agricultural land for a decade to come. Frome was highly technically skilled. He had been an instructor at the Academy of the Royal Engineers at Chatham and was the author of a textbook, *Outline of the Method of Geodesy or Trigonometrical Survey*. Not surprisingly, Governor Gawler soon reported that the department was under 'the most complete and sufficient superintendence, and is probably not inferior to the best survey establishments in the British Colonies'.[42]

Responding to the new conditions, Frome produced *Instructions for the Interior Survey of South Australia* for use in the department (and on his eventual return to England he would shape a new chapter on colonial surveying from it for a later edition of his textbook). In his instructions, he acknowledged the independent value of triangulation in remote country. A trigonometric survey was the most thorough and reliable form of survey, and triangulation was its first stage. Proceeding from a single baseline of carefully established length and using the trigonometric principle that only the length of one side of a triangle and the two angles at its end need be known to establish the lengths of the other two sides and the size of the third angle, surveyors selected prominent points in the landscape – usually hilltops which they cleared and marked with a cairn of stones supporting a pole – and began to construct their triangles. By reference to the framework of trigonometric points the land could be divided into sections; a detail survey locating significant features could then be carried out. Frome recommended that the triangulation should always proceed ahead of the sectional survey, and recognised that, in unsettled country, trigonometric stations provided pastoralists with points of reference for defining the boundaries of their unfenced runs.[43]

Frome also stressed the value of preliminary examination of the country in colonial surveying, an emphasis that Goyder would reassert and develop. Before the survey began, Frome directed, 'each Surveyor should make himself well acquainted with the nature of the ground over which his operations are to extend', recording observations and making sketches.[44] Main lines of communication were to be established and areas reserved for special purposes before the land was divided into saleable units. Boundaries which allowed purchasers to dominate resources, or features such as road and river frontages, were to be avoided. When the section boundaries were being marked out:

> all natural features crossed by the chain should be invariably noted in the field-book, on the outlines plotted from which are drawn the general character of the contours of hills, the different lines proposed for roads, directions of native paths, wells, springs, and every other object tending to mark the nature and resources of the country. Copies of the plans should always be transmitted to the principal Survey Office, accompanied by a rough diagram ... and by an explanatory report, describing the nature of the soil, description of timber, & c., upon each section, and the facilities for making and repairing roads and bridges, and peculiar geological formations of the different districts. A collection of botanical and mineralogical specimens from all parts of the province will also contribute materially to the early development of its natural resources ...[45]

Exploring was a part of the job that Frome took up readily. He investigated the Murray Lakes and the Coorong and attempted to get up into the north-east, travelling as far as Black Rock Hill. In the second half of 1843, he set out to investigate the land to the east of the Flinders Ranges, but was prevented by a salt lake. (It was eventually named after him, although it was known at the time as Lake Torrens.) Frome was possessed of the cultural accomplishments not uncommon in a man of his position, and from this journey produced a folio of watercolour sketches. One of these: *First View of the Salt Desert – called Lake Torrens* (plate 10), now in the Art Gallery of South Australia, shows a mounted figure searching the featureless, waterless horizon with a telescope. It is an image that seems to prefigure the satirical sketch, by amateur artist A.G. Ball, of Goyder 'in pursuit of the rainfall', although it seems unlikely that Ball would ever have seen Frome's work.

Because of the colony's economic difficulties, Frome took on the duties of colonial engineer, supervising the construction of bridges and buildings in addition to his duties as surveyor general and without extra pay. The Land Office, which dealt with the sale and leasing of land, was also brought under his direction in 1841 to form the Survey and Land Department (although the following year the administration of pastoral lands and leases was passed on to a newly created commissioner of crown lands). The burden of his responsibilities did not stop there. As surveyor general, he was a member of the Legislative and Executive Council until 1843, and served on the boards of the Adelaide hospital and the public cemetery. When copper was discovered in 1845, the prosperity that resulted only made things worse for him by increasing the demand for roads, and later other public works. Frome literally had to work night and day to satisfy his two roles. His reward was to have his salary cut by £100 in another economy drive and to lose the forage allowance for his horse. The Survey Department was also pruned, but the survey remained well ahead of the demand for land. Frome left South Australia because a decade of overwork had damaged his health. He went on to postings in other parts of the British Empire, retiring as a general. In a photograph taken possibly at the time of his departure, Frome's long side whiskers and moustache create the impression of a muzzle, over which, peering from under drooping eyelids, he gazes at the world like a sagacious but world-weary camel.

The rapid rise

In 1849 Frome's place as surveyor general was taken by Captain Arthur Henry Freeling, another officer of the Royal Engineers and only 29 years old. Freeling was the embodiment of a Victorian dream: he was an officer and a gentleman, handsome and of noble descent, and not surprisingly, popular. Freeling carried on in the manner of Frome, occupying a variety of roles. He was colonial engineer until the office was abolished, at times acting as colonial architect and inspector and commissioner of railways, and was chairman of the Central Board of Main Roads for most of his stay. He was one of the five commissioners of the City of Adelaide, and played a leading role in establishing a reservoir and water supply for Adelaide. As surveyor general he was a nominated member of the Legislative and Executive Councils during 1855–56, and, after the achievement of self-government, was elected a member of parliament.

At the beginning of 1853 when the position of chief clerk of the Land Office became vacant, Goyder wrote to seek recommendation to the position from the colonial secretary, B.T. Finniss, Light's former deputy. He cited Freeling's satisfaction with his performance in his first public service position and the knowledge that he had acquired about land in the colony as secretary of the Exchange as 'inducements' to employ him. He evidently had his own ideas about the responsibilities of a public servant to the public, since he also volunteered the promise that, in addition to attempting to fulfil his duties to the best of his ability, he would do his utmost to provide all the information he possibly could in response to demand.[46] (Decades later, this remarkable early commitment to the provision of information would play a role in his demise.) Goyder was recommended without hesitation and re-entered the public service on 12 January 1853. His employers very likely considered themselves lucky. One of the effects of the gold rushes in Victoria was that South Australian society had been stripped of able-bodied men at all levels, and salaries had had to be increased to entice, or retain, those who remained. As chief clerk, he was, by his own account, the only clerk, functioning with one assistant.[47] Once in the Land Office in charge of its record-keeping systems (a detailed understanding of which he retained to the end of his career), Goyder began the rapid rise to prominence which so impressed his contemporaries that no account of his career failed to mention it.

Goyder's meteoric rise was a matter of seizing the opportunities that circumstances provided. During 1853 deputy surveyor general McLaren was in the process of retiring, and Goyder evidently moved quickly to fill the void. Freeling later described Goyder as his 'right-hand man', and, towards the end of his career, Goyder described himself as having been deputy surveyor general for seven years; that is, the whole time he had been in the department prior to reaching the most senior position, although this was not officially correct.[48] Just under a year after his appointment, he was temporarily promoted to acting surveyor general. In 1854, triangulation began again to extend the surveyed area north of Adelaide up to the head of Gulf St Vincent. This was Goyder's opportunity, no doubt foreseen, to get out of the office and to begin to practise as a surveyor. On 13 September he was promoted to second assistant surveyor general. By equipping himself with two assistants, Freeling had wisely recognised that if there was enough work to require three men, then three men should do it. But the move may also have expressed his own lack of confidence, or enthusiasm, as a surveyor.

At the same time, Goyder's family began to grow. A second daughter, Florence Sarah, was born at the beginning of 1854, but in May their first-born died. In April the following year their first son, George Arthur, was born. However, there were problems when it came to having him baptised. The New Church, or the 'Adelaide Society' as its members knew it, was going through a difficult period, with no harmony between the congregation and the minister, Jacob Pitman. Goyder believed that Pitman was refusing to baptise his child, and had the baptism performed by a layman, a challenge that provoked a further round of dispute before peace was established.[49] In the same year, the young family made a move a short distance across the parklands to Medindie, close to where John and Susannah Smith lived, and, as it happened, not a great distance from surveyor general Freeling. Medindie had been subdivided by the pastoralist and sometime-commissioner of crown lands F.H. Dutton, and there was still a great deal of cleared and fenced but unbuilt land in the area. The Goyders would soon have the prominent pastoralist and parliamentarian, George Hawker, join them – in 1856 Hawker purchased 15 acres on which he would build his elaborate mansion, The Briars. The house into which the Goyders moved, although presumably larger than their residence in Margaret Street, was considered a 'neat cottage', with six rooms, a garden and a vineyard, the vineyard probably managed on the advice of his father-in-law and brother-in-law.[50] Named Hillside, the cottage was located on a private road (now Hawkers Road) close to Robe Terrace overlooking the parklands, North Adelaide, and the city. On the other side of the private road was Charles Ware's nursery, one of the leading nurseries in the colony, later to be taken over by the Smiths.

On 1 November 1855, after having served his probation as a surveyor, Goyder became first assistant surveyor general, with his duties defined as 'General Superintendence of Field Surveys'.[51] Only a short time after he had he begun, he was devising new ways to improve the efficiency of the survey parties. (One party he had reported on had literally been 'moonlighting' – rather than working on the survey, they were sleeping during the day and reaping a farmer's crop for a pound an acre at night.[52]) Goyder identified the absence of adequate pre-survey preparation as a source of inaccuracy and advised that a very thorough examination of the land should be made before the survey was planned. His recommendations imply that after Frome's departure, laxity had crept in. During the winter he had discovered that two separate surveys undertaken in the country around Kooringa overlapped. While the survey parties waited for wet weather to clear, he had taken off by himself, riding over the country and mulling over the problems as he went. He wrote to Freeling with his conclusions:

> A competent person should be sent to a locality proposed to be surveyed in sections – and that before the detailed survey is commenced he should run around a block of land, say about 10 miles long and five wide … sketching in the features of the country and chaining the distances to particular objects – such as creeks, gullies and closing the line with the utmost care – and forwarding the result for your examination; one of your assistants should then be instructed to examine this block – marking out what he considers the best lines of road, and other reserves necessary to be made – keeping in view the direction traffic is likely to take. This report would then enable you to decide correctly how the detailed survey should be conducted – and a tracing of this survey placed over the general plans would at once show if the roads had been lost sight of or reserved.
>
> It would also render the persons in charge of parties more careful in chaining – as a constant check would be kept upon them by the original distances of the person who defined the block – it would introduce the features of the country so accurately that a stranger might at once discover any particular locality – and it would save much time in supervision.[53]

This was the first formulation of the approach Goyder would bring to the South Australian survey and exhibits his characteristic broad overview, combined with a concern for detail and accuracy. Since he would be involved in putting his method into practice, Goyder was also setting up conditions which would inevitably sharpen and discipline his own powers of observation. His position ensured that he would be required to use them over vast areas of country.

At the beginning of 1856, after only three years in the department, Goyder was appointed sole assistant to the surveyor general, although the position of deputy was not made official until 1858. Another daughter, Emma Gertrude (known as Gertrude), was born in November. At Hillside, Frances settled in to a life with Ellen, who lived nearby, and the babies, while her husband was now more often in the country than at home.

CHAPTER THREE

As far as the eye could reach

Islands which have
never existed
have made their way
on to maps nonetheless.

And having done so
have held their place,
quite respectably,
sometimes for centuries.

Voyages of undiscovery, deep
into the charted wastes,
were then required
to move them off.

> – Nicholas Hasluck, from 'Islands'

Towards the end of 1856 Goyder took a small leather-bound pad with a metal clasp and modestly wrote his initials and surname, and the date, inside the front cover. So he began the first of the 27 field notebooks, which contain the records of his journeys, to his last journey in 1893. The new notebook must have lingered on his desk or in a pocket until the new year, when he took it up again – evidently in a confident, even grandiose, mood – turned a few blank pages, and inscribed 'Official Journal of the Deputy Surveyor General from January 1st 1857' in a bold round hand. The first entry – 'January 1st 1857. Holiday' – failed to live up to the promise of the title page, but it was not long before Goyder was placed in charge of a small party to carry out a triangulation of the country north of Mount Serle, in the North Flinders Ranges, where there was urgent need of trigonometric points to fix the position of runs. The project offered the opportunity to do a little real exploring.

On his way north, he was to survey the road through the Pichi Richi Pass, which leads through the Flinders Ranges and connects Port Augusta to the Willochra Plain and the country to the east. He completed this task on 2 May.[1] The entry in his notebook records his satisfaction and gives an idea of the way he went about his work:

Completed survey + examination of road – at work from daylight till dark – day showery.

Mem. 24 bullocks to draw one dray up hill – which alteration avoids.

Plotted work and wrote to Surv. General at night.

From Pichi Richi, Goyder set off to join up with the survey party already on its way to Mount Serle. This was led by J.M. Painter, a former chief clerk in the Land Office turned private surveyor and employed as such by the government. Goyder's notes show that he was moved by the appearance of the country from the time he entered the Willochra Plain, where he stopped at the head station of a run leased by the Ragless brothers. The run was formally named Balcarrie, but the head station was generally known as the Mud Hut, even on maps, where it appeared as a lonely stopping point. From the plain, Goyder moved back into the ranges on his way to the next station, Kanyaka, which he noted to be taking on the appearance of a village.[2] (The remains of the Mud Hut – a chimney and several piles of earth flecked with fragments of charcoal – can still be seen near the road that crosses the plain at some distance from an historical marker for the later Willochra Inn, while the ruins of the stone buildings which belonged to Kanyaka have become a signposted tourist attraction.)

From Kanyaka, Goyder went on to camp near Wilpena Pound, between Rawnsley Bluff and Taylors Mound. Approaching the pound, he noted that 'a very fine view presents itself to the eyes – a long rich flat bounded by picturesque hills + terminated by Rawnsley's Bluff'. Numerous photographers have since agreed: views of the bluff are among the most popular images of the Flinders Ranges. Early next morning, he reached Painter's camp, to the east of Point Bonney. The surrounding scenery was 'very beautiful'.[3]

The survey party included bullock teams, and because of the nature of the roads their progress was slow. A couple of days later, travelling over what he described as 'an extremely broken country, interspersed with nasty little creeks and gullies', one of the polers broke his horn in the course of what must have been a cruel descent.[4] ('Polers' were the bullocks harnessed to the pole directly in front of the wagon and were critical to steering.) Despite these difficulties, in his official report Goyder was to describe this country as being 'of the most pleasing character'.[5] A week later, in the Mount Hack area, they encountered a portion of road that was even worse. There was another accident; a horse and cart fell down a gully. Both turned over twice on the way down, but at the bottom the horse was 'on his feet + the cart standing properly on its wheels – and the contents – tho' scattered about – unharmed'.[6] The driver, too, was uninjured, but the next day was spent repairing the cart.

A sketch from Goyder's 1857 field notebook showing the Mud Hut with Mount Brown and the Devil's Peak in the background. Goyder evidently took care to draw the mountains accurately, but they appear large and close to the buildings rather than as they are, distant and on the horizon of a plain. [SRSA]

Point Bonney, Rawnsley Bluff (on the south-eastern side of Wilpena Pound) and the Chace Range sketched from a track north of Uenbulli, 1857. The tree-lined Wonoka Creek is in the foreground. [SRSA]

Frustrated by the slow progress, Goyder left the party in charge of Lee, the assistant surveyor, and went ahead with Painter for the last part of the way to Mount Serle, taking observations from hills along the route. By the time the main party had arrived the two had taken preliminary angles from Mount Serle and nearby hills and selected the area best suited for the measurement of the baseline. Because the country was so rocky and rough, the base had to be kept to two miles, between Mount Serle and Mount McKinlay. Each surveyor measured it separately, with little variation between results, and the mean was taken as the correct distance. The line was subsequently extended, by triangulation, to Arcoona Bluff in the north, and Mount Rowe in the south, giving a total length of 7.55 miles (just over 12 kilometres).[7]

Being in the ranges was exhilarating. Goyder's early field notebooks were essentially a record of measurements taken with various instruments and observations of land, weather and vegetation, accompanied by minimal records of his movements and the situations encountered by the surveying parties. Occasionally, he wrote notes to himself, or under severe pressure, used the notebook as a diary-like confidant, expressing strong feelings of anger, fear, or relief. But the very first notebook is filled with delight. During his stay at Mount Serle it suddenly flowered with ink drawings, apparently as much for pleasure as for work, and this is the only volume of his field notebooks to contain such sketches. Unlike Light or Frome (or Mitchell), he was not an artist, but the little sketches are arresting: they combine robust simplicity with a curious delicacy (although that effect is completely lost in reproduction) and seem to carry the great expanse of space and clear air onto the page.

In small neat writing, Goyder also attempted a thorough description of what he could see in all directions, studying the horizon through a 'glass'– a 'looking glass' or telescope. The description ran over three pages. Part of the long account of the view from the top of Mount Serle contains the most exuberant of all his descriptions of a landscape:

> To the south the Mt Hack Range stands in bold relief against the sky the Cock'scomb Mt Hack McFarlane's Hill + Mt Stuart being clearly defined. With Mt Wallace above – or behind the Mudlapina Gap – thro' which the Frome River is seen winding its sinuous way – collecting its sources into one stream and trending its way to the Nor'Nor'West until lost in the jumble of hills which intervene – with their high bluffs thrown into the air – presenting precipitous cliffs to invaders – forming into amphitheatres with the strata inclining steeply inwards absorbing in their circular pounds the surface water and numerous springs ...[8]

This grand vision merged seamlessly into an analysis that reflected on the capacity of this country to be exploited by European-style pastoral and agricultural activities. The difficulty, he explained, was that the surface and spring water were absorbed into the porous ground, 'causing rapid vegetation after a shower – which as rapidly disappears in the intense rays of the sun ...'[9]

The enchanted lake

By 26 May, with the baseline measured and the triangulation well under way, Goyder was able to begin the examination of the country to the north of Mount Serle. His purpose was to determine the extent of the survey required and to locate useful wells and springs in the vicinity of Lake Torrens, the topographical feature that made the South Australia in which he set out to do his exploring different from the South Australia of later maps. Lake Torrens, a gigantic horseshoe-shaped salt lake, a monster with arms that embraced three degrees of longitude and reached down through three degrees of latitude, was essentially the creation of the explorer Edward John Eyre.[10] Its hostile expanses had frustrated every attempt Eyre had made to reach the centre of Australia by way of the Flinders Ranges and it had halted surveyor general Frome's attempt to explore to the east. In endeavouring to skirt the lake, Frome had found himself trapped in its embrace because of errors in Eyre's bearings. While no one had followed the lake right around, it had appeared on both sides and to the north of the Flinders Ranges, glittering in the distance 'as far as the eye could reach', as Eyre put it repeatedly. Members of Sturt's expedition to the centre of the continent had seen it. Not only was the ghostly lake vast; because of its composition, it had a very particular character. Light refracted from its surface produced a disturbing array of illusions. In its vicinity, 'it was impossible to tell what to make of sensible objects, or what to believe on the evidence of vision', Eyre lamented. What the eye beheld 'partook more of enchantment than reality'.[11]

The lake's existence as one mighty, impassable barrier seemed more certain as time passed. On maps, the broken lines indicating its presence gradually became solid, until, in the year Goyder arrived in Adelaide, an elaborate decorated map of Australia was published, on which Lake Torrens appeared as the largest and most definite feature of an otherwise empty continent. But although the giant lake presented a disappointing limit to large-scale exploration and to the eventual extent of settlement, the colony's flocks and herds had continued to increase. Gradually, by means of small-scale exploration in search of suitable country and permanent water, pastoralism had spread north into the Flinders Ranges. When a new system of 14-year pastoral leases came into force in 1851, the first batch issued included leases for Arkaba, Wilpena and Aroona. By 1853, the runs had reached Mount Serle.

In 1856, in the hope of finding gold, the government had sent Benjamin Babbage, a geologist and government assayer, to search the Central Highlands. Babbage worked as far north as Mount Hopeless, the peak from which Eyre had assured himself that Lake Torrens blocked any further progress in that direction. In the northern ranges he learnt from two Aboriginal men that it was possible to cross the horseshoe lake, but his attempts to locate the crossing were unsuccessful. Although he had not found gold, or a means to cross the lake, Babbage had nevertheless found something of great value – permanent fresh water, in the form of large waterholes in the northern ranges. The colony's oldest and only daily paper, the *Register*, in an editorial titled 'Water in the

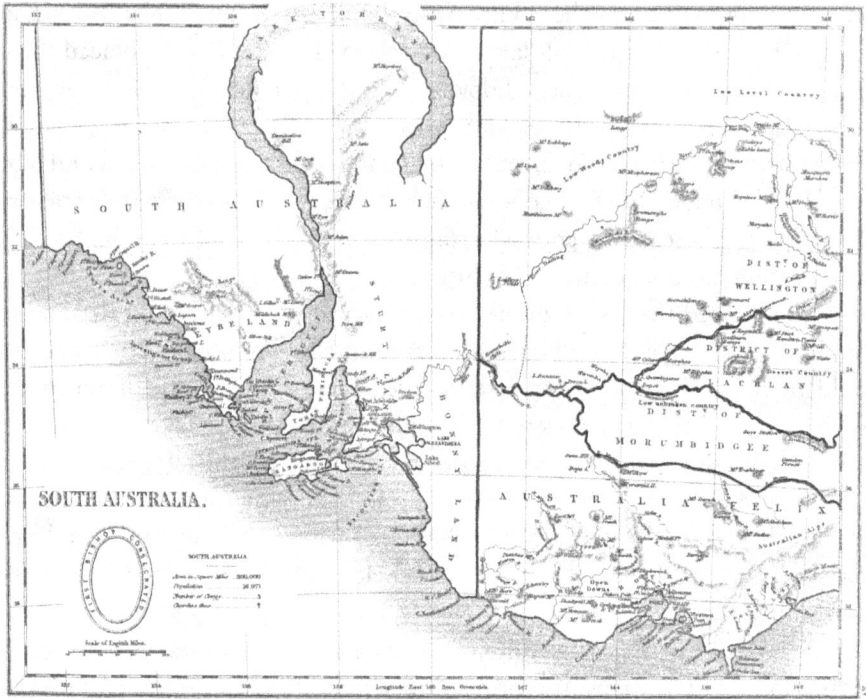

The Horseshoe Lake

This map from the early 1840s shows Lake Torrens as a horseshoe surrounding the
Northern Flinders Ranges, with Mount Serle and Mount Hopeless indicated. [SLSA]

North', proclaimed in consequence that finding 'vast tracts of fruitful country'
was still possible and observed that many districts were 'now covered with
flocks which years ago were denounced as hopelessly sterile'.[12] It was into this
world, enclosed by the baffling lake and tense with contradictory possibilities,
that Goyder sallied with the notebook that might yet receive the details of a
great discovery.

Exploration

On 27 May, Goyder left Owiendana, in the vicinity of Mount Serle in the
northern Flinders Ranges, accompanied by William Rowe, who, with his father
John Rowe, looked after the Survey Department's horses, and James Trebilcock,
a settler who had volunteered to help. The three men headed north-north-east
to Umberatana, about 40 kilometres away, and at that time the most remote
station. They were received with attentive hospitality by the Thomases, stayed
overnight, and went on the next day, crossing the plains to the north-east and

following a watercourse until it became a deep creek, winding toward the northern plains. This introduced them to another creek, the Yerelina, which was deeper, set in higher cliffs, and more difficult to follow.

Being near water and richer vegetation brought them into contact with the inhabitants of the area, who were less than pleased by their arrival. At sundown on 29 May, the party camped near the only good feed they could find for their horses. The spot was opposite some wurleys. 'Shortly afterwards,' Goyder recorded, 'I heard the voices of blacks calling to each other, as if in alarm – most probably exclamations at discovering the proximity of white people to their camp ...'[13] The people withdrew immediately. They had every reason to flee, as Goyder would have known. In 1852 a shepherd had been killed by Aborigines (possibly in retaliation for murders he had committed himself) on Aroona, then the northernmost run in the colony. In the absence of police, the pastoralist Johnson Frederick ('Fred') Hayward, an intimidating and usually well-armed man, had led a party of his men to seek revenge, and his reminiscences record that he had continued to conduct 'Occasional hunts after niggers'. A traveller passing through Aroona in 1855 recorded seeing camps in which there were only women, and was told 'whitefellows shootem alabout Blackfellow'.[14] Aroona was a long way to the south-west, but the terrible news would have travelled.

The residents of the wurleys did not return, and the next day the party continued on in the same direction for another three miles until they reached a high conical hill. The view from the top was something of a revelation, surpassing anything Goyder had ever seen 'in point of romantic scenery', a response that would later be hailed as a surprising aesthetic discovery. Although the beauty and variety of the scenery made the journey 'delightful at first', these features also rendered the way 'difficult and harrassing [sic] in the extreme'.[15] The party struggled along the bed of the MacDonnell Creek, changing course in the afternoon to begin working their way north over the ranges by bearing, crossing and re-crossing the creek as they went. Almost certainly they had been kept under surveillance most of the time, as Eyre had been when he ascended Mount Serle, his movements studied by Windawalpa, an Adnyamathanha man (or the name may have been Windhawarlpuha, an earlier form of the name, meaning 'owl bone', and the original name of Mount John).[16] Goyder recorded in his notebook that, 'at the place where we first left the Macdonnell I saw a black fellow watching us from the top of one of the hills – and was intensely delighted at his headlong speed on being discovered and my riding towards him'. The boyish satisfaction at having routed his observer foreshadows the opinion he would deliver in less than three years time, that the people of the desert country to the north were 'the most arrant cowards breathing'.[17] But

Goyder would also write of holding these people and their property – though not, of course, their land – 'sacred', and his actions would show him as good as his word.[18]

Next morning, following the same course, the party surprised two women. Goyder responded more gently:

> They were terribly afraid and gesticulated violently uttering loud sentences as we neared them, being unable to speak their language and they ours, we could make nothing of them. so gave the youngest a piece of tobacco which she crammed into her mouth whilst the older threw off her mat to defend herself from our [illegible] approach.[19]

The party carried on down the bed of the creek again, and when the cliffs on either side contracted and the bed became rocky, Rowe and Trebilcock went round to the east, while Goyder ascended a high bluff to examine the course of the creek beyond. He was rewarded with the discovery of 'a channel, from sixty to seventy feet deep ... varying in width from eighty to a hundred yards, and nearly a mile long, in which lay a magnificent sheet of water, running strongly at the south end, and increasing in depth towards the east bank'. There were fine gums on both sides of the channel, extending down to the creek and beyond the point, at which it dried up. The effect was almost overwhelming:

> This scene, so sudden and unexpected, forming so great a contrast to the arid plains and sandy-looking soil composing the bed of the creek over which we had so lately passed – the placid appearance of the waters, disturbed only by the quiet enjoyment of the water-fowl, swimming about on its surface – the rich luxuriant foliage and stately gums – afforded a feeling – a pleasure that can only be realized by persons similarly situated to ourselves.[20]

They named this channel the Freeling, made a few sketches, then continued on to camp on a grassy gumflat in the valley of the MacDonnell. From the top of a hill behind their camp – they named it Camp Hill – they could see Mounts Hopeless and Hopeful (renamed Mount Babbage by Goyder) away to the southeast. Much closer they saw another hill to the north-east, the top of which they judged would offer a good view of the surrounding countryside. View Hill, as Goyder named it, provided an uninterrupted outlook for a radius of at least 30 kilometres. In the east they could make out five creeks converging into two and flowing north-east until they were lost to sight. But the great Lake Torrens was nowhere to be seen.

The party headed east, crossing the first two creeks they had observed. When they arrived at the third, it appeared to be salt, with salt encrusted on its banks, and they were about to turn away when the horses began drinking avidly. When the men tasted the waters they were surprised to find them perfectly fresh, and apparently quite safe. 'I used them for two or three days without feeling the slightest inconvenience', Goyder reported helpfully.[21] Investigating the creeks, they travelled further east before turning back toward View Hill. The land was grassy and seemed to improve toward the north-east, and there were many recent cattle tracks. On 1 June they covered nearly 50 kilometres, reaching north to latitude 29° 20', and after zigzagging the country to the south and west, they reached Babbage's discoveries, Blanchewater and St Mary's Pool. There were 'quantities of teal, ducks, geese, cranes, cockatoos, pigeons, shell-parrots, magpies, curlews, crows, hawks, and other birds, flying about', and again, the tracks of cattle.[22]

The next day, from a hill, they saw a lagoon two miles north. The water there, emanating from fissures in rocks, also proved to be fresh. Further north there were larger springs, surrounded by the same white deposit on the ground they had already encountered and masses of reeds. These were named the Rocky and the Reedy springs. The prospect of finding more and larger sources of permanent fresh water was tantalising, but not everything they saw could be relied on. From the top of a nearby 'coronet' of eroded sandstone, which they named Weathered Hill, they saw what appeared to be a belt of gigantic gums, with a sheet of water behind them and rising land beyond. There also seemed to be a large lake about 10 miles to the east, but having travelled over that country, they knew they were seeing spectacular mirages.[23]

The three men headed north-east, back towards MacDonnell Creek, and followed its course for a while, passing several large permanent waters. The last, which was long and very deep, was 'extremely fine', and apparently an important stopping place for the Indigenous owners. There was an unoccupied camp at one end, on the eastern shore. Goyder described the wurleys as 'constructed in a similar manner to those described by Captain Sturt'; they were 'warm and comfortable, the largest capable of holding from thirty to forty people, quite round, from three to four feet high, and entered by a semi-circular opening, through which we were obliged to creep'. They named the waterhole 'the Werta-warta, from the name of the tribe frequenting the plains north of the Blanche'.[24] (The name no longer appears on maps.)

A true water horizon

On the following day, 3 June, they travelled for about 22 kilometres north-east along the bed of the creek through vegetation 'of the most luxurious kind', although the trees changed from 'lofty gums' to 'bastard peppermint', which, Goyder noted, were 'rapidly assuming a more stunted appearance'.[25] In the distance they saw what seemed to be an immense body of water, one that they judged to actually exist, and to be, most probably, part of the great horseshoe lake. They turned away from the MacDonnell and travelled about another 10 kilometres to the north-east. When they found themselves looking at 'what appeared to be Lake Torrens', Goyder determined the latitude from the meridian altitude of the sun. The trio then 'camped and had luncheon', their gentlemanly lack of haste almost certainly the result of believing that the lake would inevitably prove to be salty, or a mass of floodwater lying over a salt pan. After their meal, they made their way the remaining few kilometres to the shores of the lake, where they found that the water was fresh. Goyder noted that the bottom of the lake appeared to be composed of soil – bluish loam – and that vegetation, which included grass that was 'some of [the] finest seen since leaving Town', extended almost to the water's edge.[26] There was nothing to suggest that the water level rose or fell, and there were no flood marks. Amazingly, the water seemed to be not only fresh, but permanent. If this was so, then any youthful dreams that Goyder had nurtured of making a great discovery were being realised: the discovery of a well-watered inland with lush pastures surrounding a great lake was in the long run better than gold, especially for someone of Goyder's beliefs and outlook, absorbed as he was by the vision of establishing a thriving, enduring settlement. But Goyder had not recognised the meaning of changes in the vegetation he had recorded in his notebook. Not only had the trees dwindled and become stunted as they left the bed of the creek, but around the lake itself, there were no large, permanent forms of vegetation at all. New to the inland, and, in any case, one of the very few colonists to have visited it, he had as yet no idea what this attested about the climate. Beneath these notes he sketched a map of the lake's shores, showing the islands with perpendicular cliffs and the opposite shore, marked 'apparently land' and 'Hummocks'. Bearings to points all around the compass were included. Goyder later described how the lake:

> stretched from fifteen to twenty miles to the north-west, forming a water horizon extending from north-west-by-west to north-west; the south portion terminated by high land running south towards Weathered Hill ... An extensive bay is formed inside this promontory, extending southward to west-north-west, when the land again runs out to a point, approaching and passing us by a gentle curve to the east, and inclining gradually to the south-east, and ultimately disappearing in the distance. The north portion of the horizon is terminated by a bluff headland, round which the water appears to extend to

the north. This land passes thence to the east and forms the north boundary of the visible portion of the lake; and, from a higher elevation than that upon which we stood, appeared to extend round to the eastern wing. It is covered with vegetation, as also are several islands seen between the north and south shores, apparently about five miles distant from where we stood; their perpendicular cliffs being clearly discerned by aid of the telescope.[27]

To get a better view of the lake, they travelled about 30 kilometres due west, to the summit of higher land. On their way they crossed two creeks, one of which they named Mirage Creek 'from its forming the boundary of an imaginary lake, which we supposed we were approaching but which disappeared as we neared the elevated land'.[28] By this time Goyder considered himself knowledgeable in the ways of the land of enchantment:

> It would be perfectly useless to repeat the number of times we were deceived by mirage, and surprised by the enormous refraction peculiar to these plains ... In fact vertical angles are of little value; and the mere appearance of water no test of its existence; but this deception is only possible when away from water, the difference being so great when in its actual presence as to render deception next to impossible.[29]

Confident that his vast lake was no mirage and no mere accumulation of floodwater, Goyder also decided that he had now seen enough to enable him to give instructions on the extent of the survey. It was time to head back to town. The party returned to St Mary's Pool and Mount Freeling, taking bearings from the various hills as they went to enable the surveyors to follow and complete the triangulation, which with favourable weather Goyder hoped would extend to Weathered Hill by the end of the season. At St Mary's Pool he added a sketch map of the creek and several more illustrations to the contents of his notebook. He met up with Painter when they reached Mount Serle, and, after checking on the progress of the survey work, he left Mount Serle for Adelaide on 9 June.

The good news

For a discovery of the kind Goyder believed they had made, the entries in his notebook are almost unbelievably subdued. Without an understanding of the context, it would be impossible to tell from them alone that what was believed to be a major discovery had been made. Nevertheless, it is clear that Goyder allowed nothing to delay his return to Adelaide. The entries in his field notebook for this period are in pencil and were never overwritten in ink, as earlier entries had been. They are small and faint, and interspersed with drawings. The entry for 19 June notes: 'third day of a violent attack of dysentery', although there had been no previous mention of illness. It is a reminder that Goyder was keeping a surveyor's field notebook, not an explorer's diary, written with an eye to posterity and possible publication. His notebooks would form the basis of the reports he needed to write and would help him with work in the office, but the accounts they contained of his own experiences and responses were essentially incidental (and they are not amplified by any personal diaries or letters). It was the country that was important, not the observer.

When he reached Mount Remarkable, Goyder left his samples, plans and drawings with his luggage to follow in a cart. On 23 June, according to his notebook, he rode from Clare to the Green Water Holes (Two Wells) in a steady downpour that made the roads very heavy. The next day, in better weather and over better roads, he reached Adelaide. He jotted down reminders about waste lands, mineral leases and disputed boundaries that needed to be followed up in the office. Thursday 25 June, he stayed at home – 'Absent from duty' as the field notebook records – as he normally did on returning from a field trip.

Written many years later, in his retirement, the memoirs of Goyder's superior, surveyor general Freeling, provide a much more lively account of events:

> In … the year 1857 an officer connected with the Survey Department of South Australia came galloping into Adelaide … with the startling news that whilst employed on his duties five hundred and fifty miles north … he had discovered a large inland lake of fresh water which extended northward as far as the eye could reach, but that as he was destitute of the means of testing his discovery he had returned for the purpose of urging the Government to fit out an expedition to explore the lake. Such an announcement as the existence of a freshwater lake so far in the interior of a colony notoriously of a dry and thirsty nature caused a flutter of expectation and excitement in Adelaide, as the absence of permanent water had always been the obstacle to its successful exploration and settlement.[30]

Given that the final ride from Two Wells was less than the distance that Goyder and his mount were accustomed to travelling in a day, it is not impossible that he came down King William Street at a frantic pace, but, writing many years later and recalling events in the light of what followed, Freeling might simply have been embellishing his story with an imaginatively appropriate detail, and one that emphasised the extent of the discreetly unnamed officer's delusion. In any case, at the time there was excited talk in the government offices. The following Monday 29 June 1857, the *Register* opened with the news. The paper assured its readers that, although Goyder had refused to give any detailed information until his report was presented to the government, the preliminary information provided could be relied upon: Goyder had travelled even further north than Babbage and had found rich vegetation and extensive permanent fresh water. Apart from the obvious economic advantages, the assistant surveyor general had also reported that the ranges through which he had passed, rugged as they were reputed to be, were also so spectacular, picturesque, and inspiring that they surpassed anything he had ever seen. It was an utterly different world from the one they had previously supposed to exist, and the *Register* concluded that: 'Upon this accession of territorial wealth, we cannot be premature in congratulating the colonists of South Australia'.

By the following day, the editor was even less 'afraid of overestimating the importance of the facts now brought to light'. The discovery was as good as a gold find and equal to the major copper finds that had rescued the colony's economy in the early 1840s. The paper's understanding of inland geography had advanced apace as well:

> It is quite impossible that freshwater lakes of such magnificence as those described by Mr. Goyder can exist, as permanent waters, irrespective of vast mountain ranges, and these ranges are sure to be associated with extensive and fertile valleys.

The paper went on to voice Goyder's own desires, urging that:

> that gentleman ... should be sent out again with as little delay as possible. It is also highly important that [Goyder and his party] should be furnished with a suitable boat for exploring the newly discovered waters ... The fiction of central deserts of burning sands and boundless wildernesses of salt scrub may soon be doomed to disappear before the indubitable facts which a spirited and resolute exploration will bring to light.[31]

When Goyder's report was presented to the government, it was with a covering letter from Freeling, specially recommending Goyder and hoping that his 'praiseworthy conduct ... may be brought under the notice of his excellency the Governor-in-Chief'.[32] Governor MacDonnell was impressed by the commendable modesty with which Goyder had related his discoveries, while Charles Bonney, the commissioner of crown lands, was enthusiastic. He wrote a long letter to the chief secretary (in effect, the premier) endorsing Goyder's findings.[33] Both became convinced of the existence of some sort of inland sea or lake, and Bonney was keen for it to be explored by boat. News of the discovery was dispatched to the Royal Geographical Society in England and

The proprietors and editorial staff of the *Register* in 1857.

Standing in the centre is proprietor Anthony Forster, editor during the 'wars of the press' of 1864. On the far right is John Barrow, an editor who was soon to leave and publish his competing daily, the *Advertiser*, in 1858. Sitting between them with the walking stick is proprietor Edward William Andrews, who interviewed Goyder over the rumoured problems with the Tipara (Moonta) survey. The remaining three figures are (from the left): W.W.R. Whitridge (editorial staff), Joseph Fisher (proprietor), and William Kyffin Thomas (proprietor and printer). [SLSA]

was published in the *Proceedings* for 1857–58. After some discussion, it was Freeling who had the honour of leading the second party, becoming the second explorer, after Sturt, to have a boat – actually, two – lugged inland. On 23 July, his expedition set sail for Port Augusta. Their equipment included a specially constructed galvanised iron boat, about six metres long and weighing just over a quarter of a tonne, which they had to manoeuvre over almost impossibly difficult tracks, and a dinghy.

While the party was in the field, events continued to take their own course. News of the discovery had been received and published in other states. A discussion in the Melbourne *Argus* reviewed the whole puzzling history of the exploration of the South Australian inland in such a way as to emphasise the contradictory nature of reports about the country, while salvaging hope that it would yet turn out a rich asset. This was achieved by adopting a variation of the time-hallowed strategy of shooting the messenger: the explorers were deemed somehow personally responsible for the land they encountered. Both Sturt and Eyre were dismissed as regrettably 'short-sighted', and Eyre was found to be 'particularly so' – it was often remarked, the paper explained, 'that he was very unlucky in finding good country, and if there was any bad country he was sure to stumble upon it'. (Eyre was later recognised to have been travelling in drought years.) It was the sheer extent of the South Australian inland that was emphasised, not the climate, and what was evoked was the sense that the explorers were stumbling about in a vast realm that might contain almost anything. After all, as the *Register* had observed, reflecting on the difference between Sturt's account of the country around Lake Torrens and Goyder's, it was 'scarcely possible that so great a change can have been wrought in the physical character of the locality in a dozen years'.[34] The idea of a climate dominated by extreme changes that did not follow an annual seasonal round was beyond the colonists' conception. For journalists trying to resolve this mystery from their desks in Melbourne, their confusion was compounded by the state of the maps. In the *Argus*, it was noted that Lake Torrens 'evidently contains an immense body of water, for it extends over at least three degrees of latitude', leading to the conclusion that: 'to supply this there must be drainage of a mountainous region ...'[35]

Another report, reprinted in Adelaide from the Ballarat *Star*, documented the impact of the news. Titled 'Northward Ho!', it explained that 'some dozen or more of our adventurous and speculative pioneers' were already 'on the road for the land of promise' to secure new squatting leases in South Australia's lush north.[36] In fact, the government was being overwhelmed with claims originating from the other colonies. Goyder had sparked a land rush that generated claims for pastoral leases covering a huge area of over a million hectares. Politically, this had come at a difficult time. The first fully elected parliament had only just been assembled, and, in the absence of a party system, government had proved an unstable business of shifting alliances right from the start. The Torrens ministry, led by the son of the colonel after whom the lake had been named, attempted to alter the regulations governing pastoral leases to allow for claims made from other states that had no reference to any fixed points on the land. It was so obvious that this would result in chaos that the ministry lasted less than a month, the first to be overwhelmed by its attempts to encompass the effects of Goyder's activities.

Along with requests for leases, the government was being offered assistance with exploration. Captain Francis Cadell, a man well known to the colonists for his steamboat navigation of the Murray and whose experience in river navigation included work on the Amazon, had placed his services at the disposal of the state. 'Probably no man better adapted than Captain Cadell for an enterprise of this kind could be found,' the *Register* judged, since 'the interior ... bids fair to call for a certain amount of nautical experience to obtain complete acquaintance with it'.[37] By late August the same paper was sighing with content. 'There is truth in the old proverb – It never rains but it pours. Here we have been for years, longing for information respecting the character of the interior of the colony, and of the continent of which it forms so important a part, but longing in vain.' But now that Babbage had 'stumbled upon grassy slopes and limpid streams' and Goyder had 'discovered romantic scenery and an inland sea', it appeared likely 'that nothing will long remain secret'.[38]

Fata Morgana

It is to the credit of Goyder's honesty that nothing did remain secret for long. On 16 September, a letter from him appeared in the correspondence section of the *Register*, together with a letter he had attached. The second letter, dated 9 September, was from a pastoralist in the north and its message was desperately simple. The lake was drying up – in fact it was 'now nearly dry ... '[39] There was obviously cause for consternation, although the *Register*, suddenly less eager, did little to report it. Finally, in late September, news from Freeling, still on his way back from the lake, was to hand. Conveying it to the public, the *Register* announced that:

> we may no longer say that 'seeing is believing'. Mr. Goyder says he saw trees where Captain Freeling says there are no trees, cliffs where no cliffs exist, vegetation where all is sterility, and water where it is barren sand; and waters that were real are now, it is said, to a great extent dried-up. It is all *mirage*! So, at least, says Captain Freeling.[40]

In his letter of report Freeling explained how, after a great deal of effort, the precious boat had been lugged to the lake's edge. A party had walked over one-and-a-half kilometres through the mud, sinking up to their calves, before a few centimetres of water were found. The boat had been dragged out to it, but it could not be floated.[41]

While Freeling was away, and after the bad news had arrived, Goyder had applied to the commissioner (no longer Bonney) to be allowed to go north again himself. He acknowledged that he had been proved wrong and that the waters he had seen had been 'the mere accumulation of flood-waters from heavy rains falling in March last', but pleaded with the commissioner for an opportunity to

MR. GOYDER'S DISCOVERIES.

Early Settlers taking possession of a run, in the magnificent country discovered by Mr. Goyder, within the bight of Lake Torrens.

Cartoon from *Melbourne Punch*, 8 October 1857. [SLV]

set the matter straight himself, because he feared:

> from the tone assumed by the public journals, that an erroneous impression had been conveyed in my report, of the value, for pastoral purposes, of the country traversed by my party. I have carefully examined the statements it contains, and feel assured that, with the single exception of the permanency of the waters of the lake, its correctness may strictly be relied upon.[42]

To support his claim to have discovered fertile pastoral land with a sufficient rainfall, he pointed out that James Trebilcock, the settler who had accompanied him, had applied for a lease of about 80 square miles (over 200 square kilometres) of this land. But Freeling had failed to do anything more than march to the spot where Goyder had seen the great lake and march back again, mouthing the usual litany of disgust and horror at the country he had seen. Goyder's frustration was enormous. Regardless of the ultimate nature of the lake, he wanted to open up pastoral land, which he was convinced existed in the north-west, and he pleaded to be permitted to be allowed to investigate. Probably because he was used to getting his own way, Goyder assured the new commissioner that he felt confident of Freeling's cordial support.

When the matter was referred to Freeling, however, the surveyor general responded testily that if he had wanted the country in the north-west explored, he would have done it himself.[43] His letter implied that the prospect of a fertile area in this part of the country was no more than wishful thinking on the part of his assistant. Although Freeling had not circumnavigated Lake Torrens, he considered that there was 'a sufficiently connected chain of evidence to prove that the eastern and western wings of the lake are connected, and the lake itself is in the form of a horse-shoe'.[44] It was not possible to explore further in the north, and in any case, Goyder's presence in the office was indispensable.

Eyre's complaint about the unreliability of visual perception in the enchanted region of the lake was proving to be truer than anyone had suspected. Freeling's experiences certainly negated Goyder's claims, but if Goyder's seeing did not lead to believing, then neither did Freeling's. It was not just that, in reasserting the established view of the inland put forward by Eyre and Sturt, Freeling was the bearer of unwelcome news. The community had become restless, suspicious. There was a mystery, and Freeling had failed to solve it. The same *Register* article that had carried Freeling's report claiming that the water was 'all *mirage!*' went on to point out that Freeling had been well equipped, at the public expense. It therefore expected: 'that he would have struck out in other directions, for we cannot suppose the mirage occupies the whole interior'.[45] According to one correspondent on the letters page, the prevailing opinion was that Freeling's expedition had been very unsatisfactory, and it was understood that Bonney (probably the colony's most authoritative voice on matters connected with the land), Torrens and the commissioner were not pleased.[46] Governor MacDonnell was also dissatisfied, and in a dispatch reported that Freeling had not escaped official censure, complaining that he had not determined the limits of the lake and had added nothing to their knowledge.[47] In the six months preceding the colony's coming-of-age celebrations in early December everything had changed – and then everything had stayed unhappily the same. Another expedition was to be sent into the north and north-west, this time to be led by Babbage, who would attempt to skirt the lake.

Freeling's news had caused confusion among Adelaide dwellers and excitable newspaper editorialists, but it had not put a halt to the flow of pastoral claims, nor had it prevented the resurgence of pastorally inspired exploration in the region. Within a short period of time other parties were out looking for good grazing country, two of them led by pastoralists themselves. Regardless of the reports of Eyre and Sturt and Frome and Freeling, it was a fact that pastoralism was continuing to spread in the north and pastoralists were seeking more of this supposedly impossible country. They had discovered that salt bush made good grazing (although they were keeping this knowledge to themselves) and their behaviour lent credibility to some of Goyder's claims.

Apart from the expansion of pastoralism, Goyder's investigations had two other lasting impacts, one being the demise of the great horseshoe lake. From the time of his discoveries through to the end of 1858, the area was crowded with explorers, and eventually gaps were found. The horseshoe lake was officially gazetted out of existence at the end of 1862, to the surprise of the journalist Frederick Sinnett, who had worked as surveyor in its vicinity. Sinnett declared himself 'incredulous' on hearing of the 'non-existence of a supposed old acquaintance ...'[48] The name 'Lake Torrens' was retained by the first salt lake Eyre had seen, on the western side of the basin. When Goyder had contemplated his freshwater lake, he would have been standing on the southern shore of the western reaches of what is now Lake Blanche, named, like Blanchewater, after Governor MacDonnell's wife.

The intense delight that the Flinders Ranges had evoked in Goyder effected another lasting change. Until he described the ranges as the most inspiring landscape he had ever seen, and keyed them into the aesthetic culture of the day with his use of the terms 'romantic' and 'picturesque', they had only existed in the public imagination as Eyre's 'stepping stones', rocky outcrops containing waterholes and pockets of grass forming a pathway into a desert. It is Goyder's view that now prevails. The place these ranges hold in Australian culture as a landscape of compelling magnificence was established by later visual artists, in particular Hans Heysen, who first visited them in 1926 and in the following years produced images that startled urban Australians, most of whom had never seen the arid regions.[49] While the little sketches in Goyder's notebooks are part of the visual record of the Flinders Ranges and are the earliest made of a number of topographical features (according to Hans Mincham), they played no role in forming the now-pervasive impression. However, long before Heysen had been alerted by stories of scenery that would 'knock your eyes out', Goyder's unexpected and welcome 'discovery' had reshaped attitudes in the small world that Adelaide was at the time.[50]

Resolution and recollection

Goyder's discoveries in the north were the startling – and inauspicious – means by which he entered public life. Later accounts have remarked that he was severely criticised for his mistake, but evidence of public criticism is not easy to find. Instead the incident seems to have been followed by a silence that assumed his wounds were already so large and so painful that salt was not required. Even in an account of Freeling's career published in the *Public Service Review* in 1900, when this incident was described, no mention was made of Goyder.[51] The legacy of this silence is that the story was never revisited and re-evaluated, and throughout the twentieth century it was still presented in the simple terms in which it was first understood: Goyder was wrong about the inland sea, while Freeling was right about the 'dead heart', as it was later termed.

Nearly three-and-a-half years after encountering the flooded lake, and with more experience of the country round Lake Eyre informing his judgment, including the opportunity to see a drought broken by flood, Goyder made a frank admission of error and an apology in the process of reporting this dramatic change:

> I cannot conceive how I could possibly arrive at the conclusion, that the waters of the lake, north of Blanchewater, as seen by me in June '57, were other than flood-waters, or that they were permanent; and such an inference could only arise from erroneous premises and utter inexperience of the country; and I deeply regret being led, by the novelty of the discovery of an apparent sea of fresh water, where salt only was supposed to exist, into the line of argument then adopted: but such was my conviction at the time that nothing short of actual demonstration would have convinced me to the contrary.[52]

Goyder's criticism of himself was unnecessarily harsh (although, at the time, such harshness might have been necessary). Now that flooding in the inland is well known and documented, it is the very extent of the transformation that we marvel at. Goyder himself later admitted that even the most experienced person could be deceived.[53] In 1857, as one of the first small party of Europeans to witness the state of the country after inundation, he had stood no real chance of working out what the situation was. The first major test of his ability to examine and interpret a landscape had been one he was bound to fail.

Only a few years later he was to receive some support, although in a way that must have been at least a little galling, especially if he had harboured a youthful yearning for discovery. Julian Tenison-Woods, the Catholic pastor of Penola in the South-East and a writer in the field of the earth and life sciences, had become interested in the problem of water in the interior.[54] Tenison-Woods's interest had been aroused by Goyder's report of 1857. In 1864 he produced a paper in which he began to put together some pieces of the puzzle. He contended that the interior of the continent was wetter than was generally supposed, attributing this to its rivers undergoing rapid change, which explained why one person might see the country flooded and another as a barren desert. He also recognised that, in relation to water and drainage, a key feature of the inland was its utter flatness.[55] The following year Tenison-Woods followed up this paper with the publication of a book on the history of the discovery and exploration of Australia. Goyder's 'discoveries' in the north earned him six pages of attention, and the ultimate assessment:

The whole thing is an amusing episode in Australian exploration – a subject, which, amid its usual sombre horrors of suffering and privation, affords few such sunlit passages.[56]

In fact Tenison-Woods, far from being dismissive, had attempted a sympathetic approach to Goyder's situation, choosing to deduce from the buoyant enthusiasm of Goyder's report an objective indication of the extent of transformation in the country described. Tenison-Woods confirmed that the public in general had arrived at an angry verdict, 'in its wrath making probably too little excuse for Mr. Goyder', and concluded that it remained to be asked 'how are we going to account for the water in such quantity and so fresh …?'[57] This trust in the basic accuracy of his perceptions must have been welcome, but it can hardly have cheered Goyder to discover that, while he had indeed gained a place in the history of Australian exploration, it was by providing comic relief. Nevertheless, his grand-daughter identified the book as occupying a valued place in his office library.[58]

Unhappily for Goyder, who never admitted to being bewitched by mirage, Babbage claimed that, during his own expedition in 1858, he had seen the same mirage – a 'beautiful sheet of water, with headlands and cliffs' – that had misled the deputy surveyor general.[59] Goyder may well have been deceived about the cliffs and headlands of the opposite shore of his lake, as Freeling reported, but he had stood on its shores and seen that the water was real. However, as late as 1875, Babbage was still claiming publicly that Goyder had been misled by mirage.[60]

In 1883, in a report on water conservation, Goyder was able to express a mature understanding, not only of what he had seen, but of his status as observer:

> In 1857 the country from Mount Hopeless to the north boundary of the province, as well as the plains on either side, were mostly inundated. The waters referred to were, for the first time, seen by me whilst initiating a trigonometrical survey in the locality. Owing to inexperience, at the time they were supposed by me to be permanent; by others they were attributed to mirage; but subsequent experience and accurate geographical information has shown the region to abound in large lakes and lagoons of fresh water … [and] immense quantities of flood waters which it is known periodically inundate that country by the flooding of … rivers leading into and discharging their waters into that area.[61]

His vindication – as far as it could go – came not long after. In 1888, Andrew

Garran, who was well known at the time as the editor of the *Sydney Morning Herald* but who had also been an editor of the *Register* during the mid-1850s, published the two-volume *Picturesque Atlas of Australasia*. Together with a small portrait, it included a friendly account of Goyder's exploration, in which he was credited with finding a new path through the Flinders Ranges. The *Atlas* explained that Goyder:

> was so fortunate, or unfortunate, as to visit Lake Torrens at an exceptional time ... Mr. Goyder was not to blame for the accident of the season, though he greatly regretted the error into which he was thereby led. In 1887, through floods from the north-east, the condition of things he found was fully repeated, and there was abundance of water everywhere.[62]

It was now demonstrated that the water, all of it, had been real. But with the barrier of the horseshoe lake removed and the mystery of what lay beyond it solved, the subject was no longer a matter of fascination. No effort appears to have been made to reflect on what Goyder's experience as an explorer really meant and to modify the story accordingly: as an investigator of the continent he continued to be little more than someone who had mistakenly believed he had discovered a major topographical feature. Ironically, he had been no more mistaken than most of his contemporaries who had chosen to place their faith in the impassable lake, but the widespread willingness to dismiss his experience as largely illusory blocked public acceptance of the apparent contradiction of a freshwater lake in a region of desert and giant saltpans, and the questions that would have inevitably flowed.

CHAPTER FOUR
Systematic observation

Despite Freeling's determination to keep his deputy out of the north and in the office, the demands of Goyder's position meant that he continued to move about. In August and September of 1857 – while Freeling had been struggling through the Flinders Ranges – Goyder had made trips to Port Elliot, on the eastern side of the Fleurieu Peninsula; to Blanchetown, on the Murray; and to Truro, beyond the Barossa Valley. He had also visited the copper town of Kooringa (Burra), travelling on to the World's End and Murray Flats runs, in country he would eventually place outside the north-eastern shoulder of his line. In the following year he made six trips to Kapunda, the first copper mining centre in the colony, and three of these also extended up into the Kooringa-Burra area. In late 1858, he made six visits to the survey in progress at Myponga on the Fleurieu Peninsula.

It was not long before his mind returned to the problems of travelling in the north. One Wednesday afternoon early in March 1859, he left the office and took himself to the beach to test filtration equipment, apparently of his own design, for removing salt from water. This was a common concern among northern expeditioners: the technically minded Babbage became preoccupied with it. Goyder's equipment consisted of nothing more than a long pipe, full of sand, set up on crossbars to provide a gradient. He tested it again 10 days later but the water he collected at the bottom remained as stubbornly salty as when he had poured it in at the top, and the investigation was abandoned.[1] Later that month he left Adelaide on the journey that had probably inspired the experiment, a long trek up to Nepabunna, in the North Flinders Ranges, where mineral discoveries had been made.[2]

While he was away, an unusual opportunity to develop his skills was taking shape in Adelaide. The government, still hoping for a gold rush in their own colony, had invited the Victorian Government geologist, later the director of the Geological Survey of Victoria, Alfred Selwyn, to inspect parts of South Australia and to report on the prospects of finding gold, workable coalfields and permanent supplies of artesian water. Before taking up his post in Victoria, Selwyn had spent six years working for the Geological Survey of Great Britain. As a result, at a time when the conduct of scientific activity in general was moving away from the unsystematic investigations of enthusiastic amateurs, Selwyn was able to bring to Australia the first sharp taste of a dedicated and rigorously professional approach to the practice of geology. The South Australian Government was therefore borrowing the up-to-date knowledge and current methods that Victoria's mineral wealth had purchased and was employing their systematic geologist as a prospector.

Selwyn's inspection

The geologist arrived in Adelaide on Sunday 1 May 1859, keen not to waste any time. It was immediately decided that Goyder should be his guide, because, as Selwyn later related, 'from his intimate acquaintance with the districts to be examined he would be able to decide on the best course to pursue, at the same time to direct my attention to all points of interest on or near the route selected'.[3] Unfortunately, the proposed guide was still in the north, and no one knew when he could be expected to return.

Selwyn eventually left Adelaide with Goyder on Friday 20 May, riding out via the Tea Tree Creek to Millbrook, and then on to Mount Crawford, in the hills north-east of Adelaide. They remained in the area for about a week, journeying out several days in a row from a base at Pewsey Vale. There are outcrops of rock there that Goyder may have supposed to be of interest, but if so, they competed unsuccessfully with the Pewsey Vale winery for Selwyn's regard. The vineyards there had been established about a decade earlier by Joseph Gilbert. In his report to the government, Selwyn would write that the wines produced there were 'little, if at all inferior to the highest class of continental wine of a similar description', and an indicator of the direction in which the colony should look for its wealth.[4]

The overall plan for the journey, according to Goyder, was to traverse the roughly parallel lines of ranges, from Adelaide up to Mount Serle, zigzagging from west to east and back again and attempting to intersect 'the various formations and features' on the way.[5] From Pewsey Vale they worked north to Kapunda, and on past Anlaby, a station belonging to the prominent pastoralist Frederick Dutton, which seems to have been a regular stopping point for Goyder when travelling to the north-east. At Burra, Selwyn was piqued at being prevented from entering the mine because he lacked a written order from the directors. The pair travelled on through the Mid-North, oriented by means of the giant pastoral stations – Canowie, Booborowie, Bundaleer – until they reached Mount Remarkable at the northern end of the Mount Lofty Ranges. From there they travelled across to Port Augusta, where Selwyn began to consider the matter of artesian water, before heading back through the Flinders Ranges to the Willochra Plain.

Inevitably, they stopped at the Mud Hut, which another visitor from about that time recalled as a dust-begrimed, 'high dried', God-forsaken hole, the inhabitant of which could only offer heavily stewed tea because the water was 'so blanky bad'.[6] Selwyn had arrived during winter, when things would not have looked so 'high-dried', but nevertheless while the country was moving into drought. Writing officially, he only recorded that the course of the Willochra Creek was a wide stony channel inhabited by gum trees and containing occasional pools of water frequently so 'highly charged with salts, that it is quite unfit for domestic use'.[7] (The course of the creek has since changed, and the trees standing in its old bed are now weathered skeletons.)

From the Mud Hut they travelled on to Wilpena Pound. In the report of his journey to Lake Torrens two years earlier, Goyder had noted that there were differences between various lines of range and had speculated about the geological processes that had created them. In keeping with his romantic enthusiasm for the wildness and grandeur of the landscape, the 'perpendicular cliffs' of the 'amphitheatres' or pounds generated a vision of time when:

the earth was submerged and violently convulsed by earthquakes, acting over an immense area, and from various centres, causing the stratified rocks to separate and sink under the superincumbent mass of water into chasms beneath – while the outer portions were elevated to their present position ...[8]

This cataclysmic vision was clearly influenced by a geological theory known as 'catastrophism'. The debate over catastrophism had been at its height during Goyder's childhood, but the victory had finally gone to the opposing 'uniformitarian' approach of Charles Lyell, who had argued that the present state of the earth was the result of a constant series of changes, wrought by the same processes continuously at work. Catastrophism, on the other hand, divided the history of the earth into two phases. In the earlier one, processes were far-reaching and violent – 'universal' and catastrophic – and, in an early form of the theory, took place when the whole earth was covered in water.

The two examined Wilpena Pound on 11 June. Selwyn was as fascinated and impressed as Goyder, who must have revelled in taking on the proprietary role of host to a companion as passionate and curious as himself. Selwyn was only two years older than Goyder, and, like him, was boundlessly energetic and devoted to his work, expecting his assistants to give as much in the way of effort and dedication as he gave himself. In his account of the pound, the geologist confirmed the aesthetic assessment of the Flinders Ranges that Goyder had brought back to the surprised people of Adelaide. 'Any description I can give of this most remarkable piece of country,' Selwyn later reported, 'will fail to give an adequate idea of it – its singularity and picturesque appearance far surpassing anything I had previously seen in Australia'. Selwyn's recognition of his inability to express the magnificence of this landscape was correct: he described the pound simply as a large irregularly shaped oval area of flat or gently undulating ground, covered with grass, picturesquely timbered with clumps of pine and gum, 'and altogether presenting a most beautiful park-like scenery'.[9] In a paradoxical reversal of roles, it was Selwyn, the geologist, who responded to the pound in culturally familiar terms, while Goyder, the surveyor, had contemplated the walls of the pound and envisioned mighty events and vast geological forces.

Selwyn produced a number of drawings of the pound and areas in the ranges, one acknowledged as from a sketch by Goyder. The pair discussed the matter of the formation of the pound at length – Goyder later gratefully acknowledged Selwyn's 'liberal communications' on the subject – and Selwyn evidently succeeded in converting Goyder to his view that it had been formed by the buckling of sedimentary layers. Later Goyder would publicly repudiate the 'absurd … suppositions' about earthquakes and chasms that he had earlier put forward. His opinion had been changed, he explained, by his discussions with Selwyn, 'coupled with a little reading on the subject' – a wry way of acknowledging that his former geological notions had been outdated.[10]

After leaving Wilpena Pound, the pair travelled further north ending up at their most north-easterly point, about 50 kilometres east of Mount Serle, on the edge of the plains beside Lake Frome (or, as it was still known to them, Lake Torrens). Selwyn wanted to examine the rocks on the eastern side of the basin of the salt lake and the pair decided to cross the basin somewhere below the lake. They had to go for days without finding water, and the going must have been hard for Selwyn, who, although he was used to travelling, camping, and working out of doors, was also used to a moister climate and staff to mind his camp and horses. Selwyn was also sick and in intense pain, having nicked a finger skinning a kangaroo. The wound had infected and the infection had spread. From 23 June, they had had to ride south as fast as possible so that Selwyn could get medical advice at Burra. Selwyn was later to 'beg to express my grateful sense of the very cordial, efficient, and valuable cooperation afforded me by Mr. Goyder'.[11] This cordiality must have included a considerable effort to take care of Selwyn and to carry out his share of work about the camp. The pair reached Burra on 29 June, and arrived in Adelaide the following evening. Selwyn wrote a brief report, giving his essentially negative results and promising a longer document later.[12]

Reports

The year before, in September 1858, the South Australian Society of Architects, Engineers and Surveyors had been established, with just over 20 members, including Freeling and Goyder. After returning, Goyder attended the July meeting and promised the members a paper on colonial surveying later in the year. In the meantime, he shared the more immediate fruits of their journey together. At the August meeting he brought along the rock specimens he had collected and gave a 'very clear and lucid discourse' on the geological features of the country they had travelled, based, according to the record, on the content of his journal, but no doubt informed by knowledge gained from Selwyn. Goyder's unflagging curiosity and enthusiasm for his subject ensured that his 'discourse' was a success: the specimens, which he had presented to the association, were handed around and an 'interesting conversation followed, in which nearly the whole of the meeting took part'.[13]

When Selwyn finally produced his report, in December 1859, the conclusions he reached were unusual for a geologist. Having found no evidence of useful deposits of the gold, underground water or coal he had been asked to report upon (they had been on the opposite side of the ranges from the coal deposits at Leigh Creek), he sweetened the pill by congratulating South Australia instead on her 'many other great natural resources on which she may safely rely for future prosperity' – her iron ore, her copper and lead mines, her vineyards and her 'corn-fields'. (Selwyn was using the word 'corn' in the older English sense of grain.) On the subject of growing wheat, his comments were prophetically pertinent. Selwyn pointed out that it was a mistake to regard the soils of the northern plains as uniformly poor and sandy. The soil, he explained, had the potential to be 'exceedingly fine'. The problem was lack of moisture.[14]

Selwyn has been credited with having introduced a systematic approach to the study of geology in Australia, and with having left a generation of geologists trained in his methods and inspired by his ideals. Although not a geologist, Goyder had shown an interest in geology before Selwyn's visit – in 1857, when he was appealing for permission to return to the exploration of the north-west, he had offered to make observations of the rocks encountered – and Selwyn evidently influenced him in the same way.[15] As well as prompting Goyder to revise his views about the formation of structures in the Flinders Ranges, in the course of their travels Selwyn had imparted the method of recording systematic geological observations. Goyder's work in the next couple of years would show that he had been an enthusiastic student.

Colonial surveying

The planned paper on colonial surveying was never presented to Goyder's professional colleagues because what would have constituted its contents were put before the surveyor general and the government as a proposal for a major work of triangulation in the country south of Lake Eyre.

In 1858, when Babbage had been sent north to try to determine the real nature of the northern country, he had been misled by the high-minded public rhetoric about advancing geographical knowledge and scientific under-standing, and had failed to grasp that the reason money was being put up for the endeavour was the hope of locating good pastoral country. Encumbered by a mass of equipment, Babbage had moved slowly and devoted his time to charting the shores of what was still known as Lake Torrens. Eventually Major Peter Warburton, the commissioner of police, was sent to retrieve him. Meanwhile, Samuel Parry, a private surveyor, had made an 18-day excursion into the country north-west of Mount Serle on behalf of the government, and John McDouall Stuart had undertaken a lightly equipped, fast-paced investiga-tion of the country south of the lake for pastoralists, the Chambers brothers. Parry, Stuart and Warburton all reported finding grasslands.

In response, the government decided to continue the triangulations north of Mount Serle begun in 1857 and appointed Samuel Parry to carry out the work on contract. After a difficult beginning – the drought of 1859 meant that his horses had no food or water – Parry was lured away to Victoria by the offer of better pay.[16] For Goyder this was an opportunity to undertake a major work as well as to satisfy himself about the nature of the northern country. It was also an opportunity to increase his unsatisfactory income. When he had become chief clerk in 1853 on a salary of £300, he had been enjoying a temporary increase in salary of 50 per cent awarded to all public servants in the hope of financially lashing them to the mast against the siren call of the Victorian goldfields. This increase was eventually reduced and then abolished. For Goyder, this meant that for a brief period as first assistant, at the end of 1855, he had been on a salary of £600 per annum, although this was soon reduced to £500. From the beginning of 1858 it was cut back to £450.[17] From this he was expected to pay his own travelling expenses. But while his income had shrunk, his responsibili-ties had increased. Another daughter, Mary Ellen, had been born at the begin-ning of April 1858 and she had been followed by Isabella Agnes in August of the following year. Frances and Ellen now nurtured a flock of five.

Goyder wrote to Freeling with a detailed plan, proposing to take over the northern triangulations, and – politely – threatening to resign if his proposal was not accepted. There was nothing in the slightest degree penitential about this; Goyder was not attempting to make up for his earlier mistake, as his current entry in the *Australian Dictionary of Biography* claims (apparently on the assumption that he had been sent to triangulate the north in response to the request he had made in 1857, soon after it was reported that the water he had seen was drying up). Painstakingly conscientious as he was, Goyder never appears to have been afflicted by shame or guilt over mistakes, or accusations of mistake. His father's beliefs about child-raising had obviously served him well. If he believed the error to be real, he simply admitted it and moved on. Even while the degree of detail in his proposal indicated how deeply he was absorbed by the prospect of working in the remote northern country, in writing to Freeling he made plain that his immediate motivation was financial. The needs of his large family were pressing and he could earn up to twice his salary in private practice: 'I should be clearly standing in my own light,' he advised Freeling, 'were I not to take advantage of it …' But even though he was dissatisfied that his efforts to improve the operations of the field survey had not been rewarded, Goyder confessed that he did not want to leave the public service. He had 'a real interest in the service of the Government …' [18]

The plan as proposed was to travel in the direction of good country already located, producing accurate maps and linking the triangulated areas with lines which would follow the grid of meridians of longitude and parallels of latitude. These would be marked on the ground with mile posts. The party would include experienced well-sinkers, who would tap underground water where surface water was not available so that stock could later be brought over the waterless parts to the pastoral country. As determined to find grazing country as any pastoralist, he even noted that:

> It is also of importance that the camps of the surveyor should be at no great distance apart, as the tracks made by the drays of his party would probably become the highway in the direction where settlement was practicable and such camps would form the halting places of parties driving stock or travelling over the known country in search of new.[19]

The proposal included a detailed plan of the observations to be made and the way in which they were to be recorded in a log book. There were columns not only for readings from the theodolite, but for the outline of hills (to aid in identification) and for geological observations, which it was specified should provide sufficient information to enable a geologist to construct a map. An example was provided.[20] The report that Goyder finally prepared when the expedition approached the completion of its tasks contained detailed geological observations of a kind new to his writing, and he acknowledged Selwyn's role in making them possible.[21] A collection of astronomical observations was to be maintained so that the surveyor general could check all calculations and determine the longitude of the principal stations with reasonable accuracy. This book was also to show variations of the compass, readings of the thermometer and barometer, and the temperature at which water boiled (to determine altitude). In addition to these books of observations, the surveyor was to maintain an ordinary field notebook. As usual, the key to his plan was a preliminary survey – 'a brief but careful examination' – which would guide the direction in which the triangulation would be taken.[22]

The triangulation

The proposal was accepted, and Goyder set off on 17 October 1859. In order to do this, he had had to mortgage all his possessions. He appointed William Wadham as agent to draw his salary while he was away and to meet the interest payments and provide money for Frances. Wadham was a surveyor who had been employed by Chauncy and who later joined a prominent land agent, with whom he eventually formed the company Green and Wadham.[23]

In the hope of consolidating his financial situation, Goyder had arranged with three or four surveyors that while he was away they would send reports on small sections of land – the same reports on the value of lands that were sent to the government – to Frances, so that if something appropriate came up, Wadham could bid on Goyder's behalf. It was not a wise move. According to Goyder, Freeling had opened one of these letters, which obviously went via the office, and had attached a memo to it stating that he objected. Frances sent the memo on to Goyder, who asked the surveyors to stop sending reports to Frances and wrote back to Freeling, saying that he had mentioned his intention of doing this and had not been advised to refrain. No land had been purchased, but the incident was later used by his enemies to suggest that he had been using his access to information to benefit himself and his friend.[24]

On their way north Goyder and his party began to triangulate from the latitude of the Camel's Hump up to Mount Arden to connect the northern surveys with the original triangulation of the province begun at Mount Lofty (near Adelaide) and extending to Mount Remarkable.[25] (Freeling had wanted this done in 1857, but the governor had rejected his proposal, meaning they had had to make do with running a line to connect the surveys.[26]) On their way the surveyors triangulated 8400 square kilometres in the settled districts. As Goyder had anticipated in his plan, a line had to be laid down, running 20 miles true north of Termination Hill to Twenty Mile Hill, because the distance from hills with trigonometrical stations was too great. The line was marked by pickets roughly every 200 metres, either side of which the line was trenched for nearly a metre, and by mile posts.[27]

Drought

By going north at the end of 1859, Goyder had travelled into the end of a severe drought – the worst drought the colonists had yet experienced. As with Parry's expedition, the brunt of these conditions was borne by the horses and Goyder's field notebooks constantly note their condition and incidents concerning them – the horses were a major preoccupation. Goyder obviously felt for the animals, but he was also keenly aware that their own lives depended on them.

Goyder and his party began by making a reconnaissance of country in the Far North. Returning to Port Augusta, they found themselves crossing the Willochra Plain in the middle of a dust storm – 'a hot blast + clouds of dust', as Goyder described it in his notebook.[28] When they reached the Mud Hut, Goyder drew water from the well and had just enough for the horses when the dust cleared briefly. The hut's occupant, spotting him through the window, struggled out to prevent the party from using the water because it was needed for sheep. With access to the water refused, the most distressed of the horses, Waterman (ironically), was given all that was left in the men's canteens, but collapsed a short while later and had to be left behind with one of the men. The party had to travel a further 16 kilometres to a hut with a well where the horses could be watered. Waterman was brought up an hour after the rest had arrived, and was given more to drink and the last of the bread.

The incident prompted an anguished letter to Freeling. 'It would be impossible to convey to you the idea of the day and I bitterly felt the want of some accommodation for the poor brutes ...', Goyder lamented, going on to 'respectfully urge' upon his superior 'the propriety of surveying allotments at this inhospitable abode with the least possible delay'.[29] Freeling acted promptly, and the Ragless brothers lost 17 square kilometres of land next to the Mud Hut from their lease, but Goyder's appeal had only hastened things along. The location had been included in an earlier list of possible township sites, and had already been selected.[30] An inn was established there, now commemorated by a cairn, but no township ever eventuated. Goyder (and Freeling) were not alone in their heartfelt response to the plight of the horses. Nearly two decades later, Goyder's deputy, Willliam Christie Gosse, would curtail exploration in waterless country because 'the suffering of the poor brutes was more than he could stand', but this attitude toward working animals was far from universal.[31] Managing horses through days without water in hard conditions was a skill that had had to be learnt in Australia as part of the overall development of bushcraft. It required a great capacity to intuit how much a horse had in reserve, if anything. An entry in Goyder's notebook, in which he recorded that he thought it 'prudent' to turn back to camp because his thirsty horses would not eat, shows him exercising that intuition. The next day these animals were too tired to stand.[32]

This incident had occurred as Goyder worked his way around the shores of Lake Eyre, where parts of the countryside were so arid that five attempts to examine it had to be made 'before anything like a satisfactory result could be obtained – the party returning each time with some of the horses knocked up from want of water'.[33] He was forced by the rough and waterless conditions to abandon his investigation of the area between Lake Florence and the mouth of the Frome, something he regretted, since he had been determined to ascertain the extent and connections of Lake Eyre. There were problems with the horses throughout the summer. At the end of December, in intense heat, one was so weak that it fell into a gully and became wedged between rocks. It was pulled out and put back on its feet – supported on all sides in the process – but during the following night it collapsed again and rolled into another gully. It was retrieved once more, badly cut. Goyder nursed it to health.[34] At the end of January he recorded sending water to Captain, a young draughthorse, but 'the poor beast would not rise', or not immediately, at least.[35] Captain did survive. The labours of another of the party's draughthorses, Punch, are apparently commemorated by a place to the north of Leigh Creek named 'Punchs Rest'. The Survey Department employed horse keepers, but Goyder became increasingly involved in the care and management of these important animals. David Goyder's autobiography contains a story in which his inability to ride a horse is central, and since the family never had the means to keep one, knowledge of horses and skill in their handling were not capabilities that George Goyder is likely to have brought with him to Australia. Nevertheless, by 1860 he had enough knowledge and experience to take on the shoeing of a horse, and enough courage and self-confidence to make Punch, almost certainly a massive and powerful animal, the subject of his first attempt. (One of the strongest of the English draught breeds is named the Suffolk Punch.) Later incidents demonstrated that horses responded to him with trust, and despite the years he spent largely in the saddle, there is no mention of his ever sustaining an injury, or even having an accident, when riding or handling them.

Afloat

By September 1860, when he wrote a major report from Chambers Creek (now Stuart Creek), a further 10,300 square kilometres had been triangulated in the north, 16,615 square kilometres had been sub-divided into runs (half of this land had already been claimed) and another 7770 square kilometres had been examined prior to a more detailed survey. Plans of the runs were sent to the Survey Office so that the leases could be auctioned. It was a major work of exploration, but exploration in the form of detailed examination of the country rather than the usually linear movement that constituted the discovery work of the major explorers. As a consequence it has received little recognition. An early writer on the history of colonial exploration in South Australia, Bessie Threadgill, included these triangulations in her study, but with the rider that Goyder's work: 'was rather supplementary to, and critical of, that done by his predecessors, than a distinct exploratory performance of his own'.[36] Two works on Lake Eyre, published in the mid-1980s, one by a government department, provide histories of non-Indigenous activity in the area which appear to have been written in complete ignorance of the fact that Goyder ever ventured there.[37]

Goyder's real discovery was something he was cautious about reporting. In 1857, he had seen the country south of Lake Blanche after heavy rain and had not recognised that the lush growth he saw about him was ephemeral and that in time the country would revert to desert. In 1859 he had reached the south of Lake Eyre South at the end of a drought and, while he was there, the drought was broken by flooding rain. He was able to watch the transformation of the country as it occurred: 'In the beginning of December the country was a red stony desert. In January the country was teeming with valuable succulent vegetation.'[38] Walking on the shores of Lake Eyre, he found that 'a vast bay of salt water was before me, with hundreds of pelicans, swans, ducks, and other acquatic [sic] birds on the surface of the water, which rippled into miniature waves with every passing breeze'.[39]

This experience was later suitably commemorated by J.W. Lewis. In February 1875, Lewis was sent out under Goyder's direction on another expedition to examine Lake Eyre. When he reached an area to the north-east of the top of Lake Eyre North on the Diamantina River – much further north than Goyder was ever to travel – he too found a lagoon 'covered with wild fowl, ibis, spoonbills, and every known kind of duck in large numbers'.[40] The lagoon, of course, is ephemeral, alternating between a habitat for thousands of water birds and sandy non-existence. By calling it 'Goyders Lagoon', Lewis executed the routine courtesy of naming a discovery after a superior, but here with exquisite if unintentional appositeness, such that Goyder's name was connected not just to landscape, but also to the wildly variable rainfall that would shape his career, just as it shapes the patterns of life across the continent.

But in 1860, not surprisingly, Goyder was on his guard when he reported what he had seen. Writing in October, he admitted that he had been mistaken in 1857, and assured the lands commissioner that: 'I have paid no heed to first impressions, but have waited patiently [for] the result of a second examination of the country and waters seen during the journey'. By doing so, he saw the waters recede, turning brackish, or salty. The Frome 'contained fifteen miles of brine of a lurid red …'[41] In these conditions Goyder succeeded, at least partially, where Sturt and Freeling had failed: he was able to sail up to, if not on, the 'inland sea' – very appropriately for a man who bore the doubtful distinction of having discovered it. He was determined to test the 'extreme depth' of the waters of Lake Eyre:

> I had a rude boat of hide constructed during my absence from camp, and launched this in the Margaret, at the Walgarina Springs, and sailed with two men towards the lake, the boat – containing bedding, provisions, water, and guns – drawing but four inches. On reaching the spot previously visited by me, and where a stone had been placed at the then high water mark, I found that a fall of but six inches had taken place; and as there was still the appearance of water in the lake, from an adjacent cliff, I entertained considerable hopes of success. A few miles, however, brought us to the end of our journey – the water, which had varied in depth from four inches to six feet, gradually shallowed until, by a sudden turn, we found ourselves aground at the end of the creek and margin of the lake, which was perfectly dry. Six inches had done the whole business, and converted the apparent sea into a vast bed of mud, or rather of dry loam encrusted with salt, and surrounded by a muddy margin …[42]

After passing this muddy margin, they walked several kilometres into the lake 'on ground that horses might travel'. The ground became firmer as they advanced: 'the reaches and arms near the margin are more boggy and dangerous than the interior'. At the highest watermark Goyder gathered a handful of shells and sand, which he took to be the result of floods from the north, since he had seen none like them in Chambers Creek, to the south. He forwarded them to Adelaide. (He had also forwarded what he 'supposed to be fossils' found in some slate, preferring, in his new mode of geological humility, to submit them 'to those competent to judge, rather than saying anything of them, for it is more than probable that I may be mistaken …'[43]) In the 'salt

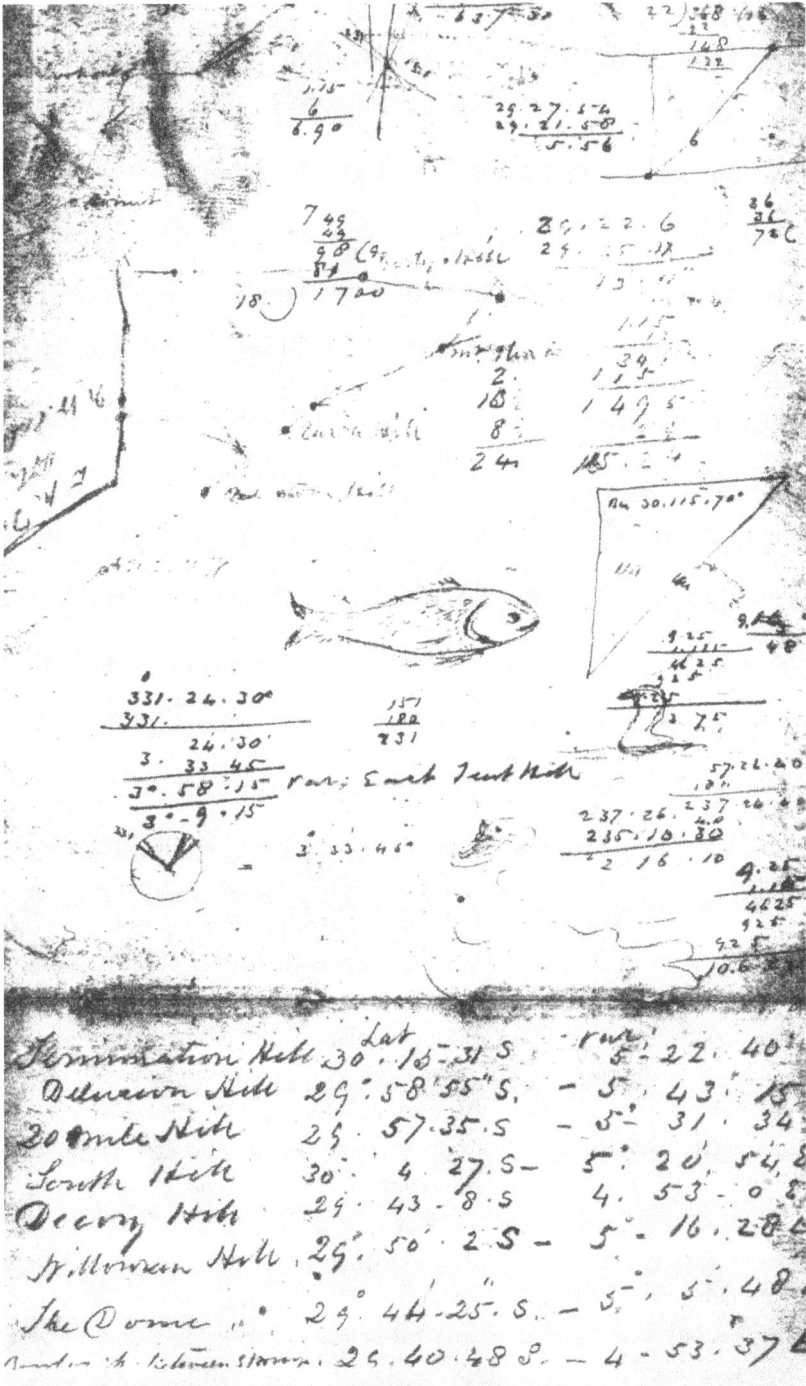

Goyder's sketch of a fish on the endpaper of one of his field notebooks for 1860. [SRSA]

reaches' of the larger creeks he reported seeing numerous small fish, from seven to 10 centimetres long. He sketched one inside the cover of his notebook, where it ended up swimming pop-eyed through a clutter of arithmetical and geometrical calculations.

The rain did not end the anguish Goyder felt at the plight of the horses. On 11 March 1860, in the area to the east of the Willouran Range, a horse named Angas became bogged. A team of men freed him but the animal was exhausted. This, Goyder complained to his notebook, was:

> the more vexatious he being one of our best horses – + there being only one spot of boggy ground about + that not more than a few feet square but his hind legs had got down in it + he must evidently have skinned himself strug-gling – This is the most untoward affair that has yet occurred vexing me more than all other mishaps since leaving town ...

The next day the 'poor brute' was 'quite dead'.[44] Angas was skinned and his hair and shoes removed. (The long hairs of a horse's mane and tail were a useful material.) In a report written a few days later Goyder vented at length the frustration and distress he'd had to keep to himself at the time: 'I could say nothing; no one was to blame ... there was nothing for it but to bear it in silence'.[45]

The country and the springs

In his final report, Goyder summarised the country he had travelled over into four types. Most of the area he had investigated was gibber plain, with grass and 'succulent herbage' in the hollows and clay-pans and saltbushes studded thinly over the rest. The next major area was the sandy country, where the vegetation was largely similar. The two other areas he identified were 'wide valleys or water-courses', full of fodder plants, the paths of which were marked by small eucalypts, acacia or mulga trees; and country 'entirely devoid of vegeta-tion suited to stock'. As prospective pastoral land, he summed the region up as country of 'red-stone plains, with one-fifth of the area covered by vegetation; tolerably grassed, sandy country; well-grassed valleys and water courses, and country liable to inundation by salt-water, or the soil of a character favourable only to production of such plants as flourish in its vicinity'. But he warned that his description 'only holds good, however, during the winter season ...' In summer, only the small trees and the saltbush survived in the intense heat. There was little else.[46]

Distinctive features of the region and of great interest to the pastoralists were the springs Goyder encountered: 'large bodies of springs ... the importance of

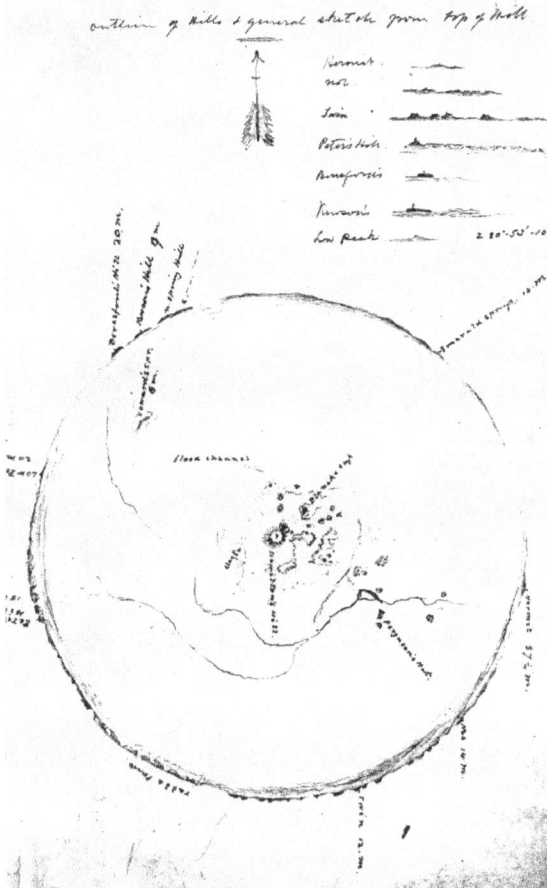

Outlines of hills and a sketch of surrounding country from the top of Hamilton's Hill, 26 July 1860. The Blanche Cup, a mound spring, is to the north-east. [SRSA]

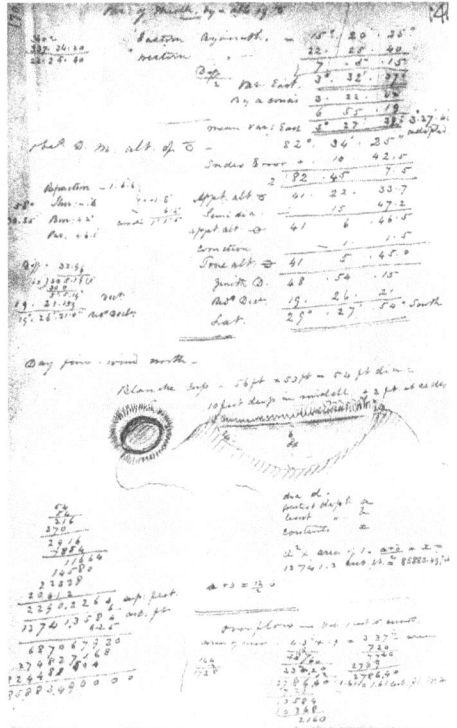

Goyder's calculations of the capacity and flow of the Blanche Cup, 26 July 1860. [SRSA]

which it is impossible at present to realize, or to award too much praise to their discoverers'.[47] Having said that, he was careful to attribute the discovery of springs to everyone but himself. He identified the groups of springs discovered by his party as 'the MacLachlan, Fred's, Gosse, Smith's, Murray's, Brackish, and one or two single springs. The first two are named after their discoverers,' he explained, 'the remainder after others of the party' (though not the Brackish Springs presumably).[48] Even so, Goyder was later acknowledged as the discoverer of the Frances and Ellen springs, as well as the Rocky and Reedy springs, which he had encountered in 1857.[49] The springs around Lake Eyre included salt- and freshwater springs, hot and cold springs, brackish and soft-water springs, springs flowing from calcified mounds and those flowing from

An illustration of the Blanche Cup from the *Australasian Sketcher* of 1883, accompanying material from Goyder's report of 1860. It shows the Overland Telegraph, completed in 1872, and the use of camels.
[SLSV]

mounds of mud. Goyder had taken wood with him specially for the purpose of constructing weirs at the springs to measure their flow. At a cooler, soft-water spring flowing from the same mound as the hot Emerald Springs (the Aboriginal name of which Goyder recorded as 'Durra-Durrina'), Babbage had previously obtained a flow measurement of the equivalent of 796,250 litres per day – a figure which had provoked mirth and derision in Adelaide.[50] Goyder obtained a measure equivalent to 20,023 litres per day but was considerate enough not to join in the attacks on Babbage, who had been made to look a fool by Warburton and who had become a ready target. Goyder reconciled the discrepancy 'by supposing the supply to be intermittent; which I am strongly inclined to believe is the case with all these springs'.[51] He had evidence of variable flow from other springs.

The survey
In his report, intended for a readership of parliamentarians and members of the public as well as the surveyor general, all Goyder recounted of the survey was that, where pastoral claims did not already exist, new runs had been drawn up in the simplest possible way. This had been done by running straight lines

between trigonometrical stations on the tops of hills. But he had been making his own discoveries. Frome had recognised the usefulness of triangulation alone in pastoral country and had insisted on the value of pre-survey examination and the recording of detailed observations of natural features. Goyder also recognised the importance of pre-survey examination and careful observation, and, influenced by Selwyn, had developed his own methods for colonial surveying. When he put his pre-survey examination routine into practice in the country south of Lake Eyre, he decided that his rough triangulation and pre-survey examination had overtaken more formal procedures in their capacity to provide a quick, cheap and adequately reliable map which could safely be used to determine the location of runs. An extraordinary amount of this work was done on foot. Early in the trip Goyder had walked with four men, leading three pack-horses, from Termination Hill to Chambers (Stuart) Creek, a distance of 120 kilometres as the crow flies – but the party had been setting up trigonometric stations on suitable hills and visiting springs on their way.[52]

Goyder presented the method he had evolved and its results a few years later when he was asked to provide advice as surveyor general to the leader of an expedition to survey a site in the Northern Territory. (The leader was B.T. Finniss, who, as colonial secretary, had recommended Goyder's entry into the Land Department in the first place.) Describing the pre-survey examination, Goyder directed that from the first station, the bearing of all prominent hills was to be observed with the theodolite and their outline shown in the observation book. The variation of the compass needles was to be carefully determined. After the hills had been named, the officer in charge and his assistant were to proceed to separate hills, noting 'the natural features and character of the country passed over, direction of water-courses & co. ...' When they arrived, the same observational routine was to be followed. The hills were also to be marked with stations, in the form of a pile of rocks with a pole showing the name of the hill. While the men were piling rocks, the surveyor was to 'sketch in his field-book a bird's eye view of the country surrounding the station', which would aid in constructing a detail map. Goyder was very proud of his creation:

This mode of survey has been so successfully adopted by me, and the information obtained whilst crossing from station to station for the purpose of observation and examining the country to obtain water for the supply of the party has been so ample, that I have been enabled to construct a map embracing 15,000 square miles of country during a two months journey, which showed discrepancy of not more than a mile between the extreme stations, a hundred and twenty miles apart, when compared with the

[handwritten field notes, largely illegible]

Water boils @ 211° ——— Wet Thm: 68°

Formation

[several lines of handwritten notes describing siliceous rocks of red stone plain with a little grass & salt bush — showing by denudation — white & red hydrates of silica — over light soil — veined with gypsum & impregnated with salt. also a good deal of ochreous iron about ... ferruginous sand stone. the hydrates of iron being aggregated like porphyry —]

[Outline of hills observed to.]

Hermit Rock
Little Hill

Cadnia

Rocky Point

cliff.

Mt. Nor' west

Potts' hill

Sandhill.

Outlines of hills in the country south of Lake Eyre, from Goyder's field notebook, Monday 20 August 1860. The hills shown include Cadnia and Mount Nor'west. [SRSA]

trigonometrical survey of the same locality, which, from the scarcity of water and other causes, required nine additional months to complete.[53]

The surprisingly detailed hand-drawn map (plate 13), which survives in the collection of the State Library of South Australia, bears out the reason for Goyder's pride. A map with such detail, he explained, would not only provide a 'satisfactory basis' for the sectional survey (if required), but in pastoral country it enabled runs to be defined and leased without fear of overlap, and even obviated the need 'for the measurement of base lines, and an expensive trigonometrical survey, until the country is in a better position to afford the expenditure'.[54]

When it came to providing names for the features encountered, Goyder continued with the approach he had quietly explored when the Frances and Ellen springs had been named. The two sisters appeared paired across the landscape, indicating how much Ellen was a part of the Goyder household, and how much home was on his mind. The 'Plan on the Country between Lake Torrens, Lake Eyre and Termination Hill', dated September 1860, shows two small lakes to the east of Lake Eyre South (one of them apparently connected to the main lake), named Lake Frances and Lake Ellen, while two creeks, 'The Frances' (as Goyder named it) and 'Nelly's Creek' (now Nelly Creek) flow side by side into the great lake's southern edge. ('Nelly' was a familiar version of Ellen (and other names) just as 'Fanny' was a familiar version of Frances.) Three daughters were commemorated by 'The Isa' (now Isa Creek), Gerty's Hill, and Lake Florence.

Jeopardy

Throughout the whole of this exercise, Goyder had been 'far from well'.[55] Some of his sufferings were the product of the extreme circumstances he'd experienced. In mid-August 1860, he and a party were in the general vicinity of Mount Nor'west in urgent need of water. Goyder had been out on his own and found his way back to camp in the evening by sighting the signal fire. There he:

> ate a little and had a pannican of tea – but too anxious to remain still on a/c of our poor horses; ascended the adjacent hills in hopes of seeing blacks fires – but none visible. horses short hobbled.[56]

The next day he 'started at grey dawn to look for Mirrabuckina', a place where he hoped to find water. He returned after walking about 27 kilometres, having found nothing but some of the horses, who had strayed in their own desperate search for something to drink. The others in the party had searched with an equal lack of result in other directions. To escape from this predicament, Goyder divided the party, sending the men off with the horses to Screech Owl Creek or to Cadnia (now known as Sliding Rock), where there would be water, with orders to give them a day's spell and then to return to collect him and a man named Tom, who was to stay with him. Goyder chose to stay himself because there was so little water left at the camp that he 'had not sufficient confidence to have two of the men alone for fear that efficient self-denial might not be exercised'. In any case his feet were 'skinned', and he was

Plan of the country between Lake Torrens Lake Eyre and Termination Hill, the map accompanying Goyder's report on the northern triangulations (SAPP 1860-61, no. 177). 'The Frances' (the name seems no longer to be in use) and Nelly Creek can be seen flowing into the bottom of Lake Eyre South, with Lake Ellen and Lake Frances to the north-east. Lake Florence, 'The Isa' and Gerty's Hill can be seen just below the title. [SLSA]

compelled to rest, and Tom was 'knocked up'. The next day, 16 August, Goyder
gave Tom a quart (just over a litre) of the remaining water, and Tom set off to
walk to some rocky holes where there had been water on previous occasions.
Goyder had been afraid to send the horses there, weak as they were, in case
the holes had dried up, and it proved a wise decision, since there was no water
there. When Tom returned, Goyder gave him most of his own tea, which he
had saved from breakfast: 'poor fellow, he was quite knocked up'.[57]

   Apart from this, there was another, even more personally threatening occasion, which was described in his obituary in the *Observer*, the most reliable of
his obituaries. Margaret Goyder Kerr has described the same incident, writing
of how:

> in a drought period in an arid part of the north-east, his stomach cramps
> forced him to dismount. At this time, he was well ahead of his party, and
> it was three days before they found him under the scorching sun, his head
> sheltered by a saltbush, without food or water.[58]

The *Observer* added that it was his strong constitution which saved his life.
There is evidence in the field notebooks of illness. On 18 July 1860, after
recording that the horses were 'too tired to go', Goyder wrote, leaving a space
on the page: 'laying down position and suffering from a return of Diarrhoea'. It
is not clear whether he is referring to the horses or himself – intestinal distur-
bances were an enduring feature of Goyder's life in the field, recorded in the
first and the last of field notebooks. But the next day he was 'up at 5 a.m.', so
if the incident in which he lay with his head under a bush for three days is
described in the field notebooks in any form at all, then it is not obvious and the
danger was entirely ignored.

   Goyder's notebooks also record his having had a swollen, infected thumb;
a problem with his eyes (common with explorers) that became so bad that it
caused him to pass a waterhole which he was attempting to reach; and pains
that he described as 'rheumatism'. These pains in his joints were most likely
symptoms of the scurvy which all of the party were suffering.

   At this time Goyder was in the process of applying for the position of
surveyor general. He feared that ill health would force him back to town, but
as it was, his application was successful and he was recalled to take up his new
post. On the way back he met up with some of Stuart's men on their way up to
Chambers Creek. They were in good spirits, but the spectacle of Goyder and
his party would have confronted them uncomfortably with what lay ahead for
them. In mid-October, scurvy had broken out among the surveying and the
well-sinking parties. Richard Paget, the head of the well-sinkers, had died. The

remainder struggled back to Adelaide. Caused by extreme conditions, this is the only death of this kind that seems to have occurred on an expedition headed by Goyder.[59]

CHAPTER FIVE

Taking charge

But all the forms of government there agree in one respect, in focusing on the public good as their objective ... This may serve to show what the officials are like. They do not make more of themselves than others, but less, for they give first priority to the good of the community and the neighbour, and lower priority to their own good ... They do nevertheless have honor and glory. They live in the centre of the community, higher up than others, and in splendid mansions. They do accept this honor and glory – not for themselves, however, but for obedience' sake.

> – Emanuel Swedenborg, on governments in heaven,
> *Heaven and Hell*, nn 217–18

When Goyder and the members of the surveying and well-sinking parties arrived back in Adelaide, they must have presented a disquieting spectacle to their families, and one likely to alarm any very young children, to whom their absent fathers had become more or less strangers. Caused by a prolonged lack of fresh vegetables and fruit in the diet, scurvy produces very sallow skin, bleeding or ulcerated gums and possibly lost teeth, foul breath, and bleeding into the muscles and joints. Goyder claimed that it took him two years to recover, despite the abundance of fresh fruit the growing town of Adelaide could provide.

The journalist Frederick Sinnett, promoting the colony in a book prepared for distribution in London at the International Exhibition of 1862, attempted to lure prospective immigrants by detailing the overflowing cornucopia of fruits, including stone fruits and citrus, along with berries, nuts and grapes, that one shop alone was offering for Christmas dessert. To reassure his English readers that living in South Australia did not mean abandoning civilisation, Sinnett took them on a walk down Rundle Street, providing an amusing commentary on the wares of each shop. There were hardware stores and saddlers and a bookseller, but the street was dominated by an overwhelming preponderance of draperies. Sinnett concluded his tour at a confectioner's, to 'have an ice and rest a little from this intolerable succession of linen-drapers', the reader, he hoped, having 'got a clear idea that people with money in their pockets need not dress themselves with sheepskins and skewers'.[1] The residents of Adelaide – nearly 45,000 of them of a total population in the colony approaching 127,000 – could not only be well dressed and fed, but clean and safe as well. Since 1860, water had been piped from reservoirs and a weir in the hills, replacing the polluted supply carted from just below the main bridge over the Torrens. A pressurised water supply was welcomed as a great advance in fighting fires.

While there was much to advertise, Sinnett was quite frank about the prospects for various types of intending colonists. For professional men as a class he observed that their situation was 'less hopeful than for capitalists – the field is full … the competition is severe and the prizes to be gained are not remarkably great'. Engineers, architects, and surveyors were not only in considerable numbers, he warned, but faced a situation in which almost all public works were carried out under government control. Surveyors, in particular, 'were in greater number than can find full and profitable employment', although there was plenty of surveying going on, both public and private.[2] There was some compensation for the more talented, in that, being a small place, any 'manifestation of special excellence that they make brings them more quickly into good reputation than it can do in a place like London …' Goyder might have been aching all over, sore and weary, but, returning to become the new surveyor general, he could at least congratulate himself on having made it to the top of the small local heap.

The succession

The competence with which Goyder so impressed his contemporaries becomes strikingly apparent when his detailed and carefully considered proposal to take over Parry's triangulations in the north is compared with the brief and sketchy proposals submitted by private surveyors. The methodical seriousness of his approach is even more evident when the report he produced at the end of the expedition is compared with a passage – an especially engaging passage – from the report by the very man he had been sent to replace. While Goyder's document is laden with descriptions of various types of rock, Parry's presentation reads like a particularly self-conscious diary. One afternoon in early August, an emu began to follow Parry and his companion and hung about the camp throughout the night, during which Parry was disturbed several times 'by a sound like the stroke of a big drum'. The next day, while Parry was alone, the emu came close to the camp. 'Very many have I seen', Parry wrote:

> but none so bold as this. It walked around me within a few yards while I
> rolled up my blankets, every now and then stopping to 'beat his drum'. His

was the noise I had heard in the night. I did not know before that the emu was provided with an apparatus so appropriate to its majestic size. I could not find it in my heart to shoot it. 'You have placed confidence in me,' I said, 'and I will not abuse it. I must be hungry indeed before I could act so meanly and cowardly as to take advantage of it.' He stopped till I had finished my blankets, and thereby learned half the art of the Australian bushman. When Smith came up and fired – but harmlessly – his pistol at him, I hope the emu observed it was not I.[3]

While it is possible that Goyder shared Parry's perception that to shoot the bird would be mean and cowardly, he is unlikely to ever have produced a whimsical account of an encounter with the wildlife and certainly not as part of a report to a superior which could be expected to appear as a parliamentary paper. (Indigenous animals and ground-dwelling birds are strikingly absent from Goyder's writings in any case.) Parry's writing indicates that the imaginative context in which he placed himself, and where his narrative functions, was partly that of the explorer story and partly the evolving form of the bush reminiscence, or bush yarn – as the reference to the 'art of the Australian bushman' indicates. Goyder, by contrast, was keen to explore, but strictly as a surveyor, demonstrating through his professional behaviour the wisdom of Mitchell's dictum that 'geographical research cannot be entrusted with advantage to amateur travellers', or as a Sydney paper put it: 'exploring altogether is a business for surveyors'.[4] Later, in the South Australian *Handbook for Government Surveyors* of 1880, this view was stated explicitly.

One leading figure in the colony who had definitely been impressed by Goyder was J.B. Neales. Neales had employed Goyder at the Adelaide Exchange and had been his superior again when he served as lands commissioner briefly in 1859. In 1863, he told the House of Assembly that, for the short time he had been in office, he had been 'surprised at the work Mr. Goyder could get through. He was then a subordinate officer, but ought to have had the salary paid to the head of the department ...'[5] In fact, it was Freeling's intention that ultimately this would be the case. In the covering letter to Goyder's proposal for the northern triangulation – the letter in which he described Goyder as his 'right-hand man' – he had expressly related the time the northern triangulations would take to the period remaining until the expiration of his tour of duty in the colony. Freeling made it clear that he had ensured that there would be two men ready to take over as surveyor general and deputy: the government, he explained, 'would have the option of rewarding Mr. Goyder's service by offering him the surveyor-generalship, and would have ... the present Senior Field

Surveyor [W.H. Christie] ... to appoint permanently to the post of Assistant Surveyor General'.[6]

Freeling had decided to resign his position in the middle of 1860. The decision took no one by surprise. He had been in the colony for over a decade and had children to educate in England. In any case, times were changing. In the years just before self-government, as surveyor general, he had been a nominated member of the Legislative and Executive councils. In 1857, he was elected to the new Legislative Council, and served in the ministry of the short-lived and transitional first government. He resigned his seat in 1859 because of the pressure of his 'numerous public avocations', but, also, perhaps, in response to opposition to his situation.[7] A year earlier, the *Advertiser* had devoted much of the editorial of its very first issue to arguing against the presence of public servants in parliament, and the position of surveyor general, although not actually named, was evidently the focus of concern. Responsible government had removed the need for an officer of the Royal Engineers, ready to take charge of surveying and engineering works as required, and had re-created the position of surveyor general as the head of a public service department, answerable to an elected minister of government. Such a figure would no doubt be expected to become embroiled in the sorts of conflicts – over the boundaries of runs and the location of roads – which Freeling found distasteful and (as Goyder later testified) was unwilling to take on.

Word of Freeling's resignation had reached Goyder while he was working south of Lake Eyre, with his health steadily deteriorating. The dispatch was carried to him by J.H. Howe, a trooper at the time, who later became a politician and embellished the legend of Goyder's vigour and toughness by recalling that he had been a hard man to catch – 'the task of overtaking him was arduous'.[8] (It probably helped that Goyder was a small and compact burden for his mounts.) Not only did the dispatch bear news of the resignation, but it contained Freeling's encouragement to his deputy to apply for the position. Goyder wrote to Freeling, enclosing an application to be forwarded to the commissioner – obviously, he hoped, with Freeling's endorsement. All he really had to do was point to his record of service and seek Freeling's support, but scurvy, exhaustion and isolation led Goyder to deal with the issue of change in the position of surveyor general in an uncharacteristically tactless and self-defeating way. The letter began predictably enough, with Goyder venturing to solicit favourable consideration for his own past service and his 'minute acquaintance' with both the activities of the office and the detail of all the surveys undertaken since settlement. Unfortunately, he was not satisfied to let the matter rest there and went on to explain that he was aware that he could not expect to receive the same salary as Freeling, since many of the responsibilities that had originally belonged to the surveyor general had now passed to the commissioner. 'I should therefore gladly accept the office at my present salary – viz. (£600–0.0) six hundred per annum,' he announced, 'and in order that the amount may not appear excessive I would endeavour to perform the duties without the aid of an Asst. Surveyor General'.[9]

Goyder offered to submit arrangements that would enable him to continue to spend half his time in the field – which he was evidently very reluctant to leave – 'without detriment to the Public Service'. He also explained how the triangulation of the northern country might be completed without him. In making this offer, especially to work without an assistant, he had evidently not paused to consider that he appeared absurdly self-confident, and he perhaps did not realise that his plan would negate Freeling's sensible scheme for achieving a smooth transition of authority in the department by having both Goyder and Christie promoted.

What Goyder would also not have known was that during 1859 Freeling had received a letter of kind that must have come as an unpleasant surprise. The letter was from William Milne, the lands commissioner, and its import was that Freeling, as surveyor general, had not adapted to the changed political circumstances and was going about his duties as if his authority was autonomous, excluding Milne from the decision-making process. According to Milne: 'the Government survey parties are moved about without any information to this Office, no proposed plan of operations is forwarded either for information or approval, the sites of Government townships are fixed, surveyed, and sold without sanction being asked or given – indeed the Commissioner of Crown Lands continues in ignorance of everything which takes place in the Survey Department until the land is settled for sale'. This state of affairs, he warned, was 'unparalleled in any other of the responsible Departments of Government and … obviously unconstitutional'.[10] While it is possible that Goyder knew of Milne's concerns, it is unlikely that he knew of the reprimand to Freeling, and that he was pressing a sore point.

Freeling passed the application on as requested. In his covering letter he wrote: 'I can speak in the highest terms of Mr. Goyder's zeal and ability; and also of his conscientious discharge of his duties …', but he refrained from commenting on Goyder's proposal and challenged his claim to be able to fill the role single-handedly.[11] He then wrote to Goyder informing him that this had been done. On 1 October, Goyder replied. After referring to other documentation accompanying the letter, he 'hastened to add a few remarks' explaining how he proposed to do all this work alone. He admitted that it had

since occurred to him that his offer 'may appear presuming or over confident'.[12]
After restating the expected change in the role of surveyor general, Goyder
referred to his previous success in carrying on both the superintendence of the
field survey and the duties of the office when Freeling had been out of town,
pointing out that Freeling had not voiced dissatisfaction.

His proposal was to leave the established procedures of the office untouched,
interfering only to the extent of 'taking care that each officer performed his
required duty in a satisfactory manner and without falling into arrears'. In
relation to extensive surveys, he explained that his knowledge was adequate
for deciding on the location of roads and reserves before beginning the survey,
and that once these had been determined, he would 'invite persons interested
in the locality to meet me on the ground on a certain date when the survey
could be pointed out and improvements or alterations at once affected ...'[13]
This was not an idle suggestion, as his later behaviour showed. Not long after
he became surveyor general, Goyder had to decide upon plans for a town at the
new mining site at Wallaroo. He recounted:

> I considered it desirable to call a meeting, and laid before them the various
> schemes that occurred to me, and I asked for information and after discussing
> various plans, a resolution was arrived at ...[14]

The letter went on to explain the way in which he planned to relate his own
work to that of the commissioner. (J.T. Bagot had taken over from Milne during
Goyder's absence.) When the district consultation had been completed, a rough
plan of the survey and an explanatory letter would be 'immediately submitted'
to the commissioner 'for his information and approval'. The plans would be
kept in the commissioner's office for reference, so that the commissioner would
be 'kept aware of the exact locality and extent of the country being surveyed
throughout the Province'. While these proposals might irritate Freeling, they
were certain to reassure any elected commissioner. This is not to imply that
Goyder was either insincere or cynical in making these promises. In his application to rejoin the service as chief clerk he had expressed his commitment to
the provision of information to the public, and his evidently democratic attitude
would have made him keen to ensure that the people's elected representative
was fully informed of the work that was proposed. It was one of the differences
that made the change from Freeling to Goyder timely and appropriate.

Goyder warned that if his application were rejected he would be forced
by frustration of his ambition to resign, although he would stay to see surveys
already in progress in other districts completed. But he reminded them that he
would be difficult and expensive to replace. In the margin of Goyder's letter,

Freeling advised the commissioner that Goyder would not be able to supervise distant parties of surveyors by being absent from the office three days a
week, and that his plan to consult those wishing to buy land in the district
before beginning the survey would lead to 'endless confusion and dissatisfaction'. Bagot added a note to his secretary, requesting a report from Freeling and
advising that Freeling needed to be aware that his successor would not have the
same burden of correspondence; at the same time he asked for recommendations which might lead to a more economical conduct of the survey. Freeling's
criticisms were not being sympathetically received, while Goyder's overtures
were clearly in the key the politicians preferred to hear.

Freeling's resignation, and his departure at the end of the year, were
announced in the papers on 18 October 1860. B.T. Finniss and Captain Douglas,
of the Marine Survey, were mentioned as rumoured successors. The *Register*,
at that time very much a part of the political process itself, chose to point out
that the government generally adhered to 'the principle of promotion where it
has been practicable; and, carrying out that principle, the office when vacant
should fall to Mr. Goyder …'[15] The news of his appointment was made public
on 23 November. The salary the government awarded was £700 – £100 more
than Goyder had asked for and £100 less than Freeling received. Christie (who
had also applied for Freeling's position) was appointed deputy. The *Register*
commented that the news of Goyder's appointment could not 'but be gratifying to his friends and satisfactory to the colony at large'.[16] In fact it meant the
colony would have its first surveyor general appointed not only for his technical
abilities but for his extensive knowledge of the country. The foundation era, in
which the colony had relied on officers of the Royal Engineers imported for
limited periods, was over. Goyder arrived back in Adelaide on 17 December,
in time to attend the official farewell for Freeling, an event at which anyone
possessing any status at all in the colony felt obliged to be present.

The final scene between superior and deputy reveals how badly wrong
things had gone, especially in view of Freeling's generally attested 'kindness and
affability', which Goyder had also acknowledged.[17] Goyder later recalled that
Freeling had written a confidential letter:

> the last thing he did in his office was to show me the document … Colonel
> Freeling sent for me and read it over; the contents, as far as I remember, were
> as follows: – He spoke highly of my ability and energy, and desire for the
> correct performance of my duty; but he stated he considered I had exhibited a
> want of judgement in advising that the District Councils should be consulted
> on the laying out of roads. He also considered it a want of judgement that I

had stated that if certain duties were taken from the office I could manage without an assistant.[18]

Freeling assured Goyder that the letter (which seems to have been identical in content to the covering letter to Goyder's application) was 'private and confidential' and that 'as soon as [it] had been read to me, it would be handed to [the commissioner's secretary] and then destroyed'.[19] Almost inevitably, this didn't happen, and not many years would pass before rumours about this letter, distorting its contents, would be used against him.

Goyder was formally appointed on 19 January 1861. His first actions were to re-organise procedures and responsibilities so that the financial aspects of land sales were passed to the commissioner's office, leaving Survey Office staff free to concentrate on technical matters. The new surveyor general also had to address the long-standing problems with the plans of the pastoral runs, inherited from early years. In relation to the sectional survey, it was only necessary to increase the frequency with which field parties were inspected to have the sectional survey proceeding to his satisfaction: as deputy – de facto and *de jure* – he had already developed and refined procedures which he considered to be sound. In 1859, applying for Parry's job, he had expressed the conviction that the system of sectional survey in use in their department was 'surpassed by none in the world', and congratulated Freeling for adopting it – in effect congratulating both Frome and himself as well.[20]

One of the first reports he was required to provide dealt with the sale, on credit, of land unsold after auction. He recommended against the practice – but for practical reasons related to the country rather than theoretical reasons to do with credit selection. The land being considered was mostly on the plains between the Light and Wakefield rivers, and north of Kapunda, well within the later border of the Line. In keeping with the attitude of caution that prevailed, Goyder expressed concern that the land north of Kapunda was not ideal, because, although the land itself was good, crops were liable to be 'cut off by causes that can neither be anticipated nor avoided'. It would later transpire that he was referring to hot winds. Always sensitive to issues of responsibility, Goyder had immediately pointed out to the commissioner that this fact raised the question of:

> whether inducements held out to cultivate such lands would not tend to injure rather than improve the condition of the labouring man; and, consequently, retard the satisfactory development of the resources of the Province, by the application of lands only fitted for pasture, to the purposes of agriculture.[21]

This was the same concern he would soon bring to the northern country and the unreliable rainfall, and his comments show that the approach he would take at that time was already part of his thinking when he came to office. In his many journeys to the north he had been confronted by the reality of an inland climate unsuited to agriculture, but although the colony had achieved responsible government, the approach to settlement remained much the same as the original plan: land was still only available for purchase after survey, and the survey continued, by and large, to radiate from the capital. Clearly, at some point, some other principle would have to be invoked. But, although the surveyor general now served an elected government minister, the need for expertise meant that the minister, and the government as a whole, was dependent upon the surveyor general for advice about the centrally important issue of land and settlement, a situation that was emphasised by the fact that Goyder was now indisputably the person who possessed the most knowledge and experience of the land and climate of the colony. While the politicians were largely city-bound, the full weight of the responsibility that Wakefield's approach placed on the surveyor general continued, effectively, to rest on his shoulders, and knowing this, Goyder had evidently been considering the impact of climate on settlement from the beginning.

Goyder's first two years as surveyor general provided him with plenty of distraction from the concerns of pastoralism, agriculture and the climate: as well as being valuator of runs, he held the second additional position of chief inspector of mines. In 1859 there had been a major copper find at Wallaroo and in 1861 another nearby at Tipara (soon known as Moonta). Apart from the mines that needed inspecting and the new settlements that needed laying out, Goyder was caught up in a convoluted dispute, involving several parties, over the claim for the main lode of the copper at Moonta, where one of the largest copper mines in the world was soon to be established.

Family gathering

In 1861 his father-in-law John Smith died at the age of 68 of consumption, like his daughter Mahala, the death reported to have taken place at 'Mr Goyder's residence in Brighton'.[22] Brighton is now a southern seaside suburb of Adelaide, but, as the name suggests, it had begun as something of a resort for the comfortably situated, the houses there, it was noted in 1851, constructed of brick and stone, with only some of pisé.[23] The benefits of sea air and bathing were seen to make the area attractive to invalids, and that is almost certainly why Goyder maintained a house there.

Walkerville Council records show that, in the year of his death, John Smith was leasing three properties – all houses with nurseries – two of which were in Fuller Street, in the ward of Walkerville, and the third described simply as being in Medindie, the area where the Goyders lived. One of the properties in Fuller Street was identified in the council assessments of 1863 as 'Clifton Nursery'. The assessed value of these properties totalled £280, a considerable sum. Goyder was the executor of John Smith's will, in which his estate was left to his daughter Ellen and his son Alfred, with the proviso that they were to provide for their mother, Susannah, who was entitled to live on the property, although only Alfred's name appears in the assessment records for all three properties for the years immediately following John's death.[24] Ellen, in any case, was drawn to the Goyders. Margaret Goyder Kerr has written that Ellen, 'lived a stone's throw from the Goyders, loved their children and often spent weeks at a time in their household'.[25]

Around 1862–63, in a move which effectively established the Goyders and Smiths in their own precinct on the edge of the parklands, Edwin Smith and his family left Findon and took over Charles Ware's nursery on Robe Terrace, across the road from the Goyders.[26] This was the Clifton Nursery that eventually occupied all the land along Northcote Terrace (parallel to Hawkers Road) from Robe Terrace to Dutton Terrace, making it a major landmark in the area.[27] The house was at the Robe Terrace end, near the Goyders.[28] At the same time, Goyder was expanding his estate, taking over a property with stables, a garden, nursery and vineyard on the edge of the parklands. He would later add pasture land.

Under the proprietorship of Charles Ware, Clifton Nursery had stocked a variety of trees, including pines, cypress, fruit trees and vines, and Australian species such as the silky oak, and the bunya and Moreton Bay pines. It had also stocked oleanders, fuchsias, potted shrubs and roses.[29] Smith, who exhibited regularly (and successfully) with the Horticultural and Floricultural Society and with the Royal Agricultural and Horticultural Society (the former he had helped to found in 1856) began to develop a rose garden on the southern slope below the house, and within a few years he had planted four hundred of what were considered the best varieties. The garden also boasted a flourishing collection of around a thousand camellias.[30] Smith was also well known for cultivating gloxinias and petunias, and a hybrid tecoma bred by him, *Tecoma x smithii* ('Orange bells'), is still cultivated as a garden plant. Not surprisingly, for a variety developed in Adelaide, it flourishes in full sunshine with comparatively little water. However, Edwin Smith's interest was not confined to his nursery. He had become a member of the Walkerville Council a year after arriving in the area and was mayor for three years during the 1860s and for another period in the late 1870s to early 1880s (but he should not be confused with the prominent politician, Edwin Thomas Smith).[31]

With the Goyders and so many of the Smiths all resident in the Medindie-Walkerville area, something of a tribe had formed. Edwin and Hannah's family, which had consisted of two on arrival, had increased by two more sons, followed by two more daughters, by 1857. Once Goyder was back in Adelaide, his own family resumed its growth. Another boy, David John, arrived in 1862.

In the middle of that year, Edwin Mitchell Smith entered the Survey Department as a field assistant. As a 15-year-old wondering what to do with himself, his decision was made after watching his uncle surveying near his home. He joined a survey party working near Lyndoch Valley in June 1862.[32]

Hopeless country

Goyder visited Wallaroo-Moonta at least twice in 1861, as well as making trips down to the South-East.[33] He was back at the copper mine at the beginning of 1862, and during the hot early months he undertook a tour of inspection of mines in the Flinders Ranges, visiting 16 mines in the areas around Wilpena and Blinman, and Yudnamutana in the far north-east. He was impressed by the indications of copper at Yudnamutana, but judged that the mines were so far from any port that, even if every mountain north of Mount Serle consisted of nothing but copper ore, without a (horse) tramway to expedite removal, the cost of cartage would still make mining unprofitable.[34]

In the middle of the year Goyder set out on a small expedition of exploration of the land behind Fowlers Bay on the Great Australian Bight. As ever, he was looking for pastoral land and water. Fowlers Bay lies between Cape Nuyts and the Nuyts Archipelago, to the east of which lie Smoky and Streaky bays, the area where Flinders had felt the hot winds off the land, which indicated to him that the aridity of the coast 'prevails to a considerable distance in the interior'.[35] Eyre's desperate trek along the coast had demonstrated how extreme that coastal aridity was. Nevertheless, the country behind the coast had attracted the attention of several leading pastoralists in the late 1850s. Goyder's notebook records the journey as having begun in a thoroughly unexcited and routine manner, despite the fact that they were planning to travel into country that was dauntingly arid, even in winter. By this time Goyder was solidly experienced in moving over dry country, and he realised that mobility was the key to their survival. That three men, lightly equipped, with only several horses, could contemplate this sort of journey confirms how far learning about how to move and survive in the arid lands of the new continent had progressed.[36]

Goyder left Adelaide on Saturday 5 July, much like a commuting public servant. His notebook records that he paid two shillings for his train fare to the port.[37] There he boarded the *Lubra*, suffering from seasickness all the way across Spencer Gulf. He was back on land at 4.00 pm on Sunday afternoon and immediately set to work, visiting two pastoralists that day. The party seem to have stayed with the second one, Murray, and the next day Goyder borrowed a double-barrelled shotgun from him, together with powder and shot-cases, although he paid another 5s 6d for more powder. The contents of this notebook also indicate his anxiety over his personal finances. Even so, he records squandering 1s 8d on 'spirits and water' – perhaps to celebrate settling two disputes over pastoral leases that day – and on two occasions he lent one of his companions, the horse keeper William Rowe, who had been with him on the shores of Lake Torrens in 1857, a pound.

Goyder and his two companions crossed the bottom of Eyre Peninsula behind Port Lincoln and moved up the west coast toward the Great Australian Bight. After travelling in the rain all day on 9 July, Goyder chose to stay at lodgings that night. The next night they camped on a station. 'Memo', he recorded. 'Fleas in thousands in hut. slept out'.[38] They travelled up through Talia, and then through Wallanippi. The day after leaving Wallanippi, 20 July, they were forced to sink a well – and fortunate enough to strike brackish water satisfactory for the horses. Goyder thought the country 'the best I have seen since our landing at Port Lincoln – the desideratum being want of permanent surface waters'.[39]

They arrived at Fowlers Bay on the evening of 23 July, and when it was night Goyder took celestial readings. From Fowlers Bay they moved inland, heading to the north-west, then north, and finally curving round to the east to arrive at a point to the north-east of Fowlers Bay. From there things did not go well. They were troubled – threatened – by lack of water, although the brief entries in Goyder's notebooks give no real indication of how badly. They do record his constant intense concern for the state of the horses, on whose wellbeing their lives depended. On Friday 1 August, although their overall situation was difficult because of the lack of surface water, he began his notebook entry with what amounts in the context to a jubilant cry: 'Night showery. Horses splendid ...'[40] In the same entry he began to record the only occasion mentioned in his notebooks when he sought the help of an Aboriginal guide:

Packed up preparatory to starting. Two emus shot + a Black came + agreed to go with us provided he got his clothes. Marchant rode with him on horseback. remainder of horses again turned out. plotted course. slight showers falling + weather still threatening. Marchant + Black returned at 6.20.

The next night they 'camped at another granite rock hole named *Metyera*. Shot a couple of very beautiful rose cockatoos. salt bush and shrubs round hole.' The waterhole was the first feature of this landscape given a name in the notebook. Ironically, the person who told them its name lost his. He was renamed 'Billy' – probably in accordance with what seems to have been standard colonial procedure, although perhaps they had been unable to determine his name in the first place. Goyder's use of the capitalised 'Black' as a sort of name before the name 'Billy' was bestowed expresses a cautious respect. Goyder did not capitalise or use titles consistently when jotting things down in his field notebooks, so his practice unselfconsciously expressed his feelings – which evidently fluctuated – about relations of power, authority and respect. For instance, despite having both a practical and obviously tender-hearted concern for his horses and other animals in his charge, he frequently did not capitalise their names – which adds to problems of deciphering the text until the names are learnt – and on some occasions he even failed to capitalise the names of his subordinates. Only rarely were men granted a title, although Frances was invariably Mrs Goyder on the odd occasions she was mentioned. By contrast, his minister was only ever a title – the 'Hon. C'sser [Commissioner]' – a formality that was also practical, given that, with the rapidly changing ministries characterising the early years of fully elected government, Goyder could ride out of Adelaide at the direction of one man and return to find himself responsible to another.

'Billy' went to visit some other people and did not return, leaving a disappointed Goyder to find his own way out of what he termed in his notebook 'hopeless country'. [41] Goyder fixed a course and they returned to Metyera, but the small supply of water there smelt so badly that the thirsty horses would not drink it, although Goyder and his companions had cleared what he described as 'filth placed by the blacks' from the hole. (Time and distance mellowed his attitude to this material. Thirty years later, describing Metyera to the Pastoral Land Commission, Goyder would explain that the supplies of water in the rocks were 'so precious' that they were covered by the Indigenous inhabitants 'with stones or branches to protect the water from birds, & c.' [42]). Despite all difficulties the group made it back to Fowlers Bay.

The reference to 'hopeless country' in this notebook provides the exception to the rule of emotional equanimity which prevails in the notebooks. In 1857, writing in relatively good spirits, Goyder had introduced the terms *beautiful*, *picturesque* and *romantic* to descriptions of the Flinders Ranges, but, on the whole, his narratives are free of emotional response and judgment, most noticeably of the standard tropes of fear, horror and rejection. (*Sublime* is the significant omission from Goyder's otherwise complete basic set of conventional nineteenth-century aesthetic terminology, although the sense of it is implicit in his theories about the formation of Wilpena Pound and similar geological formations in the Flinders Ranges.) This lack of obvious response was clearly not because he had been rendered powerless to convey the awe-inspiring vastness of an alien world, nor was it due simply to his being a practical man, as his contemporaries described him. Rather, his work and writings as a whole suggest that he was not ill at ease with the world around him and that his engagement with the country did not dramatise any personal struggles, though it cannot have been without inner significance. Within the larger framework of his commitments, he seems to have continued to be driven by the confident curiosity that his father had cultivated in his early years. On the whole, he enjoyed being on the land, even when we would not expect it of an emigrant from the north of the British Isles. Before a commission in 1887, Goyder stated that he thought the climate in the country south of Lake Eyre was 'hot, but dry and very pleasant … all that one could desire' – apart from a little unpleasantness in the rainy season. When prompted, he recalled that the thermometer had stood at the equivalent of over 42°Celsius, 'in the shade all day, for eight days during the year'.[43]

The following year he revealed the extent to which his life had been in danger in the waterless country behind Fowlers Bay in 1862. While giving evidence to the Land Laws Commission in 1888, Goyder was asked, without warning, if he knew the Nullarbor Plains country. He replied:

> I have been on the fringe of it. I went north-west from Fowler's Bay, and was myself without water three and a half days, and for five days without water for the horses. The soil in parts is good, but a great deal of the land is very scrubby, and there are sandhills, some of which rise at an angle of 45°. On the western portion I saw from the character of the vegetation that the country was not such as I should look favorably on for cultivation. In one or two places north of Fowler's Bay there are parallel islands of better country between scrubby tracts that I thought more highly of.[44]

His answer reveals not only what it took to drive him to use the word 'hopeless', but the enduring precision of his memory of the country over which he passed. It also indicates that he was already looking at the vegetation, probably as a guide to rainfall, to determine whether land was suitable for agriculture.

Apart from his investigations of the country beyond the Flinders Ranges, the expedition to Fowlers Bay was the only other journey to be classified and remembered as a work of exploration by his peers. In *The History of Australian Exploration 1788 to 1888*, Ernest Favenc recorded simply that he had paid a visit to 'the much-abused region … but found nothing to reward him but mallee scrub and spinifex'.[45] It was as far west as he was ever to go.

Resignation

In October, not long after returning, Goyder presented his first resignation. He needed more money and better security – his travelling expenses were not covered by the allowance of a pound a week, and, as he had come to realise after the 1859 expedition, 'should I be disabled or rendered unfit for office by casualty – unconditional resignation must necessarily follow without claim upon the Government for services rendered beyond remuneration already received'.[46] Recent experience could only have reinforced the lesson. He had also been receiving 'overtures' from the land agents Green and Wadham, and to these he had succumbed.

Having resolved to leave, but not wanting to be separated from his work, Goyder arrived at a bizarre solution: he proposed to take his work with him. He offered to tender for the survey at three-quarters of the present costs, and suggested that the government could appoint a 'nominal' surveyor general who would attend to other business and communicate with the government and the survey contractor – himself.[47] Commissioner Strangways forwarded this to the chief secretary with a note advising him that Goyder was a 'most efficient' officer, whom he feared they would have great difficulty in replacing, a point of view that he subsequently put to parliament, where it was greeted with cries of 'Hear, hear'. In the end Goyder was persuaded to stay with the promise that the government would increase his salary by £54 7s 6d – the amount needed to round off his allowance of £45 12s 6d for forage for his horse to an even £100 and raise his overall income to £800 per annum. There was no doubt that, with his knowledge of the country and his abilities, Goyder could have made himself a wealthy man in private practice, but the position of surveyor general in a formative stage of the colony's development offered not only unparalleled scope for his particular genius, but also an opportunity through the shaping of the pattern of settlement to contribute to the wellbeing of thousands of people in the present and untold thousands in the future, a prospect that was deeply satisfying to his Swedenborgian soul. As another public servant commented after his retirement, Goyder did not merely love his work, he had a passion for it, and a relentlessly active mind that demanded expression.[48] It was not necessary to offer very much to persuade him to stay.

The government attempted a review of civil service salaries early in 1863, through which Goyder would have become the second most highly paid officer, with a salary of £800 and another forage allowance, very slightly increased, of

£45 15s. The attempt was not successful and the government lost office in mid-year. Even the amount Goyder had accepted to stay was in jeopardy. Faithful to the agreement, Strangways moved that the sum he had promised be added to Goyder's salary, and although the new government opposed the motion on the ground that land sales were decreasing, Goyder eventually received an increase by the simple device of being granted a second forage allowance. He had also succeeded, through Strangways, in having the small travelling allowance replaced by the remittance of all expenses incurred.

CHAPTER SIX

Bird's-eye view

'Come awa' to th' hoose, mon, come. A' can see ye're a stranger to these pairts. Ye ken A' canna be fashed wi' soocial amenities, but it's right glad A' am tae see ye. Come, we'll hae a tot o' whisky an' a crack aboout your business. Ye'll need ma' help maybe …

'Na, na, mon; ye winna get hame on sic a nicht – a braw fire an' a gam' a whist and whisky toddy …'

– Surveyor Lionel Gee, recalling a Scottish welcome
in the South-East in the 1860s and 1870s

At the beginning of 1863, Goyder's duties took him back to the South-East, through country first traversed on his initial journey from Melbourne to Adelaide. A notebook shows that in 1861 he had also made two journeys in that direction to examine pastoral runs. One had taken him as far as Tarpeena, a town between Mount Gambier and Naracoorte. On a shorter trip he had gone only as far as Coomandook, to the north-east of Lake Albert.[1]

Before European settlement, the whole region constituted a giant seasonal wetland of about 17,000 square kilometres. The underlying formation is a series of dune ridges – old coastlines – running parallel to the existing coast. Before being altered by settlement, the ridges carried open forest and bushland. Grasses, sometimes under small eucalypts, covered the flats between and it was these flats which flooded. There were areas of fairly permanent swamp as well, and the region was rich with bird life. Of the Mount Gambier area in the early 1860s, Edwin Mitchell Smith, who had worked there as a young surveyor, wrote that: 'there were thousands of kangaroos in the district. These have now almost disappeared. There were also thousands of opossums in the forest and what were known as flying squirrels.'[2] He had earlier commented that, near Strathalbyn, the Wombat Plain was 'swarming' with wombats, and the Dry Plain – not dry according to him – was inhabited by numerous mallee fowl, or 'native pheasant' as he called them, which are now a threatened species.

The South-East was a wholly different world from the rest of South Australia – instead of being arid, the colonists found it too abundantly supplied with water. The rainfall was substantial and more reliable than almost anywhere else in Australia. In many places, underground water reached the surface in winter and only dropped a couple of metres below it in summer. The leader writer of the *Telegraph*, almost certainly the owner and editor, Frederick Sinnett, expressed a typical view while contemplating the region:

> If such a phrase may be pardonably used, it is among the caprices of nature so to distribute water that it is scarcely anywhere in the precise quantity which the necessities of mankind require. Everywhere there is too much or too little of it – everywhere art must step in, either to bring it from a distance or to drain it away.[3]

The roads in the region were notoriously dreadful – away from the shores of the lakes or lagoons, which were firm, they consisted of deep, loose sand, which was interspersed with large lumps of limestone and threaded with mallee roots. In winter, water on the roads could be so deep that it would reach the axles of the carts, and areas became dangerously boggy. The trip by mail coach was a legendary horror, four days of torment, and more so for the horses than the passengers.

While the view from North Terrace was that the South-East was a difficult place – boggy, mosquito-ridden and a long way away – the view from the South-East was that the politicians lacked a sense of urgency about attending to its problems. From there, money seemed to flow in one direction only, and the South-Easterners had come to feel increasingly isolated and neglected and consequently resentful. Over much better roads, Melbourne was only two days away from Mount Gambier. There was talk of secession.

Official inspection

On 5 January 1863 – in summer, when the roads were dry – a tour to demonstrate government concern for the South-East departed from the Vine Inn at eight in the morning, while it was still cool. The party was led by William Milne, a hearty and somewhat bluff-looking Scot, square-faced and squarely built, pale-haired and pale-eyed, a 'magnificent example of the healthiness of the [South Australian] climate', according to one source.[4] He was only four years older than Goyder and had been born in Glasgow. He had also attended the Glasgow High School and it is possible that the two remembered each other, but even if they did not, there was a shared point of reference in their early lives. Milne, however, had come to South Australia aged 17, travelling on a free passage as a farm worker. Having worked out his term on a pastoral station, he had gone to Tasmania before returning to Adelaide in the early

1840s. Together with his brother-in-law, he had bought up the wine and spirit business of New Church member Patrick Auld, and in a town where drunkenness was an obvious problem in the early period, the business made Milne's fortune. In 1857 he sold it and began a career as a member of the new parliament, promoting legislation encouraging agricultural settlement and establishing public works, while filling roles such as chairman of various businesses and organisations outside the parliament. He was knighted in 1876.

Milne's two companions were the colony's engineer and architect, William Hanson, and Goyder. They set out in a 'four-wheeled American conveyance with four horses', or stagecoach, but perhaps without the roof, since in the diary kept of the journey, Milne described protecting his head and shoulders from the heat of the sun with a leafy bough.[5] The purpose of the tour was to consider the need for public works, especially in relation to transport and communications. Roads and jetties had priority of concern, but the group would not be above contemplating badly functioning and smoky chimneys at the Robe gaol and the post and telegraph office, or inspecting the rebuilt privy at the Penola police station.[6]

From his perspective, Goyder later recalled that they:

> went down to look at the swamp region in the South-East. We drove down by way of the desert, and we made a short cut to Tilley's Swamp, and then making Kingston went thence to Tilley's old station, and crossing through a piece of country that for fifteen miles was covered with water, at last got to Mount Gambier. I suggested to Mr. Milne that whilst he was making his arrangements complete at Mount Gambier I should go and look at what has since been known as the drainage country. I found the land covered with waters like a sea, extending for many miles, the only parts that were dry were the ridges, which separated the swamp waters from the sea.[7]

Goyder had seen the swamps before, and he had spoken to pastoralists about runs that were half-submerged in winter. Along with the swamps, he believed, came the mysterious condition, known as 'coast disease', which caused healthy stock brought to the region to sicken, and crippled and killed in infancy animals born there. Coast disease was due to trace element deficiencies in the soil, but this was not known in the nineteenth century. In keeping with the 'bad air' theory of disease, Goyder suspected that the swamps themselves, with their 'malarious exhalations', were the cause, and eventually became convinced that they should be drained to get rid of this menace.[8]

By this time, the survey methods he had been practising and elaborating for nearly 10 years had yielded a mature and highly developed skill of observation and imagination. Since much of the South-East is flat, even a mound (such as Big Hill by the Bool Lagoon) can command an extensive view, if only of a great stretch of flat land with distant ridges. Despite this, Goyder was able to make sense, on a large scale, of his observations. As he explained after his return to Adelaide:

> After a succession of dry seasons the South-Eastern District presents a bird's-eye view of alternate low ridges of limestone covered with scrub, the ridges gradually increasing in altitude towards the south, with intervening well-grassed flats, or covered by swamp vegetation. After a moderate rain-fall the water forms into a series of lakes connected occasionally by shallow channels or divided by low transverse rises that prevent a continuous flow of water until a certain depth has been attained, after which the valleys are transformed into wide parallel sheets of water with channels in their deepest parts flowing in the direction of the outlets [in the north-west].[9]

It is an accurate account. The ability to form an understanding of the levels and lie of the land are among the ordinary skills of a surveyor, but Goyder's geographical imagination was evidently vast and sharp, his mind's eye an eagle gliding at a great height.

Goyder's unflagging interest in the world around him led to an excursion out onto Lacepede Bay while the party was at Kingston. The bay provided a reliable shelter for sailing vessels even when a severe gale was blowing, and Goyder, as keen to survey the sea as the land, wanted to investigate the secret of its calm waters. When he was offered the use of a boat named the *Swallow*, they set off on a straight line, west-south-west, with Goyder and Milne working together, taking soundings and measuring their speed, to calculate distance. The result showed a series of underwater ledges which Goyder speculated might break the force of the waves.[10]

The observer observed

Milne's diary provides glimpses of Goyder going about his work, which included attending to problems with pastoral leases and inspecting the progress of the sectional survey. It also uniquely recounts scenes in which he is shown as a traveller and a companion, and – finally – as a husband and father. Milne's straightforwardness is evident in his diary; he was a sympathetic diarist with whom it is easy to identify. In Mount Gambier, he inspected the Blue Lake on a beautiful day, and 'was insane enough to try and make a sketch … but my failure was a signal as the landscape was lovely'.[11] In the early hours of the morning, Milne and Goyder went to help when a young servant girl at the inn at which they were staying had a 'violent fit of hysteria', screaming and crying for her mother. They had helped to hold her down while the doctor was brought. All next day Milne was haunted by the 'cries of the unfortunate girl', and when they reached Penola, they telegraphed back to see how she was faring.[12]

Hanson, the engineer and architect, is virtually non-existent in Milne's diary – his most dramatic moment seems to have been when he was thrown out of his carriage and a 'good deal shaken'. Goyder, on the other hand, is present as a friend and companion. (When Hansen had his accident, Milne and Goyder were travelling together in another carriage.[13]) He appears as a lively figure moving in and out of Milne's nimbus of concerns, sometimes partnering Milne, sometimes going off on his own. His energy was both aroused and focused by the world and the problems it posed to the settlers, but he seems sociable rather than abstracted, and playful when the opportunity allowed. At Robe, they went down to the telegraph office after supper, to enjoy a long 'yabber', as Milne put it, with Charles Todd, evidently a friend to both. Goyder finished the exchange: 'saying that we parted and still we did not part, as although we were losing our Todd, we were going to our Toddy!'[14] The farewell was specially framed for Todd – 'I'd be odd without my T' – who was notoriously fond of puns. A toddy was a 'nightcap' of spirits, usually whisky, hot water and sugar, and Todd fired back that they would have to call Goyder out of bed next morning. (Although many of the Protestant denominations and sects that thrived in South Australia frowned upon the consumption of alcohol – Methodists prohibited it entirely – members of the New Church, like most Anglicans, had no such restrictions. There was no conflict for Goyder between his serious approach to his religious beliefs and drinking wine or whisky with companions.)

Two of his subordinates later remembered Goyder as an engaging and entertaining presence. He was 'capital company' as a friend and an ideal travelling companion, according to William Strawbridge, a fellow Swedenborgian who succeeded him as surveyor general.[15] Lionel Gee, a surveyor who wrote reminiscences of the South-East in the 1870s, agreed. Gee recalled that, when travelling by coach, Goyder's favourite seat was next to the driver, and that another passenger 'fortunate enough to occupy the other seat was sure of entertainment that would pass the long hours without weariness'. Goyder was:

a most versatile gentleman ... out of harness friendly, sociable, and instructive to the nth degree. He was musical, and, humming airs from Italian operas, he would suddenly break off, and with a pleasant voice that commanded attention relate a few incidents in his experiences in short stories, putting in little details that go to make the perfect story; and he was a good listener, with the marvellous intuition that with a word or two now and then would keep the narrator to the thread of his discourse.[16]

After his death, a popular paper recalled that he was possessed: 'of an abrupt manner, common with men of impulsive dispositions', but nevertheless capable of proving 'a most charming companion'.[17]

Even so, Goyder seems to have been a little more forward at times than Milne would have preferred. At the home of pastoralist Alex Cameron, 'king' of the Highland Catholic community in the South-East (and uncle of Mary McKillop), where they were staying, there were two white possum skins nailed to the wall of the house. In his diary Milne recorded that, not at all bashful, Goyder had asked for one of them, and was given it.[18] Milne had also recorded another incident earlier in the trip, which might have annoyed him – not that he expressed it – while they were at Kingston. At a celebration of the opening of the Lacepede Bay jetty – a celebration at which many toasts were drunk – Goyder had pre-empted him by proposing the health of the chairman, the man who had put the *Swallow* at their disposal. It was Milne, of course, as the senior member of the party, who should have been the one to offer the toast to the chairman.

But these were only very minor incidents and depend as much upon Milne's sensibility as anything else. The diary shows that Goyder was constantly in situations which tested his personal and social skills: throughout the journey he regularly visited squatters, or was visited by them, singly or in groups. Usually the issue was a boundary dispute, and Milne describes Goyder, 'closeted' with a group, which, 'after some two hours, broke when we understood an arrangement satisfactory to all parties was concluded'.[19]

What was later claimed to be Goyder's most widely recognised failing – his quick, fierce temper – is nowhere apparent in the diary. On one occasion early in the trip, just across the Murray, they were attempting to follow 'Goyder's new track', which had been laid out by a surveyor named Edmonds, but they had trouble locating it and then following it. When it eventually 'turned out so rough and comparatively unpracticable that we gave up the task', Goyder vowed that 'on his return he would lay it out much better himself and dismiss the surveyor'.[20] Edmonds's incompetence had first been reported in 1855 by Christie, the other assistant surveyor general at the time, and Edmonds's inattention, inaccuracy, and attempts to conceal his errors had been a constant source of trouble to the Survey Department. This incident would seem to have been the perfect instrument for lighting Goyder's short fuse, but Milne did not record any real explosion. Goyder was as good as his word, and within days of having returned to Adelaide, he had written to Edmonds to suggest that he resign.[21]

Milne's observations confirm that Goyder's tendency to impulsiveness led him to routinely place heavy demands on his own physical fitness even in circumstances where it was not demanded. On the first day of their journey to the South-East, a few miles out of Echunga, where they had stopped for a snack of bread and cheese, Goyder discovered that he had dropped his tobacco case. He left the vehicle to run back and get it. Much later in the month, when a large party went from Mosquito Plains on a trip to investigate and picnic in the Naracoorte caves, Goyder demonstrated this tendency again, making a decision which would place undue stress on his body. He also displayed a degree of pig-headedness that struck Milne as comical:

> It was now getting late – nearly five o' clock … Goyder … took it into his head, in order to save time, to strike through the bush in a north westerly direction and reach it further on. Our short cut however, turned out to be very long, as the scrub of honeysuckle [banksia] increased in density until we could scarcely move on – indeed we had to pull up repeatedly and on foot, explore a way out. Goyder showed a determination to persevere at all hazards and it was amusing to see him sometimes trudging on before with the axe over his shoulder, cutting down an obstructive tree now and then.[22]

Like his father, once he had decided upon a course, he persevered. On the whole, however, Goyder's strong impulses operated within the restraints of what was seen as a 'deliberately formed judgement'.[23] As his obituary commented, his 'restless vitality' was not seen to lead to 'waste of strength. Deliberate and systematic with a strong sense of order, he showed how well he understood the true economy of force.'[24]

Goyder's dramatic exit from the diary, as he headed to Adelaide as fast as possible on the back of a borrowed coach horse, reveals the same single-minded determination, in this instance also fuelled by intense feelings of anxiety. Returning via the Coorong on 30 January, the party had reached McGraths Flat at about midday. Milne wrote:

> Poor Goyder found a telegram waiting for him stating that his wife had been confined of twins, that one was dead, and that she was dangerously ill. This telegram had arrived the day before – he immediately notified his arrival, also that he would lose no time in getting home and wanting to know how his wife was – [25]

It was Frances's eighth confinement. The reply confirmed that she was out of danger, but during the afternoon, while travelling through the scrub, Goyder 'got impatient' and, borrowing a coach horse that had been unhitched because of exhaustion, took off, accompanied by a man who was riding with the coach. The pair was soon out of sight. Goyder's already tired horse eventually refused to go any further and he was obliged to continue on foot, walking until he reached Lake Alexandrina. There, around two in the morning, he was able to borrow another horse and was taken across the river. He still had nearly a hundred kilometres to ride.

The twins had been a girl, who had died at birth, and boy named Francis Charles, who survived for a month.

CHAPTER SEVEN

Magnum opus: the people's grass

' – Sound the trumpets, – beat the drums, – the Valuating Hero comes!'
— Eustace Mitford, *Pasquin*[1]

The year 1863 had been a difficult one. The death of the twins was followed in the middle of the year by the stress of having to attend a parliamentary inquiry that had been established to sort out the maze of competing claims and contradictory stories about the copper find at Moonta. Because of the allegations made by a malicious and desperate claimant, for Goyder the inquiry amounted to a trial. The allegations were eventually dismissed, but only after three months of hearings, and Goyder's integrity, if not the wisdom of the way in which he had handled every aspect of the various claims, was vouchsafed without reservation.

At the end of January in the following year, another baby, Alexander Woodroffe, had been born. By that time Goyder had journeyed as far into the province as he was ever to go, but his greatest enterprise of travelling and observing was yet to begin. He had been set the task of valuing a large number of leased pastoral runs located throughout the province, from the lower South-East to the Flinders Ranges and across to the west of the Eyre Peninsula. Much of this country he had already visited.

Pastoralism in South Australia had developed along different lines from that of the eastern colonies, where squatting had spread like an epidemic, and it displayed a more subdued character. Although the systematic colonisers had made provision for pastoral activity in their original plan, these provisions had been impractical, and arrangements for grazing had remained informal, until a system of annual leases was introduced for land beyond the bounds of settlement and the surveyed land. The imperial response to squatting in the eastern colonies – the introduction of longer leases in less settled and non-settled areas, with the aim of providing enough security to encourage the squatters in their wealth-generating activities without losing the option of resuming the land in the future – did not affect South Australia until 1851, when it was interpreted as a new set of lease conditions that included 14-year leases. The longer leases encouraged pastoralists to invest capital to improve their runs, and large areas were leased under this system at what were later recognised as very low rates. In 1858, when the government was short of funds, it turned its attention to these leases, but since the 'pastoral interest', as it was termed, was well represented in parliament, the government could not levy additional charges without granting concessions. An additional charge per head of stock, and the right of revaluation at the end of the 14 years, was imposed in exchange for a five-year extension of the lease. As a result, the leases taken out in 1851 required revaluation by 1865.

A most impossible task

The problem was to devise a method of assessing runs that the squatters accepted as fair. Freeling eventually came up with a scheme that classified runs into groups. He considered that it would be workable 'if the best part of one person's time were expended in travelling over the country, for the purpose of valuation'.[2] Henry Morris, the chief inspector of sheep, was selected for the new position of estimator of runs. He received his instructions in January 1860 and took 13 months to complete the task. By the time he returned, a select committee of the House of Assembly had been appointed to inquire into the whole business of assessment, and the surveyor general's office was occupied by Goyder, who had also become the new valuator of runs.

In April 1861, Goyder wrote a long report to the commissioner, suggesting how the 1858 *Assessment on Stock Act* might be amended. It was accompanied by a copy of the Act with Goyder's proposed amendments added to the text and in the margin. With a typical absence of self-doubt, 'objectionable' passages, as he considered them, had been 'struck out'.[3] His report provided the commissioner with a helpful account of the process by which the evaluating procedure had evolved, but Goyder recommended that classification be abandoned altogether and that the runs be assessed simply according to grazing capabilities and the distance from the shipping port or market where the bulk of the stock would be exported or sold. Citing with approval the system in use in New South Wales, he recommended a similar system in which, essentially, squatters would report their own stock figures. Writing of the role of Morris, Goyder's objections to the existing system might well have been penned in a clairvoyant trance. 'I may be permitted to observe,' he informed the commissioner:

> that whilst the Government have selected a gentleman whose fitness for the office is beyond dispute, and in whose impartiality there is entire confidence, it may be questioned how far any one individual, though constantly employed on that particular duty, could obtain such information, by inspection alone, as would enable the Government to value the respective run to the satisfaction of the country and the persons interested.

It was hard enough to estimate the capacities of a single new run before it had proved itself, Goyder had explained, let alone for any one person to try and estimate the carrying capacity of tens of thousands of square kilometres of country, and good rains could change the appearance of the land so dramatically that it became hard to recognise. Unless valuations were informed by the experience of the leaseholder, any such report could only be valid for the season in which it was made, with the inevitable result that 'questions will arise with every fresh assessment, and the Act, in all probability, be made a constant theme of discontent between the Government and the squatters'.[4] Morris later reported that at about this time he and Goyder had agreed that the task was 'most impossible' for one man to perform, though Goyder, as he recalled, had used 'stronger language' than this to convey his opinion.[5] It is hardly surprising that the squatters regarded Goyder as someone they could rely on.

About three months later, in mid-July, Goyder forwarded to the commissioner the report that Morris had finally completed, together with an 11-page table of valuations, based on Morris's estimations, in which the old and new assessments were presented side by side. In his covering letter, Goyder pointed out that Morris's inspection had been made during an unprecedented drought, so that, 'presuming his estimate to have been formed from actual observation, it may reasonably be increased for ordinary or more favourable seasons without injustice to the lessees'.[6] Even so, the squatters were satisfied since there were many decreases in rate.

In his report, Morris, like Goyder, recommended that the classification scheme be abandoned. It was so difficult for a single person to estimate grazing capacity correctly that an avenue of appeal would be needed, he warned, and the task was 'most laborious and fatiguing'.[7] Morris also reported on the damage to the pastures already done by overstocking. When combined with drought, the results were disastrous. On the western plains, in 1859, the state of the animals had been 'most distressing'. He recalled that:

> on many of the runs they were hardly able to crawl, and at the few permanent waters they were lying dead in hundreds, and the plains completely dotted with the carcasses of others which had managed to stagger out some distance and die. The feed round the waters, principally saltbush, was fed off for miles, and the plains about as bare as a roadway.[8]

With all the runs assessed for the purposes of the 1858 *Assessment on Stock Act*, and with rain falling again and grass growing, the whole business of pastoralism and the still-unresolved problems surrounding the valuing of runs retreated from political view – until January 1864, when the first application for the renewal of one of the original 14-year leases was received. The government was only six months old and was composed entirely of members so new to the responsibility of a portfolio that the prior experience of any of them, including the leader Henry Ayers, could be reckoned in days. Faced with the task of revaluing a large number of established leases, they sought advice from the Crown Law Office, but failed to recognise and address both the unresolved difficulties that the previous attempts had identified and the hugeness of the task.[9]

They did understand that the Act did not limit the undertaking of valuations to one individual, although this had been the practice established by Freeling. There seems to have been universal agreement in the colony at the time that there were only three men capable of valuing runs on this scale: Goyder, Morris, and the old stalwart, Charles Bonney, the former lands commissioner. Bonney had been a pastoralist himself and had made his name and reputation in Adelaide, when with his partner Hawdon he had been welcomed into the town with the first herd of cattle overlanded from New South Wales in 1838. In 1842 Bonney had been appointed commissioner of crown lands in the days before elected government. At this time the position had been largely confined to administering pastoral leases and timber licences. When the government offered him the position of valuator in 1864, Bonney declined it rather than place himself at the centre of a conflict between the pastoralists and the government. In any case, he had maintained a total and longstanding opposition to the *Assessment Act*, which he understood to provide squatters with the possibility of obtaining their lands in perpetuity. Although he did not object to these particular renewals, he considered that his position would make it impossible to be a valuator. It was not long before Bonney was expressing these opinions at public meetings, warning his fellows that wherever public lands were allowed to be held 'by a class', it had been found 'necessary always to take violent measures to wrest the land away' from that class.[10]

Morris was not a suitable candidate because he was seen as being compromised by having been permitted by the government to accept a purse of 600 sovereigns (gold coins worth a pound) from squatters for having eradicated scab, a highly contagious mite infestation that destroyed the value of the fleece, in the flocks of the South-East. Since this was a very large sum – more than five years annual salary for many – Morris was presumed to be 'mixed up' with the squatters, or, as it was later put: 'if a squatter followed him with a cheque for £5000 did they think he would throw it down? Certainly not.'[11]

With Bonney and Morris out of contention, Goyder was left to carry out the 'most impossible' task alone. By personally inspecting the runs, he was expected to estimate their carrying capacity and to determine a valuation on this basis, also taking into account factors such as ease of transport and transport costs to market. In addition to this, he would remain – apparently at his own choice – solely responsible for the work of the position of surveyor general. The agreed plan was that he would not be absent from Adelaide for longer than a month at a time.[12]

Setting out

On Saturday 27 February 1864, the commissioner of crown lands, Lavington Glyde, wrote to Goyder requesting him to proceed 'to the North, on Monday next', to begin the valuation of the nearest group of runs. Glyde deferred to Goyder's knowledge and experience in the matter, instructing only that:

> your leading principle should be, not to allow your valuations to be at all influenced by any reference to the present rentals, but, after duly considering the merits of each case, to propose a rental which shall fairly represent the present annual value of each run.[13]

He then listed five categories of information that the reports on each run were to contain. These included a plan of each run, showing the nature of the country and the location of improvements, and the estimated carrying capacity and actual stock numbers. This was almost certainly Goyder's own advice being returned to him with the authority of government, just as the startling demand that he leave in a day-and-a-half was most likely a response to the fact that he had declared himself fully prepared and ready to depart.

The first part of the pastoral 'north' that Goyder went to inspect is today called the Mid-North and is a region containing some of the best pastoral and agricultural land in the state. Pastoralists had begun to occupy these rich lands in the 1840s, and with the advent of secure long-term leases in 1851 their future had been assured. These were the squatters that Frederick Sinnett had had in mind when, about two years before Goyder set out, he produced this unflattering account:

> Squatting – that is to say, sheep and cattle farming on unsold lands, rented in large blocks, from the Government – has, especially within the last ten years, been the most remunerative pursuit carried out in the Colony. The enormous increase of their flocks and herds, the vast territory they occupy, and the national importance of what they produce are well-known. Here it need simply be remarked that many are enabled to live in grand style in England; and that our greatest nabobs, the men of leisure and fine houses and fine carriages, are almost invariably squatters. One proof of the remunerative

character of the pursuit is, that all sorts and conditions of men have thrived in it indiscriminately. Gentlemen from England, without experience ... – men who have been fortunate enough to take to the one pursuit in which they could not help making money in spite of themselves.[14]

These were also the men who in the early years had been able to purchase the best parts of their runs freehold through special surveys, and who later, as hundreds were declared around them and the lands resumed and sold for agriculture, could use the wealth they had generated while leasing these rich lands cheaply to buy back the lots. During the 1860s great stations were established in this region, and even though the leased lands were later resumed by the government, the freehold core sections of some of these giant holdings were maintained and continued as Australia's most famous merino stations and studs. The names of both the homesteads and stations of these big runs still survive, attached to townships and topographical features in the region.

Revaluing the leased sections of these runs at something approximating the real current rate meant that a body of wealthy and influential men were going to be seriously challenged. Before setting out, Goyder tried to protect himself by persuading the government to agree that his work would not be made public until it was complete.[15]

Prime country

By Wednesday 2 March, Goyder was passing through Anlaby on his way to assess Emu Flats, a large run which covered most of the land between Kapunda and Kooringa (Burra). Emu Flats was leased by Frederick Dutton (or 'Fd. Dutton Esq.' as he is identified, with unique respect, in the valuation records) and functioned as a part of Anlaby. Goyder made his base at Archy's Hut, a shepherd's hut, and on the first day of the valuations he recorded having ridden 30 miles over the run.[16] From Anlaby and Emu Flats, Goyder went north to Murray Flats, then west across to the big runs.[17] Canowie, Booborowie, Booyoolee (pronounced Beaulie or Boolie), Bundaleer, Bungaree and Anama were all included in the 18 runs that made up the first batch of valuations.

Although the official plan meant that that the colony would not be without a surveyor general for more than a month at a time, it was still necessary for Goyder to keep up with his official correspondence while in the field. On Saturday 5 March, while camped at Murray Flats, he wrote a long letter to the commissioner providing advice that had been requested on the survey of the Northern Territory.[18] In the letter he also described the method of examining the country developed during the triangulation of the region around Lake Eyre

South; he was so pleased with the systematic approach that he had begun to use a similar method for examining the runs. Like Morris before him, he had to 'cross and recross the country'. The system of examination he followed involved:

> visiting and fixing the position of huts, waters and wells; sketching the natural features of the country, and minutely noting the character and vegetation; a method that enables me to judge with tolerable accuracy, when the plot is made of the whole[,] of the country that can be fed over, and the portions omitted.[19]

Goyder was supported by a party that established and maintained camps, provided him with fresh mounts, and kept the plotting and recording of his work up to date. Since he changed horses twice on most days to enable him to travel long distances, his party was kept constantly on the move. (One of the legends about Goyder is that he wrote up his official diaries 'at times without stopping to rein in his horse', and the story probably traces its origins to this period. [20]) He lamented that his draughtsmen, whose task it was to 'lay down each day's work', had only enough 'leisure', as he put it, to provide him with the outline plans he needed to carry on.[21] His notes were transcribed neatly into large bound volumes, where the text crowds over the pages relentlessly, streaming around the numerous tiny maps, inscribed on now-yellowing tracing paper, which had served to spare the draughtsmen from the pangs of fruitless idleness.[22]

On any run, depending on its size and character, Goyder's examination might have included a series of exploratory forays from a central camp or hut, as well as point-to-point tours of wells and water sources, and of houses, huts and sheds. The entry for Emu Flats – not at all atypical, despite it being the first run examined since the procedure had been determined in advance – occupies about four pages and contains four maps. A small section from Goyder's record of observations on one of his several excursions from Archy's Hut on Thursday 3 March, displays clearly his method of recording observations along a line:

> S. at ½ m. and of plain to mallee scrub with small plains – well grassed at 1 m. thick scrub. Bushes – but little grass – at 1¼ m. well grassed plain half a mile wide. at 2 m. S. east corner of plain. followed by dense scrub & bushes with but little grass, opening & improving toward the South; at 3 m. edge of plain with clumps of scrub. plain extending east and west at 4 miles. track to bend north of water course. E. about a mile along road to old section dug water hole near proposed site for Govt. well SE. ¼ m. to clumps of pines: over well grassed plains beautifully studded with pines.[23]

These transects were plotted from bearings onto the maps. The next day was spent on a tour of wells and watercourses on several runs.

The final valuation for each run, in its printed, public form, consisted of a paragraph describing the location of the run; a long paragraph under the heading 'Estimated Grazing Capacity', describing the country, available fodder and water sources; and a paragraph of 'Remarks', which usually described and valued improvements, noted the state of the roads, and stated the relationship of the run to other runs worked by the lessee. If necessary, any comment on the estimated stocking rate of the previous assessment was also included here. The four pages of detailed notes and maps on Emu Flats yielded this summary of capacity:

30,000 sheep, or 200 per square mile

The run comprises about ninety miles of well-grassed sheoak spurs, and very well-grassed flats; fifteen miles of undulating forest land and ranges, with grass and bushes, and about forty-five miles of mallee and pine scrub, with occasional patches of grass and saltbush. There is [sic] about two miles of running water up on the run, in Brady Creek, and three springs strong enough to supply flocks, in addition to several of a smaller size, near which wells have been sunk to increase the supply. The waters are, however, all near the west boundary of the run, and [water] is only obtainable on the east side at great depths. About a third of the country, therefore – mostly scrub – is only available during the winter months. There is, also, running water upon purchased land, near Long Hill.[24]

The first batch of valuations was delivered to the commissioner at the end of May 1864.

Raising the country

A week after preparing his first report and batch of valuations, Goyder was back in the field, though further north this time.[25] In parliament, trouble was brewing. Randolph Isham Stow, the self-appointed leader of the opposition, began to raise questions and succeeded in having Glyde table his instructions concerning the valuations.[26] This prompted the *Advertiser* to bring the subject of pastoral leases and the details of the renewals back to public attention, with Goyder in the spotlight from the outset. A 'more critical and responsible duty could scarcely devolve upon any public officer', the editor, John Barrow (himself a parliamentarian) advised readers, and the failure of both the Act and the commissioner's instructions to articulate principles of procedure made the whole thing dependent upon the judgment of the sole valuator.[27] A few days later a letter appeared railing against the iniquity of the squatters' attempts to renew their occupation of 'the people's pastures' and 'the people's heritage'.[28] The 'people's grass', as it would also be known, had become part of the framework and rhetoric of the debate.

Stow launched a vague attack on Glyde, who in return accused Stow of 'coquetting with the pastoral interest' with the aim of becoming attorney-general, and describing the Pastoral Association as, 'a confederacy that would endeavour to embarrass any government that would oppose its efforts to occupy South Australia as one vast sheepwalk, at 10 s[hillings] the square mile', with the ultimate goal of reducing the government 'to the position of their most obsequious servants'.[29] Declaring himself on the side of economic and social justice, Glyde warned Stow and the Pastoral Association that any attempt to take the valuations out of his control and into their own hands would be a mistake, and, 'in solemn tones, with arm outstretched, and hand clenched', Glyde threatened that if he were dumped, he would take with him the information he already had on the valuations, and use it 'for the interests of the colony'.[30] The country 'had not yet been roused to a full sense of the importance of the question', Glyde advised them, and 'if the squatters were unwilling to pay a fair rental for the waste lands the public would be roused, and they would probably have to pay a great deal more than they contemplated at the present moment'.[31] His threats amounted to an admission that the valuations already completed had increased the rents – if the squatters were still in any doubt.

Ayers formed another ministry on 22 July 1864 with Stow as attorney-general and Milne, who had interests in a run, replacing Glyde as commissioner of crown lands. Seeing Glyde's prediction come true, even the *Register*, which had been shocked by his 'outburst of anti-squatting rage' and his threat to 'raise the country against the squatters', was compelled to admit dismay.[32] A sense that something had gone wrong in the operations of the legislature quickly gained hold. The issue was taken up by Barrow in the *Advertiser* and by the Mayor of Adelaide, who called a public meeting. By the time of the meeting, Milne had resigned, but too late to quell deep and spreading anxiety about the political intriguing of the Pastoral Association. Regardless of Glyde's threat, the country would be raised against the squatters anyway.

Meanwhile, Goyder had been hard at work. On Monday 6 June, he travelled about 125 kilometres from Adelaide to Black Springs. The next day he was on Hallett's run. (One of his daughters would eventually marry a son of the younger of the two Hallett brothers, Alfred and John.) From there Goyder and his party seem to have travelled up to the two northernmost runs they would be assessing, Wilpena and Aroona, in the central and north-central Flinders Ranges, following the usual strategy of visiting this hot dry country during the coldest and wettest season. Working their way back south, they examined a further 24 runs.

Goyder had hoped that his final report would be ready in early July, but there had been problems with receiving important mail from Adelaide. Goyder had asked for help, but no response was received from Glyde, and by the time he had returned to town to sort the matter out both Glyde and Milne were out of office.[33] Arthur Blyth had been asked to form a government, and he had taken the position of commissioner of crown lands himself. The new government was a highly irregular and unlikely arrangement of men not known to admire each other or to agree. To a constituency already on the alert, watchful for the appearance of 'combinations' and suspicious of the attempt to keep power in the hands of a few, the outcome of this latest work of parliamentary choreography was evidence that things were going from bad to worse. No sooner had Blyth finished introducing his new ministry in the Assembly than Strangways was on his feet asking for an adjournment and warning that the public could come to no other conclusion than that 'the whole thing was planned so as to place the Government in the hands of a clique', and a clique which contained some of the largest squatters in the colony at that.[34] In the *Advertiser* Barrow had already proclaimed the cure for government by cliques: dissolution.

Bringing down the House

The next day, 10 August, a motion was tabled that all valuations so far received should be presented. Blyth explained that only 18 of 80 valuations required for the northern region had been received, and that he had pressed Goyder for a finished report. Goyder had promised in writing to present a report by 10 September, and Blyth assured the House of Assembly that every document would be laid before them on that date. Until that time, however, the valuator's independence should be preserved. But the House would have none of it. That evening the mayor conducted another public meeting in White's Rooms, the largest meeting hall available, to discuss dissolving the Parliament. About 800 people showed up, including members of parliament and many leaders of the community. The line-up of speakers was familiar, and the focus soon turned to the valuations. After Thomas Reynolds, a former treasurer, had assured the crowd of Goyder's honesty and integrity and his capability to value the runs – though he believed that two or three valuators would have provided a better safeguard against error – Barrow spoke. Raising laughter, cheers and applause, Barrow worked the crowd with a mixture of satire directed at his opponents, a finely tuned presentation of himself as the put-upon but upright defender of the people's cause, and regular displays of modesty carefully calculated to call forth roars of encouragement and approval from his supporters. Justice should be done to the community, Barrow exhorted, 'and it was certainly not just to require the farmer to pay £1 per acre, or £640 per square mile, and only demand at the rate of 1d. per acre from the squatter'. He concluded by defending his use of his newspaper to promote his views, which was followed by tremendous cheering. Barrow was followed by Bonney, who placed the valuations and the battle they had engendered in the grand context of world history and class oppression.[35]

What was not generally appreciated in Adelaide was that the situation in the Far North was very different from that in the Mid-North, where, as Sinnett had commented, even those with no experience or ability could grow rich. While Goyder was still working his way assiduously across the runs in the north, but after the furore over the rich leases had begun, a 'Voice from the North-East' wrote to the *Advertiser* to warn that people in Adelaide:

> hear of sheep-farmers (a few) going home with large fortunes and see others doing the 'heavy swell' in town, and naturally enough form their opinions from what comes under their eyes. Forgetting that for one squatter who can afford to live in Adelaide, there are fifty who cannot: forgetting, too, that these men are the early birds of the colony who picked up the worms. We, therefore, who came later in the day and had to work up the refuse of the country, reasonably object to being placed in the same boat as our more fortunate brethren. Before many weeks are over the popular delusion that all squatters are coining money will have exploded.[36]

The delusion would be exploded by the continuing lack of rain in the north.

The first 18 valuations of the prime country in the Mid-North were tabled two days later, on 12 August, in a summarised, tabular form. They seem to have been received like a short sharp blow that left the squatters and their supporters temporarily winded. The historian Edwin Hodder, writing only 30 years later, recorded that at first they were 'dumbfounded'.[37] The *Register* introduced these first valuations with the sort of solemn directness appropriate to reporting a death or disaster, the awfulness of which words could neither truly convey nor in any way ameliorate. The absence of any developed response betrayed real shock. The valuations amounted to 'confiscation' according to the writer, who could only muster in response a request that Goyder be called upon to explain his enormous increases.[38] At the *Advertiser*, Barrow crowed that the valuations proved, 'what we have been contending for – namely, that down to the present time the people's estate has been literally thrown away. The public will now see what they have hitherto lost, and what they may hereafter gain. But the utmost vigilance will be requisite.'[39]

For some runs the estimated grazing capacity had also been increased dramatically. At Anlaby, Frederick Dutton Esq. found himself assessed as capable of carrying an extra 5000 sheep on Emu Flats. His annual liability for this run was to explode from £250 to £2700 – almost an elevenfold increase – if not to £3000, if the value of improvements was not deducted. Emu Flats sustained by far the greatest increase of any valuation, but the annual rents on all the big runs of the Mid-North were increased by between five and ten times. As one wit later summarised the situation, in a ditty to the tune of 'Green Grow the Rushes O!':

> One mornin' blue came Goyder true,
> And turned things tapsaltearie o!
> Four hundred odd, says he, won't do,
> Four thousand is more cheerie o![40]

The *Register* soon recovered its poise, warning of such profound business insecurity that foreign investment would be driven away. Squatters could find no escape in going to auction, because buyers from other colonies would be deterred.[41] At the *Advertiser*, Barrow totted up the tens of thousands the government would gain in revenue for the public good, while warning that, 'the public must be watchful, or the prize will even now be snatched from their grasp'.[42]

Goyder was fortunate to be out of town during the uproar of the following weeks. In what were acknowledged as 'wars of the press', the two papers, both with weekly subsidiaries and editors who were also members of parliament, argued figures and cases.[43] Both agreed in calling for Goyder's statement of principle or a more detailed report. Impressed by the degree of commotion, the government produced a series of surprising decisions: it accepted the validity of Goyder's appointment as valuator, which the squatters had questioned; it declared that no statement would be delivered until after Goyder's report due on 10 September had been received; and it announced that, unless the valuations were 'manifestly unfair', the government would recommend the governor to accept them.[44] It even exceeded Goyder's zeal for making the squatters pay by choosing, against his advice, to include the value of improvements in the rental calculations.

The squatters and their supporters were dismayed, but the anti-squatters were not mollified, suspecting that the whole thing was a diversion designed to allow their opponents time to formulate their complaints in detail and demonstrate that the valuations were indeed 'manifestly unfair'. As a result of their pressure, the complete version of the original 18 valuations was tabled. In both houses, pressure was brought to bear on the ministry to accept or reject the valuations immediately, to prove that no cards from the Pastoral Association were hidden up their collective sleeve. At the same time, petitions calling for additional valuators to be appointed were being prepared for presentation to both houses. 'We understand that an attempt is about to be made to induce the Government to set aside Mr. Goyder's valuations, and to institute an entirely new system of valuing', Barrow raged. 'Should the Government agree to such a course it would prove that they are mad ... The eyes of the whole country are now fixed upon Mr. Goyder, upon the Parliament, and upon the Government, and little quarter will be given to those who attempt to defraud the people of their just due.' Again Barrow advised the people to beware, exhorting them to: 'shake off all apathy, and rouse themselves to the really great question before them'. It was clear what should be done:

> let the people meet, in town, in country; in large meetings and in small meetings. Let them petition His Excellency to dissolve the present Parliament, for unless this is done there is little hope and no security.[45]

And meet they did. All over the province people met in public halls and public houses. Hardly a day passed without a meeting, and on many days there were more than one. Transcripts of the proceedings of many of these meetings were reproduced in full in the papers, and even the outcomes of the smaller and more remote meetings were presented in brief. These reports filled columns alongside the transcripts of debates in parliament, where members of both houses debated the valuations and pastoral leases almost exclusively. The valuations and the leases were the most frequent subject of newspaper editorials and leading articles, and through their correspondence columns, others joined in. When the discussion touched on the valuator, as it constantly did, it was usually the adequacy of a sole valuator that was questioned: Goyder's competence was acknowledged by most, and his integrity acclaimed by all. Barrow dismissed all the squatters' complaints by pointing out that it had been well known before the appointment was made that Goyder was 'the squatters' favourite', that they had urged the government to appoint him, and that there was evidence to prove this. Henry Ayers, the chief secretary, agreed, telling the Legislative Council that not a single word of protest or rejection had been heard from the holders of pastoral leases when Goyder had been appointed sole valuator, and that the leaseholders had 'professed every confidence in him'.[46] One of the leading opponents of the valuations, Samuel Davenport, claimed to admire Goyder's 'honesty and intelligence', and admitted that he was an 'excellent judge of water capabilities and feed for sheep and cattle'. Nevertheless, Davenport argued, Goyder was necessarily 'incapable of knowing the influences of climate and soil on stock', since this was always particular, local knowledge.[47] The squatters as a whole pursued this theme, expanding upon the difficulties of anyone valuing any run other than their own – a position that Goyder himself had championed in 1861. It was his recognition of the importance of local conditions that had driven him to examine the runs with such exhausting care.

In the end, all the squatters' complaints came to nothing. In the prevailing political climate, any suggestion that they value their own runs was inconceivable, while the only other possible valuators were the unavailable Bonney, who wanted to send them all to auction in any case, and the compromised Morris. The colonists were stuck with Goyder, just as much as he was stuck with the task. As a correspondent to the *Advertiser* advised, they could not even come up with a suitably qualified and impartial assistant for him, let alone a replacement, and, they were 'much indebted to the Surveyor-General for his moral courage and evident desire to do his duty in this thankless office. He has undertaken a task in the performance of which he is sure to displease either the lessee or the public.'[48]

Eventually, the members of the Assembly, unable to bear each other and the constant wrangling that occupied them, and incapable of dealing with other issues as they waited for Goyder's return, decided for a second time to adjourn, this time for a fortnight. Both papers agreed that this was for the best since continuing the debate would do nothing but generate more bad feeling. Yet another 10 days were added to Goyder's submission date of 10 September – to give the government time to discuss the submitted valuations with Goyder and to prepare their decision.

A wholesome precedent

Goyder arrived back on the evening of Saturday 10 September, weary and sick. On Monday, he remained at home. The press and the government waited anxiously. On September 19, the last day left to him, Goyder wrote to the commissioner. The reports on the runs he had examined still had to be laid out and formally recorded, and Yorke Peninsula remained unvisited. He regretted the delay, but was unrepentant:

The principle upon which the valuations are framed admits no undue haste – certain elements have to be considered in minute detail, and these form the data from which the result is obtained: their compilation is necessarily a work of time; and, as any omission would render the documents incomplete, I cannot take blame for the delay, although I much regret that the time taken has been longer than was anticipated.[49]

He attributed his delay and the fact that he had not been able to examine the runs on Yorke Peninsula as intended to both the 'rough nature of much of the country under inspection' and the 'unusual dryness of the season'.[50] On the same day, in another report, he presented exactly the view put forward by the anonymous squatter from the north-east:

an exaggerated idea appears to be gaining ground as to the total amount likely to be realized upon the whole of the leases falling in. Many of these in the north are of the roughest possible character, and managed at only greatly enhanced cost compared with those nearer town: and, from the uncertainty of the seasons, do not realize profitable returns, at most, more than five seasons out of nine.[51]

Goyder's letter and the 15 completed valuations were submitted with a two-page report providing an answer to requests for a statement of the principle guiding his work. At the same time, Goyder returned a swag of letters of complaint from squatters, some accompanied by replies, others with only his 'endorsement' – the terse comment he had scratched onto the outer sheet of the folded correspondence for the information of the commissioner. On H.B. Hughes's letter challenging the stated number of animals on Booyoolee, Goyder noted: 'Mr. Hughes objects to the quantity of stock stated to be on his run; Mr. Hughes was my informant' – and, in any case, he had made an independent assessment as well.[52] He advised that Abraham Scott of Canowie could test 'the justice, indeed the liberality' of the valuation he had received by going to auction.[53] Since Goyder refused to accept any error and considered all his valuations liberal, his brief replies seemed to some to constitute an arrogant dismissal of the petitioners. Certainly such an old pastoral pillar of the colony as George French Angas regarded them as an unwarranted insult to men of dignity.[54]

Goyder's account of his guiding principles did not please his critics either. Those arguing the squatters' case seemed to imply that only some sort of mathematical formula (probably involving a value per head of stock) could avoid reliance on the valuator's fallible judgment and be truly fair – though no doubt if such a formula had been produced it would have been deemed arbitrary and impractical. What they received was a common-sense description of what had been done. Goyder explained that the runs were carefully inspected, so that the area of different qualities of land could be calculated and the carrying capacity estimated, and listed other factors which were taken into consideration: seasonal availability of water, trespass, vicinity to stock routes, nearness to markets or ports, different weight of clips, and the value of improvements.[55] In determining the final figure, a reduction was then made in each year's rental, of up to one-fifth of the estimated value of the improvements. The expectation of renewal, the impact of stock travelling over the run, and the understanding that the runs would provide basic hospitality to travellers had been taken into account as well. Although the precise way in which these factors were brought together was never disclosed, Goyder seems to have played along with the prevailing view that some kind of formula would be used. The tabular presentation of the first valuations seems to imply that a final rental was calculated from the figures for the various columns, while Goyder's own words suggest a thoroughly objective, quantified approach. In an unidentified autobiographical account, he declared that:

> my journals were so complete that the valuations were made by clerks reading the records and without reference to the names of the leases until the valuation was complete, and I was placed in possession of data that thoroughly qualified me to perform the duty required of me ...[56]

Even so, in every one of the first 18 valuations of the prime Mid-North runs (though not in later valuations), the total annual rent, both with and without deductions for improvements, divided by the number of square miles leased, results in a whole figure (which was not included in the valuation). It can only be inferred that the real, but unstated, valuation was Goyder's estimation of the number of pounds per square mile each run was worth, informed, although very conservatively, by an idea of what these huge runs might bring per mile, if divided up and auctioned. At a meeting at Port Gawler, G.S. Kingston, in support of Goyder's methods, explained the valuations in terms of pounds per square mile.[57] The valuations were not the product of a crude formula, but of a complex assessment – as Goyder tried to show. What he would not explicitly admit, for fear of playing into the squatters' hands, was that they were also ultimately the product of his own perception and judgment. Because no formula was ever produced, people remained curious: nearly a quarter of a century later, when C.J. Sanders, a surveyor who had been with Goyder during his inspection of the runs, was a witness before the Lands Laws Commission of 1888, he was asked how the valuations had been calculated.[58]

Concluding his report, Goyder warned about anticipating the same level of income from the northernmost runs and finally took the opportunity to respond – very directly – to his critics:

It has been alleged, since the publication of my report of the eighteen runs ... that my valuation amounts very nearly to confiscation. I have carefully perused all that has been printed, and listened to much that has been far from agreeable or complimentary upon the subject, but I have seen or read nothing to alter my conviction that the valuation made is liberal. Indeed, it is difficult to understand the outcry that has been made in the face of sums paid to private persons and companies for grazing land of even inferior quality.[59]

In fact, on 19 out of the 39 of the northern runs visited, Goyder had *lowered* the estimated grazing capacity and explicitly acknowledged that the previous estimate had been too high. For many of the remainder, the estimate stayed unchanged, but the rents were increased all the same, though he did not increase rents on the far northern country to the extent he had done for the prime country further south.

Once the valuations were accepted (which did not mean that the struggle was over), the treasurer, John Hart, told an Oddfellows anniversary dinner in late September that the ministry had examined Goyder on the valuations, 'but his answers were so discreet, so straightforward and honest, they could only say he had come to a just decision'. Hart claimed that he did not know:

a more wonderful thing than the amount of work and intelligence brought to bear on these valuations in the short time by Mr. Goyder, who deserved well of the colony – he was second to no officer that that colony had ever had.[60]

The day after these assurances were published, Hart repeated them in the House of Assembly, assuring the members that 'no one could imagine the care and labour which Mr. Goyder had bestowed upon his work'.[61]

When the government accepted the valuations and all of Goyder's related recommendations, Barrow, naturally, was overflowing with satisfaction:

Ministers have examined and re-examined Mr. Goyder; they have questioned him and cross-questioned him with regard to everything in his figures appearing in the slightest degree doubtful; but his explanations have been so complete, and his justification so conclusive, that all hesitation has vanished, and the Government unitedly pledge themselves to support the valuations.

Indeed, it would have been strange if they had decided differently:

Mr. Goyder's valuations are most trustworthy. Successive Governments have reposed full confidence in that officer; confidence in his knowledge, confidence in his judgment, confidence in his independent honesty ... The people have confidence; and we say it would have been strange indeed had the valuations of so able and upright a man been rejected.[62]

If anything, it seemed Goyder had been too kind. Barrow reported that the Honourable Messrs Baker and Davenport (who led the opposition to the valuations in parliament) were even now transferring the Tungkillo run from one to another at nearly twice the price per square mile assigned by Goyder. The mid-October issue of the *Australasian* devoted its second leading article to the valuations in South Australia, finding them 'full of moment' for Victoria and other colonies. South Australia was providing a 'wholesome precedent' in the business of revaluing runs, and the paper, referring to the letters of complaint written by prominent squatters, supported Goyder in his blunt response.[63]

Mumbo jumbo

Barrow's warning that the people would need to be watchful proved correct: the squatters twisted and turned in their attempts to evade the valuations, even after they had been accepted. Another challenge was made to the validity of Goyder's appointment, this time through the law. One squatter who was seen to have succeeded in his evasion was George Hawker, a wealthy parliamentarian with runs in the Mid-North and beyond. Hawker requested the declaration of two hundreds around his runs in the Mid-North, despite this process being known as 'killing a squatter' because it involved the resumption of leased pastoral land for surveying into agricultural lots. Since he had already bought up all the best land on Bungaree and Anama, he could not be 'killed', and once inside the hundreds his leased land was subject to different regulations: Goyder's valuation would no longer apply. But Goyder, who despite his intransigence did not share the polarised anti-squatter perspective of the supporters of his valuations, recommended the declaration of these hundreds anyway. The *Advertiser* was appalled, leading the *Register* to remark that:

> When Goyder is in favour of high valuations, then Goyder is a little wonder; but when he recommends anything that does not accord with the creed of his worshippers then his recommendations are not to be trusted. Like the Mumbo Jumbo of the fanatics, he must be reviled because he has not prophesied correctly.[64]

Another squatter who continued to resist was John Baker in the Legislative Council. Baker moved the establishment of a select committee, through which individual valuations could be declared unjust and therefore invalid. The Legislative Council agreed with alacrity – but it presented Baker with a committee composed of only two squatters, himself and Samuel Davenport, and three anti-squatters, including Barrow. This result was so little anticipated by Baker that he believed that there must have been a mistake. Baker's misjudgment was greater than at first appeared. Squatters who were called before the committee and given an opportunity to protest made fools of themselves and their cause by refusing to divulge anything of their financial situation. This made it impossible to prove that the valuations were unfair, but at the same time confirmed the perception that they were literally embarrassed by their wealth. Let down by his friends, Baker was frustrated by his enemies. Since the committee as a whole would not recall Goyder from the field to attend it, Baker attempted, acting on his own authority as chairman, to have a trooper sent to inform the valuator that his presence was required, but the attempt failed.[65] The committee never returned a final report.

The squatters finally brought about their own defeat by engineering the passage of new land Regulations containing a clause that would function, in the long term, as a loophole to evade the valuations. This resulted in a deadlock in the legislature, where members could not pass the Regulations or the Budget. Eventually, they could not even put together a ministry. Encouraged by the governor, an arrangement was made whereby temporary supply was granted until elections for both houses could be held in the new year. The subject of the valuations drifted into exhausted abeyance and the citizens of South Australia were able to spend summer thinking about something else, although one undaunted correspondent to the *Register* still had enough energy to complain that the 'enormous power' vested in the valuator, while possibly 'very well suited to a grand Russian General or Governor of a Province', was 'utterly preposterous in the hands of a little surveyor in an English colony'.[66] Everyone else was sick of the subject.

Barrow applied himself to influencing the way in which voters assessed candidates in the coming election. Since it was possible that the squatters and their supporters could initiate fresh legislation to overturn the valuations, Barrow decided that candidates merely confirming that they accepted the valuations was not good enough: they should be required to express positive support for Goyder's valuations and endorse a determination to see them *not* overthrown.[67] In a continuation of the spirit of the previous year, Barrow's advice significantly influenced the popular response and the course of events. The valuations were already the stumbling block over which one government had tripped, and now they had become a sort of gate to divide parliamentary candidates into the fit and the unfit, the 'goyderites', as they were sometimes termed, and those who opposed the valuations. Goyder, of course, was not aligned with either side. While he was no doubt keen to ensure that the public resource of the best pastoral country was not exploited for the exclusive benefit of a small number of people, he was certainly not an anti-squatter, as his enthusiasm for locating and surveying land for pastoralism made clear. For him, as a public servant, there was no inherent conflict between wanting to see pastoralism prosper and grow and wanting to see the pastoralists pay a fair price to the rest of the community for the land they used. The extraordinary labours he was imposing on himself were intended to ensure fairness for the squatters as much as for the government – although the squatters didn't see it that way. But although he was neither for nor against the squatters, it was not mere accident that he was the figurehead at the centre of this major political crisis. In the eight years of responsible government there had been 12 ministries. The South Australian Parliament was not dominated by opposing parties – parties did not come until much later – and the government of the day consisted of shifting alliances of parliamentarians. Although some politicians occasionally managed to hold onto a portfolio from one government to another, the situation was hardly conducive to the development and execution of consistent public policy and inevitably meant that the influence of senior public servants was enhanced. Since land was the central concern of government during the early decades, and Goyder possessed both considerable administrative ability and an unrivalled knowledge of the land, which rendered any government largely dependent on his advice, it is not surprising to find him at the centre of a major clash of interests over the use of the land. But Goyder's character, too, placed him personally at the centre of the dispute. In his willingness to revalue runs dramatically where it was appropriate, even though he knew he would be attacked, and in refusing to make any concessions when he was, Goyder left the parliamentarians and society as a whole with no option but to deal with the issues of pastoral power and of proper returns for the use of public assets. His moral courage was acknowledged at the time and for the rest of his life. As South Australians prepared for an election, another round of public meetings addressing the vexatious issues of the valuations were held around the town and the province as a whole.

West and east

Following his belated return from the north, Goyder had promised that he would press on with the task of valuing. After 'an indispensable visit to the Murray, I propose visiting Yorke's Peninsula and Port Lincoln, and finally the South-East District,' he assured the commissioner, 'and, should no unexpected casualty occur, I still hope to complete the whole by the first week in January, 1865'.[68] It was an absurd hope. He was reported to be in the South-East in October, and in mid-November the whole party steamed across Spencer Gulf from Wallaroo to Port Lincoln, to continue valuing runs there.[69] He arrived back in Adelaide on Friday 30 December, having valued the runs in Streaky Bay and Port Lincoln.[70] These journeys resulted in a further 33 valuations.[71] In two-thirds of these, Goyder judged that the grazing capacity had been over-assessed or slightly over-assessed, and reduced the estimation accordingly. Even so, as elsewhere, the rentals were increased.

As the process of valuing proceeded, the business of recording the work had developed its own routines and refinements. Diary-like elements, scarcely present in the first place, vanished altogether, and a system of recording movements between points on each run, identified by letters of the alphabet, was used for ease of both preparing and comparing the text and the accompanying maps. The names given to features, usually huts and wells, if available, continued to be used. (On one run, trips were recorded from points A4 and A7 to what were described as 'oak huts', drolly named Windsor Castle and Temperance Hall.[72]) The recording of improvements also seems to have become more thorough and systematic, even extending to include abandoned pig pens, that were noted, not because Goyder was obsessive, but in case of future claims for 'improvements' that had been overlooked. Other details (such as the number of windows and doors in the head station) provided indications of the profitability of a run and of the attitudes of its proprietors. In the key matter of recording the quality of pasture, standard phrases became standard abbreviations: 'very fairly grassed', 'pretty well grassed' and 'very well grassed' became 'VFG', 'PWG' and 'VWG'. And Goyder did not confine himself to recording pasture grasses. On a run in the Avenue Range, in the South-East, he observed:

P. W. limestone, oak, Honey-suckle [banksia], a few gums, grass tree, ferns – bushes herbs & a little grass – Swamp cutting grass, rushes – a few useless bushes & weak herbs & patches of weak grass in places.[73]

Like the uniquely 'hopeless' country behind Fowlers Bay, these weak and useless bushes and patches of grass seem to have been very nearly the only ones he chose to dismiss in this way.

Examining the country and recording its vegetation in the systematic way he did during the valuations seems to have had an enduring impact on Goyder. Lionel Gee's experience of travelling with Goyder in South-East in the 1870s led him to remember Goyder as a keen observer:

nothing escaped him. Plants, trees, shrubs he would name in passing, giving the botanical name that betrayed a most retentive memory. On one of these journeys along the Coorong he pointed to the remains of a burnt tree with the simple remark, 'Sheoak burns well'.

There had been a grass fire, and lying on the blackened ground was the perfect outline of a large sheoak which had been blown down some time previously and caught alight. There had been no wind, and every limb to the tiniest twig, was outlined in white powdery ash, most delicate tracery, as though drawn in chalk.

No charred wood, no charcoal. Detail, yes ...[74]

Goyder and his support team had left Adelaide for the South-East on Sunday 5 February 1865. They were to remain there for several months. His final valuations under the 1858 *Assessment on Stock Act* were forwarded to the commissioner, together with a report, on 30 October 1865.[75] (More valuations were to be required, but under the conditions of a later Act.) Altogether, Goyder had carried out 175 valuations for the purposes of the 1858 *Assessment on Stock Act*, producing a detailed description of pastoral country – modified by grazing, but not cleared – across the entire area of the province occupied by the colonists; he'd also compiled a partial Domesday Book of South Australian squatting, covering leases extant in 1864. In the process, he claimed to have ridden 30,000 miles (over 48,000 kilometres) which would mean an average ride of 50 miles every day for 600 days. His field notebooks indicate that he commonly kept tallies of distances travelled, but these were based on his daily estimates, and he had no way of knowing exactly how far he had gone each day. As startling a figure as it is, Goyder must have been convinced that it was accurate, because he used it in his own public account of the events and again in a letter to the commissioner, written on 26 January 1871, in which he described having been 'required to ride over thirty-thousand miles in less than twenty months'.[76]

Whatever the actual distance covered in his surveys of vast areas of South Australia, Goyder had spent so much time in the saddle that his health was affected. When he finally withdrew from valuing runs in late 1866, it was because his health was immediately under threat, and for the rest of his life he regarded the valuations as the chief cause of his broken health – and as indisputable proof of his willingness to be a loyal servant of the government. In the *Australian Dictionary of Biography*, Goyder's estimate of the distance he rode has been pruned with a bold hand (or misprint) by a third. It is still a remarkable 'over 20,000' miles. If he had been a bushranger instead of a public servant, he would have entered folklore.

CHAPTER EIGHT
In search of the rainfall

The Adnyamathanha people of the northern Flinders Ranges possessed stories that prepared their hearers for the exigencies of life under extreme drought conditions. One of the tales stressed that knowledge of available water must be shared between groups and not kept secret. Another warned of the temptation and horror of cannibalism when the supply of food was exhausted. Hot springs in the northern Flinders Ranges (Vadaardlanha, or the Paralana Hot Springs) constantly testified to the reality of drought. The springs were understood to have become hot after 'drought' sticks, traditionally driven into the ground to end drought, were thrown into water, to 'cool the country down'. As a result, the springs themselves became permanently hot.[1] Further inland, in the central desert, where climate is not experienced as an essentially predictable round of weather changes, but as observable patterns that recur (however infrequently and irregularly, as with floods), the indigenous languages of the region do not contain words equivalent to our 'summer', 'winter', 'spring'. Instead, there are terms that identify recurring features of the weather and its effects, such as the period when vegetation grows rapidly after rain.[2] The term applies when the events which characterise the 'season' take place.

The pastoralists knew nothing of this in the early 1860s, inclined as they were to drive Indigenous people away from the waterholes rather than to listen to their stories, but consideration was being given to the problems of aridity and dry seasons. Frederick Sinnett, in his account of the colony, had noted that farmers on the Adelaide Plains:

occasionally suffer from the effects of hot winds and want of rain; and it appears as if good and bad seasons succeed one another in cycles of considerable duration; but the laws governing them, and the rotation in which they may be expected to succeed one another have yet to be ascertained. The past two seasons are considered by many persons to be the first two seasons of a cycle of comparatively rainy weather; the previous seven years were injuriously dry.[3]

Introducing the north and what was understood of the climate there, Sinnett provided a description of the 'injuriously dry seasons' that had reached their extreme in 1859. His account shows that he was already well aware of the different sources and seasonality of rain in the north, and had recognised that the rainfall was unpredictable as a matter of course:

> From Port Augusta, for perhaps three hundred miles northwards, the climate is of a most uncertain kind. It appears to be too far inland and too far north to secure with regularity the winter rains that visit the southern portions of the Colony, while on the other hand, it is not far enough north to secure with regularity the summer rains of the tropics. Sometimes it comes in for one, sometimes for the other, and sometimes for neither, experiencing droughts of two or three years' duration …[4]

Annus memorabilis

The beginning of 1865 brought another new baby to the Goyder household. Francis Etherington – to be known as Frank – would begin his career as a cadet surveyor in a survey department still headed by his father. At the same time, news of another drought was reaching Adelaide from the 'Far North', as it was termed. Failure of the rains had been reported since the middle of 1864 and mentioned by Goyder after his long-awaited return to Adelaide in the second half of September of that year.

The *Register* began to publish regular reports and letters from the stations, recording the constant loss of sheep by the tens of thousands, the deaths of the draught bullocks, and the steady desertification and desertion of the north. Kites, gathering to feed, circled in the sky, as they had been seen to do in the previous drought. Even where there was still water in springs and wells, it was impossible to remain on the country once the vegetation had gone. Without bullocks, the movement of goods had ceased; without forage, the movement of people on horseback came to an end as well. Specially arranged services of horse-drawn carts had to be organised to bring fodder for the horses that remained.

Nevertheless, when the northern squatters declared that drought was ruining them, their complaints were not received sympathetically. After six months of angry debate, people in Adelaide thought they were hearing yet another verse of the now-familiar lament that the pastoralists were being destroyed by the valuations and that they would have to throw in their runs, sell their 'handsome buggies that make Adelaide look so aristocratical' (as Barrow had teased in his papers), and flee, literally, to other pastures.[5] Even when the government did succeed in getting a team of commissioners into the north to investigate, it was on the condition that the members neither visit nor report upon any run that had been visited by Goyder and included in his valuations. To do so would be seen as attacking the valuations, rendering their report effectively useless in the highly polarised political environment that had developed in Adelaide.

The three commissioners inevitably included Charles Bonney, along with an anti-squatting member of the House of Assembly, Wentworth Cavenagh, and the new inspector of sheep, Charles Valentine. They confirmed that the situation was as grim as the squatters were claiming. The drought seemed to be of unprecedented severity, and the country had been heavily overstocked. Few squatters had been seriously deterred by the previous drought, which they had chosen to regard as a freak event. Runs had changed hands, and, according to the commissioners, in the process had become seriously overstocked and overvalued. Purchasers had paid no attention to past experience of the climate.[6] But rather than accept any responsibility, the squatters attempted to blame Goyder for everything. He had been misled by the good seasons, they claimed, and made the valuations too high, and that had resulted in overstocking – despite the fact that his valuations had not been given until three-quarters of the way through 1864. Since Goyder had been hindered in his work by 'unusual dryness', as he had termed it, the claim that he had been misled by the good conditions in the north was also hollow. But when it came to the impact of the drought, the squatters had been speaking plainly, neither twisting nor exaggerating the truth.

Late in the year, a 'pictorial representation of the death and desolation consequent upon the drought in the Far North' would be exhibited at the Adelaide Exchange.[7] It showed, near the ruined huts of an abandoned station, a faithful old riding horse, left to fend for itself, sinking to the ground, while a starved cat wailed miserably. Stock – referred to in the newspaper account of the picture as the 'proper' occupants of the run – lay dead among the bodies of what were curiously described as 'trespassers from the forest' – presumably the indigenous animals. (The impact of pastoralism meant that the 1865 drought permanently altered the ecology of the Flinders Ranges and contributed to the local extinction of several species of animals.[8]) Long letters to the *Register* from pastoralists Thomas Elder and J.B. Hughes reported the terrible plight of the Indigenous people in country stripped bare by overstocking. (Of the two daily papers, it was the *Register* that was most concerned with the situation of Indigenous people.

The *Advertiser* was inclined to be populist and dismissive.) In December, a correspondent writing from Nuccaleena reported that: 'I saw 40 sitting down the other night with nothing but a bush wallaby between them'.[9] In Adelaide, parliament was petitioned to extend aid, and a regional sub-protector was appointed in 1866 to supply what help he could.[10]

The idea of a line

Late in March, the government led by Arthur Blyth, the lands commissioner, was replaced. The new commissioner of crown lands and immigration was Henry Bull Templar Strangways, commonly known as Bull Strangways, as a slip by a Hansard writer reveals.[11] Strangways had been trained in the law and admitted to the Bar in England in 1856, a year later returning to settle in the colony he had visited as a boy. He had entered the parliament in 1858 and was soon one of the players in the game of government. Strangways was a difficult man, with an 'unconciliating sourness' of manner. In parliament his tactic was to cling to office at all costs and to regain office, when it was lost, by de-stabilising any government of which he was not a member, regardless of its performance. Despite all of this, as Sinnett had advised readers of the *Telegraph*, Strangways was also 'a man of energy and ability, and a member doing his departmental work well … far above the average of our members, both as to culture and ability'.[12]

The new government was soon receiving petitions and requests for special consideration from the northern squatters, but an unsympathetic Strangways dismissed their demands as unreasonable. On a Thursday morning in late June, the northern squatters held another meeting in the office of Levi and Watts. There were about 30 of them present or represented, and the meeting was chaired by former government leader George Waterhouse. The outcome was that:

> A line was agreed upon showing the country affected by the drought, and a resolution was passed recommending the Government to grant within [beyond] such line an extension of tenure.[13]

By their own admission the line that the squatters had agreed on was a crude administrative measure, perhaps something like the line already described by George Hawker several years before, in 1858, when he had attempted to impress on the members of the House of Assembly that squatting was not all riding about town in smart carriages. As well as his runs in the prime country, which formed the foundation of his wealth, Hawker had runs further north, and it was especially hard there, he advised. 'Let them draw a line from Mount Bryan to Mount Remarkable, and another from Mount Remarkable to Port Lincoln, and if honourable gentlemen would go and see the country beyond these lines they would learn the hardships to be endured.'[14] Since those assembled in the Adelaide office in 1865 thought that 'without actual survey no boundary could be fixed which should be sure to meet the justice of every particular case', they also agreed that the government should be empowered to consider individual claims for drought relief from those not covered by such a clumsy tool.

Six days after the squatters held their meeting, Goyder was called before a select committee of the Assembly which was gathering evidence for a report on the land sales system. The questioning had only just begun, when a remark Goyder made – that he had never imagined that land was to be resumed and sold for pastoral purposes only – prompted the chairman, C.H. Goode, to move inquiries in a very particular direction. He began by establishing Goyder's credentials in the customary way, by stating that he believed Goyder to be 'intimately acquainted with all parts of the Colony'.[15] Goyder, still involved with the valuations after more than a year and probably savouring the novelty of sitting on something that didn't move, replied with dry restraint that he had a 'tolerable acquaintance'. Goode then set about trying to determine the amount of time left to the province before the supply of new agricultural land was exhausted, requesting a sketch of the current agricultural boundaries from the surveyor general.

Using Mount Brown (north of Mount Remarkable and to the east of Port Augusta) as his point of reference in the north, Goyder explained that, as far as agriculture was concerned, the land beyond it was 'available for anything, but it is the uncertain rainfall which renders it comparatively useless for tillage. It is strong enough in wet seasons to grow anything, but it is the climate and the aridity of the soil that decides its character'.[16] The 10,000 acres that remained unsurveyed on the Yorke Peninsula he considered as suitable for growing hay, but not wheat, again because of the 'uncertainty of the seasons'. He added that 'some cereal, not yet known in the Colony, might be found suitable for the higher latitudes – and in that case very much more land would be available', and advised that they still had 'a great deal to learn so far as the agricultural and pastoral resources of the Colony are concerned'.[17] However, on the basis of current understanding, and continuing at the same rate of expansion, he estimated that the supply of agricultural land would be exhausted in five years.

The discussion drifted to other areas, but Lavington Glyde, the former commissioner of lands who had threatened to 'raise the country against the squatters', steered it back to the topic. Glyde remarked that a suggestion for dividing the country into pastoral and agricultural areas had been made and enquired: 'I suppose you would have no difficulty in preparing a map that would show that?' Goyder – almost certainly the source of the suggestion in the first place – replied he would have no difficulty at all, and that such a map was 'desirable'.[18] The land designated pastoral should never be sold, he believed, but only leased. The squatters naturally had an interest in discovering where this boundary might lie, since, if this approach was taken, those to its north would be able to rest comfortably on their leases, confident of not having to face resumption. John Williams, a northern squatter whose runs were being badly affected by the drought, then asked Goyder directly if he could fix the boundaries of the agricultural lands. It was at this point that Goyder replied: '*I think I can fix the boundaries by what would be the reliable rainfall*' [my emphasis].[19]

Assisted by Walter Duffield, a member of the committee who was not so much impartial as fully committed on both the agricultural and pastoral fronts – he was the proprietor of both wheat mills and pastoral runs – Williams extracted a little more detail about what is now the central section of the Line. As if testing Goyder, Williams asked whether he was aware that those boundaries had been found to be 'pretty much the same, year after year'. 'I am quite aware of that', Goyder snapped back,

> and I am quite within the limits in the rough estimate I gave. I take the apex of the cone; Mount Brown forms the apex, and the line comes down south, and very nearly south-east – spreading out to the eastward, and then dropping to the south again.

Duffield asked if it defined the low lands on the Murray. 'Quite so,' Goyder agreed, 'and … [there] is less rainfall at Ulooloo than at One Tree Hill and a greater at Mount Lock than at Black Rock'.[20] Goyder had already expressed this understanding in the valuations. In his comments on the Gottlieb's Wells run (in the area of what is now Terowie) in particular, he had remarked that the 'rainfall upon this run is not so reliable as upon those to the west and south, a circumstance allowed for accordingly'.[21] The importance of what Goyder had said for the long-term development of the colony was recognised by the *Advertiser*, when the parliamentary paper recording this committee hearing was released. In early August, the paper made an attempt to resurrect Goyder's ideas from entombment in the sepulchre of the official record, but no response was forthcoming.[22] Drought had shrunken the scope of the community's attention.

Goyder versus the goyderites

It was not until the end of October that Goyder's desire to map a line of reliable rainfall for use as an agricultural boundary and the demands of the northern

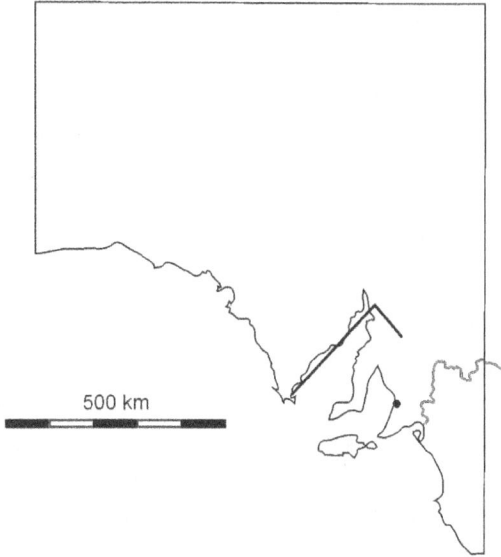

The line described by George Hawker, dividing the highly profitable grazing country in the south from the more difficult country in the north

The line described by Goyder in June 1865 as the boundary of the reliable rainfall

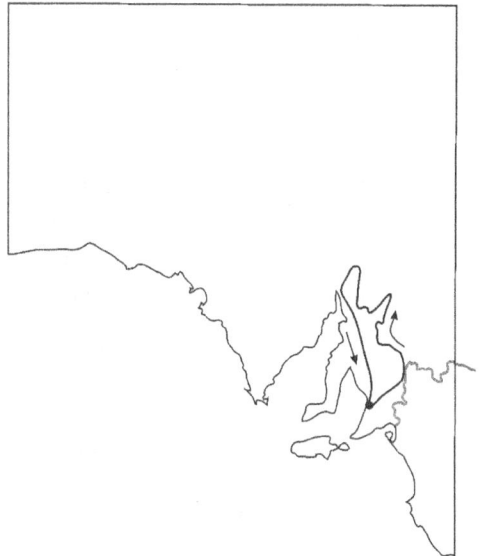

The general course of Goyder's journey to inspect the drought

Hawker's grazing line (1858), Goyder's proto-line (June 1865), and the main route of Goyder's journey, November 1865.

squatters for drought relief and a definition of the drought-affected areas coalesced. In a climate of instability, the government fell yet again and once again the portfolios changed hands. The new government was led by the chief secretary, Captain John Hart, the 'Ancient Mariner' as he was known, who, as treasurer, had been so impressed by Goyder's work on the valuations. Hart was an old hand in the colony's politics, and was regarded as having a taste for secret manoeuvring and little stratagems.[23] Accompanying him was the political new arrival and now attorney-general, J.P. Boucaut, one of Adelaide's leading barristers, who had been deployed against Goyder during the inquiry into the Moonta leases and who was to go on to become premier himself. The lands commissioner was Lavington Glyde, still intransigently anti-squatter in his attitudes and regarded as an incongruous member of a Hart ministry, and the treasurer was Walter Duffield.

Hart introduced his government on 24 October, a Tuesday. He promised that relief measures for the squatters would be implemented, but not until the northern runs commissioners had returned from the north and presented their report. On Thursday he met George Waterhouse in King William Street. Waterhouse, still pursuing the goal of a line for the squatters, suggested 'that Mr. Goyder should be sent to the North to report upon the runs suffering from drought which had not been visited by the Northern commissioners'. Hart agreed that Goyder should be sent out, and claimed to have already decided upon such a course. On Friday 27 October, he formally proposed that Goyder be sent north 'to ascertain the line to which the rain has extended and which defines the country suffering from drought ...'[24] There was unanimous agreement until he suggested that they draw up instructions and send Goyder out immediately. Glyde refused to agree to this, and the discussion was postponed. Cabinet met again on Saturday morning, with Glyde still against Goyder's immediate dispatch. He had made inquiries and learnt that the northern runs commissioners were expected the following Tuesday, and he was adamant that Goyder should not be sent out until their views had been heard. If the instructions to Goyder were drawn up, he would not sign the letter, he warned.[25]

On Monday 30 October, Goyder delivered his last instalment of valuations under the 1858 Act. They were accompanied by valuations under an 1865 Act and a report to the commissioner, the contents of which suggest that he had a good idea of what was going on. In the course of the report he found himself 'compelled to state' that his valuations of some of the runs in the Murray lands had been excessive, especially those on the droving route for stock from the Darling and northern districts, which had had to bear the passage of the equivalent of 350,000 sheep in the previous year. At the end of the report he explained that the drought afflicting the north was an unprecedented calamity and that even lessees who had not objected to his valuations, which were justified at the time, now doubted that their runs were worth holding. Given that the northern runs commissioners had been instructed not to report on runs he had valued, Goyder could see only one way for justice to be obtained for these squatters. 'I would,' he offered:

> with the approval of the Government, although suffering from fatigue, willingly undertake a special journey with the view of placing the Government in possession of a statement, by which to compare the country as it now exists and as it appears in ordinary seasons, in terms of which my valuations were made.[26]

Cabinet met again at 2.30 pm that day, but Glyde, who had not been to the Crown Lands Office, had not seen Goyder's offer to go north.[27] Hart presented the case for his proposal and Glyde agreed to its fairness. Nevertheless, he refused to withdraw from his previous position, because, according to Hart, his pride would not let him, and he left the meeting.[28] This was effectively a resignation. He was replaced as commissioner by J.B. Neales.

When it became known that Goyder was going north again there were cries of displeasure, doubt, and suspicion on all sides. 'The country will have need to be vigilant … Secret agencies are busy at work …', the *Advertiser* proclaimed under the heading 'Another crisis!' as it announced Glyde's resignation.[29] The anti-squatters feared that Goyder was being sent north to undermine his own valuations, while 'the pastoral interest', through the *Register*, feared that he was being sent to provide evidence with which to overturn the findings of the runs commissioners.

Goyder received his brief paragraph of instructions on 3 November. 'I was sent up specially to see where the rainfall ceased', he later summarised succinctly, and finally set out on 13 November.[30] The report of the northern runs commissioners was released after his departure. It had been received a week before, but withheld to dispel the idea that Goyder was going to amend its evidence and conclusions. Even so, the tone of the squatters and their supporters remained aggressive.

On the same day that Goyder received his instructions, G.S. Kingston wrote to the *Advertiser* with an 'analytical digest' of the rainfall in Adelaide for the past 27 years. Kingston had maintained rainfall records for Adelaide from the early years of settlement until Todd had taken over and arranged for records of the rainfall to be 'treasured up' across the colony. Kingston's letter was received so enthusiastically that, instead of being included in the correspondence column, it was included in one of the leading articles, which began by stressing the importance of rainfall records. 'The climate of South Australia, like that of all other countries, results from and is subject to certain laws,' the writer (presumably Barrow) intoned, 'but at present these laws are all but unknown'. Nevertheless, the article displayed a strong awareness of the distinction between average annual quantity of rainfall and reliability, and looked forward to the day – not to be realised until near the end of the twentieth century – when drought could be predicted.[31]

While Goyder was away, the editors of the *Register*, entirely unaware of Goyder's view of the situation and assuming that he would remain committed to his valuations regardless of circumstances, began a campaign of editorial prophylaxis against the work it supposed he would do on his return. 'Where is the rainfall which he is to measure?', one editorial sneered:

> A few years ago that gentleman went to the Far North and discovered an inland sea. There in the basin of Lake Torrens was a vast expanse of blue water – so Mr. Goyder said. But when Colonel Freeling, with a boat, arrived at the spot, mirage and mud were all that remained. The inland sea, the glittering cliffs, and the tall trees in the distance had passed away like the baseless fabric of a vision.[32]

The rest of the editorial was directed to the plight of the northern squatters and also included a condemnation of Goyder for the overstocked runs. The phrase: 'in search of the rainfall', became the heading used for articles and letters referring to Goyder's journey.[33] A similar phrase, 'in pursuit of the rainfall', became part of the caption given to what is now the most well-known image of Goyder, A.G. Ball's cartoon of him at 'Squatters [*sic*] Paradise' – another *Register* phrase – on the north-east plains.

In pursuit

In his field notebook Goyder used the heading 'drought line', and that was clearly what he had been sent to define. A substantial section of pages of the notebook for this period have been torn out – unusually – and the only references to the journey are on the pages since numbered 1, 43 and 44. Page 1 contains a list of locations that probably represent the proposed route; page 43 contains notes on the various categories of drought relief; page 44 contains a description of a line and some very brief notes. At the back of the book are Goyder's notes on his route and travel expenses from 13 November through to 19 November. (These included 2s. for a pint of ale for the ever-present Rowe. Goyder spent a more modest sum on his own consumption of lemonade and soda water.)[34]

His account of his route on his return confirms that he had traversed the country in the same way as during the valuations and when examining country prior to survey, 'passing in various directions over the intermediate country' between the key points of his route.[35] The preliminary part of the journey took him from Adelaide through Mount Pleasant and to Blanchetown on the Murray, and then up to Von Rieben's hotel, near the Murray bend.[36] On his return he described his whole route as having been:

> from Adelaide north-easterly to the Murray, northerly and north-westerly to Kooringa, north to Gottleib's [sic] Wells, east to Ketchoula [Ketchowla], northerly to Tetulpa [Teetulpa], southerly to east of Mount Lock, north-west by Boniah Creek, Rocky Gully and Hogshead to Pekina, south-west and north-west through Boolooroo [Booleroo] to Mount Remarkable, north-easterly over plain to north of Kanyaka Run, south-westerly by Western Plains to Port Augusta, and by Baroota and Telowie to Crystal Brook and Broughton, and thence by Clare to town.[37]

The final route seems to have been influenced by his observations en route. Before he left, the *Register* reported him as intending to travel 'first to the Murray by Tungkillo, then to work northward and westward, so as to finish the journey at Port Lincoln'.[38] In a letter to the office three days after his departure he reported that he intended to take a different route.[39] He would now travel down the west coast of Spencer Gulf as far as Franklin Harbour, returning via Port Augusta and down through Crystal Brook and Wakefield to Adelaide, although both these areas were eventually omitted from the tour. In his letter Goyder predicted that he would not be in Adelaide until 16 December, but wrote again at the end of November to announce his early return.[40] The going must have been very hard.

Mapping the line

He returned on 4 December and had a long talk with Hart on the same day. Hart recorded in his diary that he had been assured that Goyder's report would fully justify his having been sent, and that he would be able to support a fair measure of relief. Presumably Goyder took his usual rest the following day, and on 6 December he prepared his report. At the beginning he admitted what had motivated him all along. While he was no doubt genuinely concerned about the northern squatters, who were now stuck with his pre-drought valuations, his underlying goal had been to collect the information with which to adjust his line of reliable rainfall. If the drought had been 'of an ordinary nature,' he explained:

there had been no necessity for my leaving town upon this duty, as the line of demarcation might have been shown from information previously in my possession, and specially referred to in my report on the valuation of some of the northern runs. The drought, however, being of an unusually severe nature, and extending more generally than any previously known, it became indispensable to add to my previous experience the knowledge of the state of the country as it now exists.[41]

Goyder repeated this in an account given many years later. He described himself as having been 'not satisfied' until seeing again some of the land he had travelled over during the valuations, and explained how he had, 'with the sanction of the Government … made a further examination which involved a 3,000-miles ride when the drought was at its worst'. In this account he also made plain that he had done this because he had hoped that, in addition to the line being used to determine which pastoralists should be granted drought relief, 'a new Act would have made it illegal to lease or sell for agricultural purposes without [beyond] "Goyder's line of rainfall"'. But in reality, the government had not sanctioned Goyder's journey to investigate the drought with a view to clarifying his understanding of the climate and its meaning for future land use in the colony, nor had it directed him to define the line 'to the south of which … rainfall might be considered fairly reliable' – not officially at least.[42] Perhaps, in conversation, Hart had appeared to support Goyder's enthusiasm for this venture, but Hart made no mention of the subject in his diaries. He and the Cabinet were really only interested in the squatters and the political exigencies of the moment.

Having gone out with a definite understanding of where the reliable rainfall ended and a knowledge of past drought-affected areas, Goyder was in no doubt about what he observed:

The result of my investigation shows the line of demarcation extending considerably further south than I had anticipated. The change from the country suffering from excessive drought to that where its effect had only been slightly experienced, being palpable to the eye from the nature of the country itself, and may be described as bare ground, destitute of grass and herbage, the surface soil dried by the intense heat, in places broken and pulverized by the passage of stock, and formed by the action of the wind into miniature hummocks, surrounding the closely cropped stumps of salt, blue, and other dwarf bushes, whilst those of greater elevation are denuded of their leaves and smaller branches as far as the stock can reach. This description generally holds good of all country upon which stock has been depastured and where the drought obtains. The change from that to where the drought has had a less serious effect being shown by the fresher and more leafy appearance of the bushes, gradually improving to those in their ordinary state, and the gradual increase of other vegetation from bare ground to well-grassed country.[43]

Country which had been unaffected by the 1859 drought was now parched. In response to seeing the drought Goyder dropped the peak of his line about 40 kilometres, from Mount Brown to Mount Remarkable, and to an equivalent point in the east, among hills in the South Flinders Ranges, while excluding the Willowie Plain between them. He described the new boundary he had drawn:

> The line of demarcation I found to extend from Swan Reach, on the River Murray, in a north-westerly direction to the Burra Hill; and thence north to the Oak Rises, east of Ulooloo, and by the last named Hill to Mount Sly; and in a northerly and westerly direction … by the Hogshead and Tarcowie, to Mount Remarkable; thence southerly by the Bluff and Ferguson's Range to the Broughton; and south-westerly to the east shore of Spencer's Gulf, crossing the Gulf to Franklin Harbour; and thence north-westerly to the west end of the Gawler Ranges.[44]

Goyder accompanied this with a map, signed and dated on the same day, 6 December 1865. It showed a single line, hand-drawn onto one of the standard and not completely accurate office maps of 1862, with the note: 'NB. Firm black line shewing demarcation between country suffering drought and that to which drought has not extended'.[45]

Given that his investigation of the drought was confined to the north-central area, with an excursion into the north-east, and that Goyder would later state explicitly that he had drawn upon knowledge gained over years, it is reasonable to ask how Goyder's Line was actually mapped. Unfortunately, documents relating to Goyder's work consist of raw observations in the field notebooks and finished conclusions presented to government. The middle phase of interpretation and reflection is missing. In the case of the Line, his observations in the field are missing as well, so there is no straightforward account on which to draw. But his activities in the years preceding the drought make the basis of an answer reasonably plain.

As well as taking note of the rainfall over the years, Goyder had also been systematically observing the vegetation of large areas of the country. Australia is exceptionally rich in plant species and these form complex and changing communities. In areas where the rainfall is sparse and variable, there are plants adapted not just to aridity, but to drought. What seems most likely is that, primed by his experience of the extremes of the inland and having spent month after month doing nothing but record the presence and condition of grasses, trees and shrubs, Goyder had formed some ideas about species to be found on drought-prone country, although there is nothing in his writings to confirm this, other than a warning, made in 1876, that the different vegetation of the saltbush country (well north of the Line) indicated a different climate.[46]

In the report which accompanied the first presentation of the Line, Goyder referred to the condition of vegetation, or its absence, as showing clearly where the extreme drought conditions prevailed.[47] In the same report he also presented an observation relating to long-term climate patterns. In his field notebook, under a description of the Line, there is the comment: 'Dying Timber Far N. Evidence of drought at long periods'.[48] In the report, he had elaborated:

> Thinking over the appearance of the whole country during the various times I have passed over it recalls to mind undulating lands, covered with forest trees from fifteen to thirty feet high, all dead – showing that, whilst there are successions of seasons sufficiently good to ensure for a series of years the growth of timber of considerable size, droughts have before prevailed sufficient to cause the trees to perish from want of moisture; and as such may prevail again ...[49]

There is evidence given by others who had been close to him. In 1902, four years after Goyder's death, his nephew, Edwin Mitchell Smith, by that time deputy surveyor general, confirmed to an inquiry that the feature by which Goyder had determined the Line was the 'herbage and the vegetation'.[50] Smith and his superior, surveyor general William Strawbridge, had been in the Survey Office all their working lives, both having joined the year after Goyder became surveyor general. Strawbridge elaborated on the point to the members of the inquiry, explaining that Goyder had:

> divided the grass or herbage land from the bluebush and saltbush. He considered the line of bluebush and saltbush was the line of demarcation between fair and poor rainfall, and I think he defined it in the original plan correctly, because there is saltbush north of that line, and I have not seen any south of it.[51]

The agricultural expert, W.S. Kelly, who could have been drawing on hearsay, claimed that Goyder 'continually sought the guidance of surveyors as to the southern edge of the saltbush', although this was probably in the 1870s, since in 1865, except in the area around and below Mount Bryan, the survey was nowhere near the Line.[52] Lionel Gee's recollection of Goyder identifying species as they passed and giving their botanical names while travelling in the 1870s, and the fact that the field notebooks from his later career, up to the very last in 1893, contain records of the kind of vegetation he was passing through, indicate that noting the presence of species had become habitual with him, and was a constant part of his observational practice.[53]

Saltbush and bluebush are not merely adapted to arid conditions, but to unpredictably arid conditions, and Goyder evidently recognised them as key indicators of unreliable rainfall, but knowing that he looked to the presence of saltbush and bluebush does not mean that they were the only plants which served as guides – Smith's broad reference to 'herbage and vegetation' certainly suggests otherwise. At the time it was drawn, much of the Line passed through mallee, but his does not preclude the possibility of discerning changes in vegetation. Mallee trees are small eucalypts with a large root or lignotuber and very often have multiple branches rather than a single trunk supporting their umbrella-like canopy of leaves, but the term 'mallee' also refers to plant communities dominated by these trees. The composition of mallee communities varies, and can include saltbush.

In considering how Goyder could have mapped the limit of reliable rainfall, it is useful to consider the Line in sections. The central section follows the spine of hilly country running up through the Mid-North to join the Flinders Ranges. This was country Goyder had ridden over many times, and about which he had definite ideas, as his remarks from earlier in the year indicate. He redefined the top of the Line by siting it much further south, and gave it its distinct two-horned shape by excluding the Willowie Plain, which lies between lines of hills. In the 1870s, the south-eastern section contained large areas of mallee with patches of native pines. There was bluebush on the north-eastern portion and saltbush to the west. It can be assumed that, in 1865, the condition of this country would have been desperate. When this land was settled and cleared, life there was characterised by unpredictability and extreme variability of rainfall: seasons could be wonderful or disastrous.[54] To add to the burden of the bad years, the plain when cleared of its mallee and other vegetation became subject to severe soil loss through dust storms. A local poet joked that the unhappy farmers could not leave because: 'If we were to leave the country / The country would follow us'.[55]

In the east the Line tracked down beside the range of hills, moving further east to form a noticeable shoulder when it encountered Mount Bryan. In this upper north-eastern section, Goyder is reported to have been assisted by pastoralist Alexander McCulloch. Unlike many other squatters, McCulloch would probably have been a sheep farmer if he had remained in his native Scotland, and was known as 'a good man among sheep'. He was first recorded as the proprietor of a flock in the colony in 1841. In 1856 he had moved into the more northerly area, becoming the proprietor of Gottlieb's Wells, Eldrotrilla, an associated run, and Nackara, which contained saltbush and bluebush plains, even further to the north.[56] In the collection of portraits, *Pastoral Pioneers of South Australia*, written when the squatters described were still the objects of living memory, it is stated that McCulloch had made a close study of local rainfall conditions, which had enabled him to 'materially assist' Goyder in describing the Line.[57] Gottlieb's Wells was one of the planned destinations on Goyder's journey, but it was evident from his valuation of the run that detailed discussions had already taken place.

Apart from being a committed and successful permanent settler with a serious interest in the local rainfall, McCulloch had other qualities that would have impressed Goyder and prompted communication between them: he was renowned for his generous hospitality (possibly of the same whisky-flavoured variety as dispensed by the Scottish settlers of the South-East), and he was respected for his fairness and honesty in dealing with his employees. He was also the father-in-law of J.P. Boucaut, the attorney-general. (That Boucaut had a squatter in the family did not escape the attention of other politicians at the time.) Over the years it is likely that the ever-curious Goyder had gathered information about rainfall from other pastoralists as well. As was usual at the time, he stayed at homesteads when travelling, and, being good company when not preoccupied with work, he was well placed to acquire whatever information the squatters had to impart, especially as his reputation as a hardy and expert bushman would have eliminated any tendency to dismiss him as a bureaucrat from the city.

Below Mount Bryan the Line drops south, still following the hills. This section of the Line was never disputed because the country to the east is not suited to agriculture, due as much to the soil as to the rainfall, and it is here that the Line appears as a road running north–south just to the east of the hills, with wheat cropped right to the western edge of the road and saltbush stretching away in the east.[58] This is also the one area where the issue of the extent of the agricultural land had already been addressed. By 1865, the land along the hills had been surveyed and sold, and here the Line was effectively formed by the eastern edge of the surveyed area. From this point the Line continued south, before touching the tips of other surveyed agricultural areas as it curved east into mallee country toward the River Murray.[59]

The western side of the central section of the Line swept down from Mount Remarkable to the coast, passing through mallee as it curved out to the coast. In 1865 the techniques for clearing mallee had not yet been developed, so this country was unsuitable both for agriculture and, except in the north and on the coast, for pastoralism as well. This was one of the areas passed over when the journey to examine the drought was cut short. Nevertheless, Goyder felt confident enough to draw his line through the mallee (which extended much further south in the central region and on the Yorke Peninsula), and when the Line was redrawn onto a new map in 1882 after this country had been cleared, he shifted it only a little to the north-west. Later, wheat growing was successfully established north of the Line here, although much of the area proved prone to soil drift.

Without the central section, the two wings of the Line – the eastern section, from the River Murray to the Victorian border, and the western section, from the Eyre Peninsula across to country behind the bight – could be joined to form one long curve, and both wings passed through vast tracts of mallee. Despite having planned to go to the Eyre Peninsula to investigate the impact of the drought, this part of Goyder's trip had also been abandoned, and his earlier visits had only taken him into the southern tip of the peninsula and around the western coast. His notebook, or what remains of it, contains only one cryptic note about the western region – 'Gawler Ranges Eastern plains'.[60] There is no evidence, then, that Goyder ever visited the part of Eyre Peninsula where the Line runs, so this section must have been drawn on the basis of information obtained from other sources. Pastoralists are one possible source of information about the drought and vegetation, since the Line crossed the southern tip of a large section of leased pastoral country behind Franklin Harbour, under which it passed.[61] Wheat farming has since been successfully established beyond the Line here, too, through the hills above Cleve, but it is not based on the expectation of a reliable crop every year, which Goyder had in mind as a necessity for small farmers in the 1860s.

In the east, the Line curved away neatly from the River Murray at Swan Reach to meet the border with Victoria at 35° S, passing through another vast region of mallee which, at that time, the settlers found impassable. As with the Eyre Peninsula, Goyder's earlier visits to the South-East had not taken him as far north as this, but he had not even planned to visit this region in 1865. He did, however, have his knowledge of the country in the central section to guide him in taking the Line through the mallee to the river, from where it began its sweep to the border.

Drought relief

On Saturday 9 December Hart arrived at Goyder's house at 9.00 am, as arranged. Goyder produced his map (plate 18) and they had 'a very long conversation'. The pair agreed (according to Hart's diary) that the valuations would be left as they were, and the whole of the northern pastoral area would be divided into scheduled areas for which extensions of tenure would be granted. They left open the issue of remission of rents (although the final measures included rent relief). Goyder's notebook shows that he had been considering a plan that involved extending the leases – a clever solution to the political situation, since it provided compensation while leaving the valuations untouched.

They had just arrived at this point when Boucaut appeared, 'evidently very nervous', according to Hart. Boucaut was afraid that if the government did anything that affected the valuations they would fail, but he was reassured by Hart. Despite his trepidation, it was Boucaut who had provided the initial impetus for drought-relief measures, in Goyder's mind at least. In a confidential reply to a letter written by Boucaut nearly 20 years later, Goyder asserted that 'the squatters' relief measures were initiated by you – and that prior to [my receiving] my instructions. You came to my house at Medindee with the late Cap[n] Hart, and before you left that day the [Regulations?] for relief had been framed.' This also applied to the Line and everything related to relief, as far as Goyder could remember 'at this distant date'.[62] In any case, the discussions were prolonged. Their immediate outcome was the drafting of a letter to Goyder asking him to furnish the commissioner – Neales, who was not present – with whatever suggestions for relief the circumstances seemed to require.

On 11 December Goyder presented the report containing his suggestions for the northern runs to parliament.[63] He proposed that the area north of the Line should be divided into four regions, for which extensions of three, six, 10 and 14 years would be granted. The report was accompanied by a coloured map (plate 19) showing the scheduled areas. The area south of the Line (now the southernmost of several lines) was coloured green. The map which had accompanied the first report, showing only the rainfall line, had not been shown to the parliamentarians and was not included in the papers of the parliament at the time when the first report was ordered to be printed.

Unexpectedly, it wasn't the suggestions for relief that caused an outcry. In the first sentence of the report Goyder had referred to the earlier report he had presented on 30 October, and quoted it at length. But nobody had seen this report. The suspicion that there had been secret plotting and intriguing going on was compounded when the missing report was discovered to have been languishing in a pigeon hole in the Crown Lands Office for over a month.

Hart demanded to know what had happened. The secretary of the Crown
Lands Office, E.T. Wildman, offered the excuse – bizarre in the context, as
was noted at the time – that he had been overwhelmed with work and had no
time to attend to a report on valuations from the surveyor general. Wildman
claimed to have handed the report, unread, to the departing commissioner, who
put it aside unmarked, after which it was supposed to have been mentioned to
the new commissioner, Neales, and bundled, still unread, into the pigeon hole
and forgotten. (Neales claimed that his pigeon hole had been emptied several
times.) Strangways, who had been the commissioner succeeded by Glyde on
23 October, told the parliament that not only was it strange that a document
which so influenced Glyde's situation had been missing for so long; moreover,
it was somewhat suspicious that it had been written in the first place and that
a journey to investigate the drought made. Strangways claimed that Goyder,
in a long conversation with him on the last day he was in office, had ridiculed
Waterhouse's idea of drawing a line.[64] If there were any truth in Strangways's
claim, it is more likely that Goyder was ridiculing the idea of a blunt administrative instrument such as a straight line between two points.

There had been concern, too, about secret instructions having been issued to
Goyder. Cavenagh, the 'anti-squatter' among the northern runs commissioners,
had earlier astutely deduced that:

> since to determine the line of demarcation between the drought and the rain-
> fall was in fact the difference between the pastoral and agricultural country,
> surely Mr. Goyder must have received private instructions in addition to those
> published.[65]

Cavenagh's were not the only suspicions. Along with the questions raised by
the doubtful circumstances surrounding the pigeon-holed report, there were
hints of collusion between Goyder and Waterhouse. Boucaut defended Goyder
by simply asserting that the Assembly and the country had too high an opinion
of him to believe that he could be a traitor.[66] What no-one suspected was that
the surveyor general had 'secretly' instructed himself.

When the *Register* learnt the part Goyder's search for the rainfall was to
play in delivering drought relief to the squatters, its assessment of him changed
suddenly and absolutely. When he and his map were attacked in parliament,
it defended him, claiming that he was being 'found fault with because he has
mapped-out the drought in colours too distinct for the taste of honourable
members'. Now that there was 'great difference' between the views of Goyder
and those of his former enthusiastic supporters, the anti-squatter so-called
'goyderites', the 'Mumbo-Jumbo whom they worshipped must now be kicked

and cuffed ...' Of course, the paper was oblivious to its own equally dramatic reversal of attitude.[67]

Since the lines defining the various scheduled areas could hardly have represented what Goyder had seen on his tour – they went into areas well away from the route he had taken – the Cabinet requested that Goyder explain the principles on which they were based. In the memorandum he returned in reply, Goyder made plain that the lines did not just map the present drought: rather he had 'the honor to state' that the lines 'were the result of personal experience and knowledge of the country, obtained by frequent visits extending over a series of years and variety of seasons'.[68] Modifications had been introduced in some places to match the borders of affected runs, some of which had been made in response to prompting from politicians (seeking assistance for certain runs), others on the basis of information provided by 'reliable persons'.[69]

One feature of the map which mystified some of the politicians was the way in which the lines ran much closer together in some areas. Goyder explained that this expressed the reality, shown to him by experience, that: 'the reliable rainfall more rapidly decreased to the eastward and westward of the main range than along its course from north to south'. The simplicity of his answer conceals how novel a map expressing changes in a hitherto undefined, unfamiliar climatic variable (seasonal rainfall reliability) actually was, especially to people who had possibly never seen a map indicating change in a natural variable of any kind. If they had, it could only have been a naval chart showing water depth or a contour map, and their experience cannot have been enough for them to have recognised that lines close together express rapid change. What was to become known as Goyder's Line did not make its first solo appearance in public until early in 1866 when *Map of the Northern Runs* was presented to the parliament. On this map (plate 20) the Line was accompanied only by a red line showing the route taken by the northern runs commissioners.[70]

As the drought was ending, Goyder made another tour of investigation. In 1865 he had believed that the country was in as bad a state as it could be, but when he returned early in 1867, he discovered that things were a lot worse:

> even the surface of the ground was gone – blown away; and all the vegetation had disappeared, except in places – leaving the sheep and cattle tracks standing up like tessellated pavements three or four inches high.[71]

It would take years to recover, he warned.

Understanding the climate and the country extended to understanding the people and their plight. Although he had been criticised and blamed by the squatters and their supporters and from 1857 had been taken to task (by armchair critics especially) for failing to accurately interpret the salt-lakes country in flood, when Goyder was given an opportunity to give his opinion on the northern squatters and the way in which they had managed their affairs, he only commented: 'I could scarcely blame them, knowing the country as I do ... I could scarcely blame them.'[72]

Part two: THE DARK DIVIDE

CHAPTER NINE

Colonial morality: 1861–63

The story of Goyder's rapid rise to a position of great responsibility and influence that appeared in every nineteenth-century account of his career overlooked his unpromising beginnings as an explorer, when his report of permanent water in the inland led to his being regarded as the victim of mirage and floodwater. A far greater challenge, one that threatened his future and which dogged him until the end of his first decade as surveyor general, was also ignored by the record books. Given the repeated testimonies to his honesty, fairness and impartiality, it comes as a surprise that this threat took the form of very serious allegations of corruption.

The workload of the Survey Department and the Land Office had increased dramatically when massive resources of copper were discovered at the head of the Yorke Peninsula, first on a run named Walla-Waroo in 1859, then, in 1861, a little further south on the limestone plains at Moonta-Moonterra, near the Tipara Springs. Three new townships were spawned: Wallaroo, the port on the coast; Kadina, a little inland; and Moonta, to the south. By 1861, smelters had been built at Wallaroo. They were fired by coal brought by ship from Newcastle, and operated by skilled Welshmen, although the actual mining was largely carried out by miners from Cornwall. By the 1870s, South Australia had replaced Cornwall as the largest copper region in the British Empire and had taken in enough Cornish families to have its own 'Little Cornwall' in the 'copper triangle' of the three towns.[1] The Moonta mines would not close until 1923.

The new townships had to be surveyed – although the miners preferred to cluster in little cottages on the mining leases – but the waterless flat country covered with mallee and native pine (*Callitris*) posed problems. The settlements suffered from a lack of clean drinking water, resulting in outbreaks of typhoid, while the water pumped from the mines did not drain and lay about in dark stagnant sheets, poisoning the soil and posing a serious threat to the prospects for the new district. Goyder was called upon to address these problems, and only a week after he had returned from the Lake Eyre country as the new surveyor general, it was reported that he was about to visit Wallaroo to inspect the area for a town site.[2] The visit took place in January 1861. Since he also held the position of inspector of mines, he examined the copper mines as well. The number of shafts being sunk meant it was no small job, but Goyder tackled it with his usual close attention.

Even before mining had begun, the discoveries had created problems in the Land Office and added to the workload. After word had got out about the first find, there had been a rush to place speculative claims around the site, straight from the map. However, the department's map of the 'waste lands' (as the unsurveyed lands were termed) had been put together on the basis of information provided by the applicants for pastoral leases. Even if the lessees or their employees had established the bearings correctly, the distances between points of reference were only estimations. In places, the map was wildly inaccurate. Across the top of the Yorke Peninsula seven miles (over 11 kilometres) had been added to the country, as Goyder discovered on his visit at the beginning of 1861.[3] Nevertheless, there had been no choice but to mark the secondary, dependent claims on this map, and to sort things out later.

Claiming the prize

The problem of the maps, coupled with financially straitened circumstances, not only added to Goyder's workload, they made him unexpectedly vulnerable. One morning late in May, only months after his official appointment as surveyor general, Goyder entered the busy front office to encounter an arresting tableau. Waiting for attention was Captain Walter Watson Hughes, the former sea captain turned pastoralist who held the Walla-Waroo lease on which one of his shepherds had found indications of copper in 1859. He was now a wealthy mine owner. With Hughes was a clerk employed by his agents, Elder, Stirling and Company, specifically to aid clients in lodging mineral claims.

Goyder offered his assistance, and learnt that another of Hughes's shepherds, Patrick Ryan, had discovered more copper at a location identified, very vaguely, as 'eight or ten miles north of Tipara, and south of the Bald Hill'.[4] He pressed for a better description but Hughes assured him that this was the best he could do. On that basis, and keen to avoid a repetition of the rush to place claims around an original claim placed on an inaccurate map, Goyder indicated a spot and instructed the chief clerk to mark the claims, but not to draw attention to them until a correct survey had been received. It was government policy to protect the original discoverers, so time was allowed for Hughes to have the location accurately established by a private surveyor. To further protect the discovery, Elder Stirling followed the more-or-less standard procedure of taking out the four initial claims in the name of an associate, W.S. Peter.

Hughes, however, was not being open with Goyder. In his pocket he had a letter containing a more precise description of the location of the find, but to show it would reveal that he was acting on second-hand information, and Hughes wanted time to ensure the claims were as well sited as possible.[5] Showing the letter would also draw attention to the fact that, while he considered he had an implicit agreement with all of his shepherds about copper finds, he did not have the signed agreement the law required. But what even Hughes did not know was that the shepherd, Ryan, who was inclined to be drunk, had already signed an agreement with the publican of the Traveller's Rest in Port Wakefield, and another in Adelaide with Samuel Mills, the former city surveyor, in the Black Bull in Hindley Street. Both men had enlisted the support of friends, and competition ensued for what amounted to possession of Ryan, who was so drunk that all attempts to have him lodge a claim in the Land Office failed. In an attempt to define the actual site of the claim, the struggle shifted to the top of the Yorke Peninsula, where angry exchanges took place between the two groups. The find was eventually located, but in the end neither party could prevail against Hughes, who also turned up in the area. Ryan fell into his employer's arms wailing that he had been plied with drink and led astray, and yet another agreement was signed.

A dramatic competition followed, as representatives of both Mills and Hughes raced back to Adelaide to be at the Land Office when it opened on Monday morning 3 June. Hughes made use of the young son of a friend, William Austin Horn, who later became a politician and influential public figure whose wealth was founded in mining and pastoralism. Horn took a circuitous route through the scrub, riding over 260 kilometres in 22 hours. The information he brought was given to one of the members of Elder Stirling, John Taylor, a wealthy pastoral pioneer and brother-in-law of Edward Stirling. After a team of clerks had copied out multiple applications, Taylor took them to the Land Office via a short cut from the Elder Stirling office in Grenfell Street across what was known as the 'Corporation acre' (where the Town Hall now stands).[6] He entered through a back door opening off the courtyard in the centre of the building. Mills, who had been waiting outside for the office to open at 10 am, came in through the front door. When the chief clerk arrived, 10 minutes late, he attended to Taylor, who lodged a block of 30 claims on behalf of Hughes.

At the same time, Goyder was helping Mills, who seemed to him to be diffident and unsure. Unaware of the claims Taylor was placing, but with Hughes's

The quadrangle within the Government Offices in the early twentieth century – a view from the verandah behind the Land Office. The Treasury is to the right, and the back door through which John Taylor entered to lodge the Moonta claims is out of sight to the left. [SLSA]

King William Street, the Town Hall and the Government Offices from Victoria Square. The Land Office was on the south-west corner of the building (shown) with an entrance from King William Street. Goyder's office looked south over the square. [SLSA]

application in mind, Goyder warned Mills that his claim might fail because of an existing application. Mills did not approach the chief clerk with his finished claim until another half an hour had passed.[7] He was later to claim that Taylor, whom he remembered as having been in the office when he entered, had been let in early, but at the time neither side was aware of the other's activities. Hughes had men working the obviously rich lode almost immediately, while Ryan, with the income from his agreement, began steadily to drink himself to death.

The first dispute

It wasn't long before a complaint that Hughes had shifted the location of his claim was reported in the *Register*. Goyder attempted to dispel the rumour with an interview, but in the end, he was compelled to go to the top of the peninsula to settle the claim at the site.[8] While there he unwisely followed his usual procedure of staying at Hughes's homestead when travelling in the area.

Just as Goyder had trusted Hughes, and had resorted without hesitation to an off-the-cuff solution to the looming problem of another rush of claims to be resolved using misleading maps, so his method of settling this dispute was also personal and straightforward. Reasoning that if he 'went to the locality by survey alone and without being guided there, it would be more satisfactory than by going in the first instance and hearing conflicting statements, and being biassed [*sic*] by knowing the position beforehand', he took a chain and theodolite and set out from a point at Wallaroo early one morning with only the directions of Gavin Young, Hughes's surveyor, to guide him.[9] (Along with the theodolite, the tool most associated with surveyors at the time was Gunther's chain of 100 long rectangular links (plate 12), the whole chain being 22 yards, or a little more than 20 metres long.) If Young's work was accurate, Goyder would finish on the disputed claim. Word had got about that this was what he was doing, and surveyors representing mining associations with claims that would be affected by the outcome gathered on the site with Hughes to see where the tie-line would end.

Goyder emerged from the bush at about midday, chaining towards the waiting group. He completed his work just a short distance away, turned to Hughes, and in an ironic tone, regretted to have to tell him that the survey was wrong – by a few links. He then unambiguously confirmed that the claim belonged to Hughes.[10] Some of the surveyors remained unhappy, and went with him to carry on the discussion in one of their tents, where they referred to a copy of the application which Goyder had brought, bushman-style, tucked inside his hat.[11] Eventually he told everyone he was 'sick of the matter', or 'sick of the plans', refused to discuss it any further, and walked away.[12] Grumbling continued for some time after, but Goyder's demonstration finally ended that aspect of the dispute.

Eustace Reveley Mitford

Quite apart from Hughes, Mills, and the other groups squabbling over Ryan's discovery, there was another person who considered himself the discoverer of the Moonta lode. He, too, had backers. This was an older man named Eustace Reveley Mitford, one of the Mitfords of Northumberland, an old landed family hailing back to companions of William the Conqueror and members of the Saxon aristocracy, as Eustace was well aware. His grandfather, William Mitford, was the author of a monumental, multi-volumed *History of Greece*, which he had begun at the instigation of his friend, Edward Gibbon. The history, which took a definite Tory stance, was celebrated in its day, but unlike Gibbon's famous work, *The History of the Decline and Fall of the Roman Empire*, its acclaim has not endured. While his relation to William Mitford distinguished him to people of his day, he is now best identified as the great-grand uncle of the famous 'Mitford girls' of the early twentieth century, to whom he was strikingly similar in appearance. His later writings also foreshadowed those of the sisters in their distinctive voice, inventive playfulness and sheer joy of expression – along with a brutal capacity for making fun at the expense of others.

Despite his background, Eustace Mitford seems to have brought nothing with him on his journey across the world to South Australia, apart from a burgeoning family. Formerly a sailor, in South Australia he tried his hand as a carrier or teamster, and as a farmer, both without success. He also tried his luck on the Victorian goldfields.[13] But by 1856 he had been imprisoned for debt. His attempt to declare himself bankrupt was rejected by the court and his debtors continued to pursue him for some years more.

Mitford claimed that as early as 1853 he had seen indications of copper not far from where the main Moonta shaft would eventually be, but he had not been able to get the needed backers. (Because copper cannot be picked up in nuggets, but must be smelted from ore, a large investment is required to obtain it.) In 1861, a businessman (and former bankrupt) named Montague Phillipson approached him to search for copper on behalf of a group of associates. The result was the archly named Mitford and Wallaroo Adventure. Mitford later claimed to have relocated the spot he had previously seen in early May, before Ryan had made any discovery, but was unable to take any action because his horse and trap had been seized to pay his debts. In any case, he did not write to his backers until 4 June – 10 days after Hughes had visited the Land Office, and two days after Ryan had signed an agreement with him. In that letter, Mitford warned that Hughes had taken out claims 'a few miles off', and urged his associates in Adelaide to acquire the area he had identified if it was still possible.[14]

Mitford identified his find by reference both to the coast and to the Bald Hill, but on the waste lands map these were not in correct relation to each other, a fact that was not disclosed to the public. Goyder gave Mitford time to fix his claim properly, but warned him that it might turn out to be on an area for which a claim had already been lodged. When Mitford and his associates were eventually told that this was the case, they repeatedly demanded to be shown the prior claim. According to Mitford, this was so that he could verify with his own eyes that the descriptions matched and that it did indeed claim the same ground. When Goyder refused to reveal the details of the prior claim, which he considered himself bound to defend, Mitford interpreted this (or claimed to) as evidence of false dealing and attempted to mount a case to obtain the Moonta mine. Mitford claimed that he had been misled and had faced deliberate obstruction in the Land Office, and that he had been shown a map which later disappeared, while other maps were altered. (Surveyors and Land Office clerical staff later testified that a map corresponding to the description given by Mitford had never existed, and changes to the other maps were coherently explained by Goyder.) What was at first implied, and later alleged explicitly, was that the information revealed on Mitford's claims had been stolen – by Goyder – and given to Hughes, who was then able to shift the location of his claim.

Mitford's allegations extended to include Strangways, who had become commissioner of crown lands on 20 May, just five days before Hughes had arrived to place his claim. Strangways had already had a year's experience in government as attorney-general, but Mitford portrayed him as Goyder's puppet. The attack on the pair was backed up by an all-encompassing conspiracy theory: anyone who denied Mitford's claims was accused of having been bought. Since Hughes had distributed shares to a small crowd of influential men, not all of whom were his personal friends, this strategy had some appeal. Regardless of his tactics, it can at least be said for Mitford that, of all the Moonta disputants, he was the only one to have actually been on the Yorke Peninsula prospecting for copper.

Colonial morality

From mid-1861 to the end of the year, a torrent of earnest, angry and demanding correspondence had poured into the Land Office from the various mining associations, as each vied for the Moonta lease. Visits were made to the surveyor general, several a day at times. The constant demands – made not just by Mitford but also by the representatives of other mining associations – to see the original claim appear to have inflamed Goyder's naturally quick temper, already aggravated by the irritability that is a characteristic symptom of scurvy, from which he was still recovering.[15] More than one such 'visitor' later reported being ordered (with formal politeness) out of his office.[16] But his behaviour also shows how strongly Goyder, who was generally regarded as courteous and helpful, reacted to what he perceived as attempts to fraudulently obtain information.

Mitford and his co-conspirator Phillipson took the step of employing a Supreme Court barrister to attempt to obtain access to Hughes's claim on their behalf and increased the implied threats and charges in their letters. Strangways warned that any charges should be put in writing so that they could be investigated, and on 26 January 1862, Mitford responded with a long letter containing a numbered list of complaints, ultimately summarised as:

> a chaos of contradictory statements, distorted facts, impudent assertions, and absurd blunders – a hill moved ten miles – two hills knocked into one – that one finally ignored – claims granted – claims denied – claims without forfeiture, and forfeiture without claims – maps made to suit all purposes, and documents altered to suit any requirements.

Mitford then declared that he could not reasonably dare to hope that:

> when Government officers are made things of sale, as sworn in the Insolvent Court; and the Treasury mulcted to augment a fixed salary, as declared in Parliament; the mere connivance of an interested official with a cunning interloper to defraud an original discoverer of valuable claims, would be thought worthy of particular notice, or very nice inquiry.

Although he had had to mortgage his house to enable him to embark on the journey to undertake the triangulations in the north in 1859, Goyder's name, unlike Mitford's, does not appear in the index of Insolvency Court activities, as implied. If this smear was aimed at Goyder, as it appears to have been, Mitford was mistaken. The whole business of his failed claim, Mitford concluded, was an attempt to protect the interests of 'money exalted offenders' and was 'a brilliant sample of colonial morality, that I trust did not originate in the Crown Lands Office'.[17] As well as implying that it was Hughes who was behind the scheme to cheat him, by referring to the Crown Lands Office, Mitford was implicating both Goyder and Strangways. The Land Office (where sales, leases and claims were handled) and the Survey Department were the responsibility of the surveyor general, but the office of his elected superior, the commissioner of crown lands and immigration, was the Crown Lands Office.

It is not clear how much of Mitford's fulminating was pretence and how much was delusion fed by frustration and disappointment. In any case, adopting the position of a victim enabled him to locate the source of his unhappiness in the moral failings and hypocrisies of his more materially successful fellow South Australians, and having found this comfortable perch, he never left it. In later writings, in which he constantly vilified Goyder, he identified the path to financial and social success in the colony as: 'turn Methodist – get into Parliament – borrow money in Europe, and rob right and left', and wailed that: 'no honest man can live in South Australia. I have tried the experiment myself, just for a change, and believe me, it can't be done – for it don't pay'.[18] He attacked what he labelled 'colonial morality' as the inevitable outcome of self-government and democracy, while proclaiming the belief (put forward by Gibbon) that democracies must inevitably descend into corruption, disorder and tyranny: 'Universal suffrage and lynch law are inseparable'.[19]

The person in Mitford's universe who most perfectly embodied what was wrong with the whole state of affairs was the surveyor general, whose colonial voyage of discovery had travelled a route quite different from his own, riding a steady breeze from genteel poverty and obscurity to prominence and responsibility. Goyder's rapid rise to a position of central importance on the basis of his ability and dedication and his democratic attitude made him an embodiment of the values South Australians proclaimed. But through the eyes of Mitford's bitter frustration and outraged sense of self, Goyder was someone 'brought up (and only fit) to sweep an office', someone who'd been fished from the 'squalid dregs of universal suffrage'.[20]

Out of doors

In 1862 Mitford approached Anthony Forster, one of the editors at the *Register* and member of the Legislative Council (and a recipient of shares from Hughes), and Frederick Sinnett at the *Telegraph* with his version of events concerning the mining claims. At both offices, his attempts to have his grievances made public were ignored – he was 'repulsed'.[21] It was perhaps this failure that led to his publishing a pamphlet, probably sometime in 1862, in which he claimed that the owner of the original four claims in the name of 'Peter' was Goyder himself.[22] It was possibly around this time that, on some occasions when he left the office, Goyder was followed and spied on.[23]

The crescendo was reached in September, by which time, as the *Telegraph* reported, 'the persons said to have been corrupted, and the price of their alleged corruption, are almost as freely discussed on 'Change [the Adelaide Exchange] as the price of flour ...'[24] Debate in parliament led to some of the correspondence being tabled, including Mitford's letter of 26 January. Goyder replied to questions made in the house, while Strangways asked members, 'not to be swayed by the statements made by certain persons out of doors, by persons who had formed themselves into a kind of committee, with a view to endeavour to obtain a large sum as compensation from the government'.[25] A petition from the Mitford Mining Adventure (as it had become) followed, and Goyder was asked to respond with a report, which he did on 26 September. So urgent did the situation seem that the commissioner had telegraphed Goyder at Kadina to ask him to come down to prepare it.

The petition was a rehash of the usual claims. Goyder had by now read Mitford's defamatory letter and he was furious. He began by pointing out that most of the claims to which he was required to respond were erroneous in the first place, and addressed the 15 allegations he could recognise in the letter in order. His response made plain his contempt. It was true as alleged, he explained, that the disputed claim was being worked when the disputants wanted to take it up, 'as occupation had been continued by the original discoverers from the first'. It was also 'perfectly true' that access to the details of Peter's claim had been constantly denied to the petitioners, 'and it is to be regretted that so stringent a precaution is indispensably necessary to defeat the machinations of persons ... who do not hesitate to take advantage of misdescriptions to deprive the original claimants of the benefits arising from their discoveries'. 'Finally,' he concluded:

> I do not believe that the petitioners suffer any loss whatever beyond what is justly entailed upon persons attempting to obtain the property of others; and I utterly repudiate the possibility of the petitioners proving the truth of the allegations set forth, and beg respectfully to state, that had I seen Mr. Mitford's letter ... on an earlier date, I should have urged upon the government the propriety of prosecuting that individual, or have left the service in order that I could have done so myself.[26]

The South-East

Unaware of the allegations, Goyder had been occupied for at least some of the time during 1862 by the journey to the waterless country behind Fowlers Bay.

On his return he had presented his report, which was apparently not interesting enough to be published as a parliamentary paper. However, the tour of the South-East made with Milne, the commissioner of public works, and the engineer Hanson in January 1863 was to have a result that would engage him, with some interruptions, for the next 20 years, and again in his final years. In addition to the survey and the affairs of the pastoralists and the miners, he would be absorbed in realising the dream of draining the vast seasonal swamps of the region, and engineering a new landscape. This massive disruption of a natural environment was at the time universally and unquestionably accepted; the only criticism of this project was that it might not succeed.

The stated purpose of the tour in 1863 was to find ways to improve traffic and communications, but at the same time, the subject of drainage had been 'a matter of special observation and remark'.[27] As they travelled, Goyder had pointed out to Milne how the inundated areas, where the water was often both extensive and deep, could be drained cheaply and effectively by simple cuttings to create new pastoral and agricultural land, and at Milne's request he examined and marked out sites where these could best be made.[28] Once back in Adelaide, Milne requested a report from Hanson and also from Goyder, and it was in this report that Goyder accurately described the lie of the land of the South-East.

Goyder had grasped that the flooding was the result of fairly high rainfall, a groundwater level that reached the surface during winter, and raised ridges of land running parallel to the coast, resulting in water gathering in the channels between them, which then slowly drained to the north – at Maria Creek, which enters the sea at Kingston on Lacepede Bay, and further north through the Salt Creek. Salt Creek drains into the Coorong, a long strip of water that begins just north of Kingston and is separated from the sea by a narrow ridge of land, the Younghusband Peninsula. Further south, from Robe to Port MacDonnell, water also gathered in coastal lakes.

To prevent the flats from being flooded in winter, Goyder proposed four cuttings, two connecting nearby swamps to Salt Creek and Maria Creek, and two much further south, through narrow coastal ranges, to connect German Swamp and Mount Muirhead flat to Lake Bonney and Lake Frome.[29] This was intended as a safe beginning. Goyder had a larger scheme in mind, but advised that no more work should be done until 'a careful survey of the whole of the swamps and low ridges' was carried out to determine their levels in relation to the sea. He knew the results would make plain to others what he, with his powerful geographical imagination, could already see:

it would at once expose the practicability of the formation of drainage channels at right angles to the direction of the valleys; that is, in a direct course
towards the sea – giving an increased fall and consequent scouring effect ...[30]

As well as proposing to simply cut across the ridges and flats directly to the
sea, he envisaged forming lakes with no natural outlet in the basins. These
might be drained by sinking shafts 'to conduct the waters from the surface into
the cavernous hollows beneath'.[31] Given the thoughtful attention Goyder paid
the landscape, this plan was probably inspired by his observation of those places
where water could be seen draining naturally into large holes in the limestone.
The strange sucking, rumbling sound made by the water being drawn down
into the caverns was enough to attract attention and reveal the location of the
holes.[32]

The Public Works Department suggested doing no more than enlarging the
outfalls of Maria Creek and Salt Creek. There was another government tour of
inspection to consider the matter, but the focus of official concern had shrunk
again to improving roads and maintaining the telegraph line, and drainage was
seen as a little more than a cheap means of keeping the existing roads passable.
Two cuts were made according to Goyder's suggestion – an outfall at Narrow
Neck draining Cootel Swamp into Lake Frome and a cut draining Tilley's
Swamp into the Salt Creek. (The Lake Frome referred to here is an intermittent coastal lake located behind Cape Buffon, not to be confused with the better
known salt lake to the east of the Flinders Ranges.) This work was so immediately successful that surveys were undertaken to identify areas where it could be
continued, and land was held back from sale in anticipation of the higher price
that would be received if it were drained.[33] But the government's attention was
distracted by the effects of the drought in the north. For the next few years the
Public Works Department would carry on with the drainage work, although
in a perfunctory manner.[34]

Odium mineralogicum

Meanwhile, the problem of the Moonta had not gone away. At the end of February
1863, Mitford had been back in the Land Office attempting to make another
claim. The battle over points of reference continued. According to Goyder:

> I explained to them ... that the land they referred to would be in the sea.
> Mitford stated that he had camped there very often. I stated that if he wished
> to mine in the sea I had no objection. I left the office then ... and went into
> my own room. I had, a few days before, received from the surveyor who was
> laying out the Moonta township, a correct survey in detail, with the Nalyappa

and Warburty Points; and with that I returned to the office and laid down the position of the coast line from Port Hughes to about the place where they wished to make the application.[35]

This only provided Mitford with more ammunition, and Goyder, realising that the affair would eventually be investigated, ordered the chief clerk to restore the maps to their original state – lest he be accused of tampering with them. A select committee was appointed to inquire into the allegations surrounding the claims at Tipara (Moonta). It consisted of Arthur Blyth, the treasurer, and his brother Neville Blyth ('the spouting ironmonger' as Mitford was later to dub him), J.B. Neales, G.S. Kingston, Walter Duffield, Alan McFarlane and Lavington Glyde, all prominent figures in pastoralism, agriculture, mining and business. Mitford, complaining pitifully that his character had been unjustly attacked by the surveyor general, asked to be represented by counsel, as did the other members of the association connected with his claim. (Goyder's 'unjust' assault on Mitford's character seems to have consisted of nothing more than his having visited Mitford's new partner to warn him against being mixed up with a 'notorious scoundrel'. This 'strong language' had been provoked by the pamphlet Mitford had published accusing Goyder of being the owner of the four leases in Peter's name.[36]) It was agreed that all parties – including Goyder – could be represented. By the end of June the committee had heard from over 60 witnesses. At this point it became apparent that more sympathy had been generated for the claims of Mills than those of Mitford, and Mills petitioned to have his complaint considered by the committee as well. By this time Patrick Ryan was dead.

It then only remained for Goyder to give his evidence. Over six sessions in July he was presented with nearly 1500 questions. He was questioned first by Rupert Ingleby, the government counsel, and then J.P. Boucaut, who had taken over as Mitford's counsel. Boucaut's technique was to ask a series of technical questions, interposed with inquiries designed to undermine Goyder's integrity, leading to complaints in parliament about the use of courtroom tactics. Without warning, Boucaut suddenly inquired if Goyder was aware that Freeling had written to the commissioner of crown lands, 'setting forth that you were a person not at all fit to hold the office of Surveyor-General?' Goyder answered simply, 'No'. The questioning continued:

'Did you never hear of such a letter?'
    'No.'
    'Did you know Mr. Stevenson, a land surveyor?'
    'Yes, I did.'

After it was established that Stevenson was dead, Boucaut asked if he had been in the habit of sending reports of the value of lands he surveyed to the government. 'Every diagram is accompanied by such a memorandum,' Goyder replied, and then added, obviously addressing the committee, 'I know what he wants, and will give all information'.[37]

In answer to the clearly implied charge that Freeling had considered him not fit to be surveyor general because of his earlier involvement in corrupt land dealings, Goyder explained the circumstances surrounding his employment as a trigonometrical surveyor for the northern triangulations, the mortgaging of his possessions, and the arrangement between his agent Wadham, Frances, and the surveyors who were to advise her. Freeling had known of this arrangement, Goyder explained, although he had apparently forgotten when he was told of it again and had objected. Kingston asked if the information to be given to Frances was of a 'fuller nature' than is usually noted on the government documents. 'Certainly not', Goyder replied.[38]

At Goyder's next appearance Kingston had taken over in the chair, and he began by immediately returning to the matter of Goyder's appointment, giving him a free hand to defend himself and inviting him to give any evidence he thought appropriate to the committee. What Boucaut had done was imply that Freeling's displeasure at the arrangement with Stevenson was the basis for the concerns he had expressed about Goyder's fitness to become surveyor general – that Freeling had been concerned about a willingness on Goyder's part to exploit his position for personal gain. To clear his name, Goyder had to remove that confusion, and to do that he would have to admit to knowledge of the contents of what he been assured would remain a confidential letter. He had previously claimed to know nothing of this letter, but with the gravity of his situation clear, he was less reticent. As 'the contents of the private and confidential letter appear to have been made public,' he explained, 'I will state what they were'.[39] Once Goyder had divulged the real reasons Freeling had given for opposing his appointment – his misjudgment in wanting to consult district councils about roads and in claiming that he could manage without an assistant – Kingston questioned him about the arrangements that had been made with Stevenson in a way that was usefully investigative and direct without being insinuating or brutal. By the end of this exchange, Goyder had begun to look like a put-upon and poor-but-honest individual whose privacy had been violated.

Boucaut responded by presenting the claim that in his letter Freeling had described Goyder as 'very deficient in judgement, rash, and hasty'.[40] He went on to hint that the information sent to Frances was given to Wadham for use in the land agency of Green and Wadham, the business which Goyder had been invited to join in 1862. Goyder explained that Frances was to use the information to provide Wadham with a bidding ceiling, if instructing him to purchase. Kingston again clarified the situation by interrupting to ask if Wadham had been made acquainted with the contents of those reports, even though no land was purchased. 'No,' Goyder replied, 'I do not think Mr. Wadham has purchased a single acre of land in the colony on my recommendation, or any information got from me'.[41] Boucaut then put a question to which Goyder objected. The room was cleared while the committee considered the objection. When the doors were re-opened, Kingston informed them that Boucaut could put the question if he wanted, although the committee would not insist on it, nor would they insist on Goyder being required to answer. Boucaut put his question – which, like a question to which Mitford objected, was not recorded in the minutes – and Goyder declined to reply. Almost certainly the question would have pursued the same line of questioning and may have broached the nature of Goyder's connection with the land agents.

The inquisition into the surveyor general's probity and administrative competence continued for another four days of the hearing, on 10, 13 and 15 July, and 6 August. Along with the maps and the issue of 'vagueness', questioning constantly returned to Goyder's instruction to the chief clerk not to draw attention to Hughes's claim while it remained unsurveyed. Boucaut also returned to the matter of Goyder's involvement in the inquiry itself, suggesting that he been showing Hughes the evidence, that he had been in constant correspondence with Hughes since the inquiry began, and that he had been providing information and instructing the government counsel in his cross-examination. Goyder denied all of this, although in one of his replies he observed that he had been taking notes on the proceedings and it was possible that Hughes had read them in his room while he was being examined by Ingleby – which seems to raise the question of what Hughes was doing in his office in his absence, although no eyebrows were raised at the time.[42] That Goyder did instruct Ingleby, the government counsel, seems also to have been admitted by implication on a day when Ingleby was ill. On behalf of Ingleby, Boucaut asked Goyder a question that Goyder revealed he had wanted put.[43]

Goyder also had to face another insinuating line of questioning from Boucaut, who repeatedly implied that he controlled Strangways. Both Mitford and Phillipson had already attempted to score points off the relationship between Goyder and Strangways by recounting an exchange in which they told Goyder that Strangways had admitted that the maps were all wrong. According to Phillipson, Goyder was supposed to have exclaimed that, 'the Commissioner knew nothing at all about it; and that it was a piece of ex-official information, which he had no right to give us; and that the Commissioner ought to talk less', or, as Mitford put it, to 'mind his own business'.[44]

In response to another line of inquiry, Goyder acknowledged that there had been a possibility of his being given shares in the mine. Stirling had told Goyder that he and Taylor had been considering 'the propriety' of allotting him some shares. 'I stopped him at once,' Goyder related, 'and told him on no consideration whatever would I accept any interest in any mine in the Colony whilst I held my position as Surveyor-General of the Colony'.[45] He repeated that statement later, stressing that shares had not been offered to him, only the possibility had been considered.[46] Towards the end of the inquiry, Kingston again tried to discover when the 'offer' had been made. By this time, his question was prefaced: 'I do not mean to offend you …' Goyder answered that he knew nothing of the division of the shares and added that:

> if my impression had been that there was any attempt to influence me by what was said, I should have made an indignant rejoinder, but I considered they were only offering me a friendly gift; but I could not receive a gift of that kind, therefore I stopped them.[47]

No evidence was found in the records that Goyder had ever owned shares.

Goyder was eventually granted the opportunity to make a general statement. The committee heard at considerable length the history of the waste lands maps, on the state of which Goyder thought they may feel compelled to remark. (They did.) The point he wanted to impress upon them was that he had been ignorant of important information that had been revealed during the hearings – obviously alluding to Hughes's evidence in which Hughes had admitted to withholding information when his claim was first lodged – and he concluded by requesting that once again they bear in mind the fact that 'my whole knowledge, up to the time of coming into this Committee room, was obtained from documents which came officially before me, and from applications made personally to me'.[48]

The committee's report, together with the minutes, appeared early in September. Containing over 7000 questions and responses, the minutes were baffling in size and complexity. As the historian Edwin Hodder recalled in 1893, all the circumstances connected with the claims 'would fill a moderately sized volume'.[49] Since all the members had dissociated themselves from the report, principally because of legal aspects, responsibility was borne solely by the chairman, Kingston, who decided in favour of Ryan as the person who had made the original discovery and Mills as entitled to the claim after Ryan's death. Mitford was scarcely mentioned. (Both his legal representatives, Charles Fenn and later Boucaut, eventually admitted that he had no claim.) Great care was taken to stress Goyder's innocence, but problems remained. The committee deeply regretted Goyder's 'extraordinary want of judgement and discretion' in accepting Hughes's claim without investigating the situation of the other claimants or interviewing Ryan, but they had 'much pleasure in expressing their conviction that, in doing so, he was in no way influenced by pecuniary or improper motives, as has been insinuated'.[50] Because of the defective state of the maps, it was decided that payment of costs would not be demanded of Mitford and his associates, despite their having no case. They had only to pay their own costs. The report concluded with the expression of:

> the fullest conviction that there is not the slightest ground for the accusation that any maps and papers have been destroyed, improperly altered, or tampered with by the Surveyor-General or any of the officers of his department, or that there are any grounds for accusing that officer, or those under him, of being influenced by any other motive than a sincere desire to discharge their duty faithfully to the public ... [51]

Goyder had been caught in a difficult situation, administering what were acknowledged to be unsatisfactory regulations that were certain to generate problems with inadequate tools inherited from his predecessor. But the real issue was that the situation had arisen at a time when the colony was undergoing transformation and the small world of gentlemen's agreements and transactions based on public reputation was disappearing. (Apart from trusting Hughes, Goyder described accepting an application from one of Mitford's associates, despite his misgivings, because the person who had presented it 'was a respectable man, and I was bound to receive his statement'.[52]) The 1860s were a time of transition, and in the early 1860s the rough dwellings of the pioneers were being superseded and piped water and gas lighting introduced. Similarly, personal responses and ad hoc solutions were no longer an appropriate administrative approach, even under pressure. Goyder's down-to-earth approach, while it reflected his own willingness to take responsibility for his decisions, did not combine well with his habit of conducting himself, as John Western, the Elder's clerk who had assisted Hughes, put it, like 'any business man about town', making personal visits to the offices of land and pastoral agents. To those who had taken to 'dogging' Goyder when he left his office, it looked suspiciously like partiality – although, as Western also remarked, he could recall that on two or three occasions when Goyder had visited Elder Stirling, he had been 'anything but friendly', and 'had not done what we wanted him to do'.[53]

Western summarised a great deal when he noted that: 'generally in putting in mineral claims at that time, there was not that particularity required that there has been since'.[54]

Not so black

At the *Telegraph*, Sinnett judged that Goyder had been exonerated, 'very properly, we think, according to the evidence'.[55] But Mitford and the adventurers were not willing to give up, and not long after there was an encounter in the streets – Sinnett reported it with relish – between Strangways and Phillipson.[56] In parliament Strangways had accused Mitford of perjury in declaring himself bankrupt, described Phillipson as 'a swindler, as anyone could see who referred to the proceedings of the Court of Insolvency', and accused both of them of making 'vile and unfounded charges' against himself and Goyder.[57] Phillipson then took to stalking Strangways with a horsewhip hidden in his coat, and eventually confronted him in the street, but the infuriated politician fended him off with a good shove that knocked him against a wall. After that, interest in the whole affair finally fizzled out, although Goyder had not escaped criticism. Apart from Kingston's remarks about his rashness, Sinnett held that, while innocent of corruption, he was guilty of being 'not strong-minded enough to be a fair arbitrator between a few insignificant people and the great gods of our commercial Olympus'.[58] Neville Blyth gave his assessment of the affair in parliament, accusing Goyder of having one law for the rich and another for the poor – an expression that, as it happened, had originally been coined by John Mitford, Lord Redesdale, Eustace Mitford's great-uncle.[59]

Fortunately for Goyder, John Barrow rose to his defence in the *Advertiser*. Barrow agreed that the question was important: they could not have a surveyor general who favoured the rich over the poor. But Barrow distinguished between the principles that had governed Goyder's approach, which he held did not favour Hughes (Goyder, after all, had been trying to defend what he believed to be rights related to an original discovery), and the eventual outcome, which certainly did. Barrow dismissed Mitford's complaints, noting in Goyder's defence that, 'as Mr. Blyth considers that Mr. Mitford has no valid case at all, he will not find fault with the Surveyor-General for taking the same view of the matter'. That left Mills, who, as Barrow pointed out, had said himself that Goyder had been as helpful as possible when attending to his application. Barrow acquitted Goyder of 'morbid reverence for wealth', and challenged Blyth to make a clear case, if he had one.[60] Goyder, who had been out of town and unaware of the debate taking place about him, came back to find Barrow's defence in the morning's paper. He wrote immediately to thank him. 'I am the more obliged,' he assured Barrow, 'as the paper will be read at home and my friends will have the opportunity of knowing that I am not quite so black as I have been painted ...'[61] Barrow's article did its work. The affair blew over and all was quiet for a while.

On the same day Goyder also wrote to thank Thomas Reynolds, one of the five commissioners of crown lands who knew of the letter that Freeling had written and which had been presented at the Tipara inquiry as evidence that he had opposed Goyder's appointment as surveyor general. When the minutes of the inquiry were made public, both former commissioner John Bagot and the current commissioner, Frederick Dutton, had been asked for an explanation of how the contents of this letter had been available to Mitford's counsel (who had refused to explain). Dutton denied all knowledge of the letter, while Bagot claimed he'd put it in a private deed box – where it remained – and forgotten it. Strangways explained that he had been informed about it when he became commissioner in 1861, and told the parliament that the only others who knew of it were former commissioners Waterhouse, Reynolds, and Hay, and he was able to exonerate Waterhouse and Hay, leaving only Reynolds, whom he was sure could proffer a firm denial. Bagot wondered if an official had been guilty of a gross breach of trust and an act of personal antipathy towards Goyder – could Freeling be to blame?[62] Reynolds immediately penned his denial, assuring Goyder that he had: 'too much regard for your conscientiousness, industry and ability to do anything that was calculated to damage you in the estimation of the public ...'[63] In reply, Goyder criticised Freeling for not having shown him the letter earlier and claimed that, had he known of the letter before his appointment, he would have withdrawn his application for the position. He assured Reynolds that he had never believed him responsible – 'I blamed Mr. Bagot – but he has stated that he has not divulged the fact and this on his honour ...'[64] When Bagot addressed the matter in parliament, he made the point that had been so far overlooked: that while there were 'one or two other remarks', Freeling's letter 'gave the highest character to Mr. Goyder' and 'bore testimony to his being one of the ablest men in the colony'.[65]

The case was noted outside the colony. In Melbourne, the *Age* had been scandalised that £60 000 worth of shares had been distributed among people of influence, and although it was acknowledged that Goyder had not taken any, he was seen as a lackey of the rich and powerful.[66] In the following year, 1864, Goyder would confound that view by substantially increasing the valuations of the richest pastoral runs, while in Adelaide, where the surveyor general's 'price' had been freely discussed only two years before, the public voice would be solidly united in its assertion of his unshakeable integrity. Through their patent falsity, Mitford's allegations seem to have almost embarrassed the community that entertained them: during the turbulent period of the revaluations, the surveyor general's probity remained unquestioned. Unfortunately for Goyder, the now somewhat unbalanced Mitford remained determined that the incident would not be forgotten. While the members of the closely connected core of wealthy and influential people at the centre of South Australian society diligently feathered their own and each other's nests, Mitford continued, very publicly, to focus his rage on the one person who invariably gained nothing but his salary. But this challenge to Goyder's career would never become part of his public curriculum vitae – for the good reason that he was found only to have made an error of judgment – and it would eventually be completely forgotten.

The Murray

Inevitably, Goyder's responsibilities and interests also brought him to the banks of the Murray, Australia's largest river, which collects the waters of the continent's greatest river system, the Murray-Darling Basin. Before European settlement, the river had both a seasonal cycle, which reflected winter rainfalls in the surrounding areas, so that it peaked in spring and was at its lowest at the end of summer and in autumn, and a variation in flow, reflecting the extreme variability of the continent's rainfall. In dry years it was in danger of becoming little more than a 'chain of waterholes', as Goyder remarked, but in wet years it stretched across its flood plains.[67] This flooding flushed away the shallow saline water (the salt rising with the ground water) and freshened and revitalised billabongs and backwaters.

The notion of a large river without a navigable mouth was baffling, and early in the colony's history attempts had been made to find an entrance suitable to seafaring traffic. The Murray emerges into the sea through a narrow gap in sandhills and its shallow mouth and surrounding waters are in constant flux as variations in the river's flow redistribute huge volumes of sand and redefine the river's entry point to the sea. In 1856, while still an assistant to Freeling, Goyder had been sent to the mouth of the Murray to assess the possibility of connecting Hindmarsh and other islands to Pelican Point, on the Lake Albert Peninsula, after the governor had been petitioned for a road to follow the Aboriginal route that ran along a rocky ridge between the two points. Goyder tested the entire length of the limestone ridge and proposed a route that would connect the islands and at the same time shorten the journey to the South-East. But while contemplating the proposed road, the ever-curious Goyder found himself 'naturally led to consider the direction of the river current, and its influence upon the Murray mouth'.[68] The problem for navigation was the fluctuation in flow and the shallow, shifting mouth and bars. Goyder proposed bringing the main channel of the Murray, once it reached the sandy, islanded area near the coast, down Holmes Creek (or Channel) to the east of Hindmarsh Island, which he believed to be the original main channel, rather than down the Goolwa channel, the present main channel to the south-west. He believed that the course had been changed when the river encountered the rocky bar that blocked the Holmes Creek channel. Cutting through it would restore Holmes Creek as the main channel and would place the mouth at right angles to the coast again, and so stabilise it. Vessels would gain a direct entrance from the sea into a safe channel between Hindmarsh and Mundoo islands, he explained, 'nor is it likely that the Goolwa channel would be injuriously affected by the altera-tion'.[69] His proposal was never implemented because of the cost and because, as he recalled at the end of his career, it 'would have resulted in affecting vested interests'.[70]

In September 1863 he produced a report on routes to the Murray – or at least that was what he had been asked to report upon and how the report was titled.[71] The contents, though, were the product of Goyder's circumspect and logical, independent approach. Rather than beginning by comparing the virtues of various routes, as expected, his approach was to consider the state of the river and the extent of its navigability. He reported that there were banks across the river at several points, and these were believed to be increasing. The location of shallows and bars that were impassable during times of low flow were also identified. To deal with them, he suggested having the steamship *Grappler*, sitting idle at Blanchetown for half the year, converted to a dredge. The conver-sion would be cheap and would enable the river to be navigated all year round, capturing the flow of products from the north which would otherwise go to Victoria. When he finally came to consider roads Goyder simply declared that the five road routes he had been expected to compare would all be built in due course and should all be accepted: the critical task was dredging the Murray. This advice took everyone by surprise, and cut the Gordian knot, as the *Register* put it – the weighing and measuring of the merits of the various routes had been going on for some time without reaching a conclusion.[72] The *Grappler* was converted and, at some point, dredging of the Murray began.[73]

Modes comiques

That year – 1863 – was also the year in which it was decided that ministers of government, along with their under-secretaries and heads of department, would all be required to wear an official uniform on important occasions of state and in the presence of the governor on official occasions. The uniforms were a ready

Goyder in his official uniform. The photograph was probably taken in 1863, when uniforms for senior public service officials were introduced. Goyder would have been about thirty-seven. 'For behold, I will show unto you a little bloke in a cocked hat and a big sword' crowed Mitford, who also characterised Goyder as 'South Australian drought in a cocked hat and pigsticker'. [SLSA]

target for wits. The most common joke was that the wearers of these outfits looked not so much like important public figures as liveried footmen. The *Register* published a comic poem in which the narrator is a servant or former servant:

> Agin to show my loyalty
> To hour most grashious Queen,
> And also fur to watch the swells,
> So splendid to be seen,
> Hi took my way to Guvmint-ouse
> Hand stood amongst the best,
> And sor wunse mor wot jeered my art,
> Those men sooperbly drest.
> Around the Guvnor there they stood,
> While pride wos in their hi;
> They looked so like my friends hov hold,
> They almost made me cry.[74]

Nevertheless, the introduction of uniforms certainly transformed vice-regal levees – events which were already important in publicly defining social position – into literally more glittering affairs. An early photo of Goyder shows him in his uniform and was probably taken to record the full glory of his appearance effected by this expensive official addition to his wardrobe.

One of those to join in the fun that was being had at the expense of senior public servants was Eustace Mitford. Pamphleteering and maintaining a correspondence with the Land Office had given Mitford a taste for wielding the pen, and in 1863 an unsigned article appeared in the *Telegraph*, satirising the new uniforms. In the 1760s, a literary hoax involving the discovery of ancient Gaelic writings made popular what were known as the 'Songs of Ossian'. The songs, the authenticity of which did not go unquestioned for long, had a distinct, readily parodied style. Having noted that the *Government Gazette* had appeared as 'a sort of *Magazine des Modes Comiques*', the writer ('X') suddenly exclaims, in Ossianic style: 'Weep, ye gods, and rejoice ye tailors, at the change. JURY [a tailor] lift up thy head; for thy shop in Hindley-street will be the resort of Ministers and other high functionaries. Poet of MOSES & SONS indite a pæan fitted for a lofty theme!'[75] And so on. Mitford would later prove that he could churn out parody like this effortlessly.

Mitford had also submitted a satirical article to the Gawler Institute, which had offered a prize for a history of the colony. His history, this time in a mock Biblical style, did not win, but it was published in the *Critic* and the *Border*

Watch.[76] For Goyder, this modest success was an ominous development: one of the targets of Mitford's satire was the Survey Department. Later, when he had his own weekly paper, Mitford would describe the officially costumed surveyor general as 'South Australian drought in a cocked hat and pig-sticker'.[77] He intended, of course, to demean Goyder, but in a way he could not have comprehended or foreseen, his jibe was profoundly apt.

CHAPTER TEN

Transition: 1866–68

The years 1864 and 1865 mark the end of the first phase of Goyder's career. That beginning – full of energy, exploration and the search for understanding, and characterised by unshaken confidence in his own ability to devise and effect appropriate strategies for settlement – was brought to a close by the valuations, the massive labour of which gradually wore him down, and the overwhelming drought. But it was a fruitful end. Central to his vision of a settled landscape was the line of reliable rainfall that divided the country into pastoral and agricultural zones according to the realities of the climate. That line was now drawn and had been checked against conditions far worse than anything the colonists had previously experienced and had already been put into use, if only to benefit drought-stricken pastoralists. However exhausted he was, Goyder had good reason to believe he was ready to deal with what lay ahead.

But the years that followed, from the beginning of 1866, when the Line emerged, until 1872, when it became enshrined in law, run through Goyder's life like a dark valley. Between 1866 and the end of 1868, when he left to lead an expedition to select a site for a settlement and to survey a large area of land in the Northern Territory, Goyder continued to be occupied by a range of concerns apart from the survey.

While the need to value runs ensured that Goyder was 'not much in the office', it meant equally that he was unable to supervise parties in the field, and he finally had to accept the need for a deputy.[1] In the brief portrait of Goyder that Lionel Gee included in his recollections of the South-East in the 1870s, Gee commented that Goyder was 'jealous of rivalry', and the fact that he was able to resist having a deputy for so long during his early years, despite the evident need, bears this out.[2] The hiatus of power that this created around him ensured not only that he was entirely in control, but also that he was indispensable – no one else was able to gain the sort of knowledge and understanding that might challenge his. But there was only so much even Goyder could do, and on 1 September 1865, Arthur Bevan Cooper was promoted to deputy surveyor general in charge of field staff.

Goyder had finished valuing the large group of runs that had come up for renewal in 1865 (and for which he had been paid £1400 in addition to his normal salary), but he was still expected to value other runs as the leases expired, and for much of 1866 he was still undertaking this task, during the first part of the year tabling valuations of runs in the South-East. In early May the government, now led by Boucaut, with Milne as lands commissioner, initiated a process aimed at avoiding the uproar of 1864 by appointing additional valuators. To make up a trio, the government chose one valuator and the squatters another. The chief inspector of sheep, now Charles Valentine, and former commissioner Charles Bonney – the only two people capable of performing the task – were duly chosen.

On 10 September, Goyder wrote to the commissioner, Milne, asking to be relieved of the duty of having to visit and evaluate runs himself. 'I find that my health will not permit me to continue much longer the tax entailed upon body and mind by the extra duty which I have to perform as Valuator of Runs', Goyder warned, repeating that he needed to be free of 'duties that I cannot much longer undertake without serious injury to myself', and that it was 'indispensably necessary that other and more efficient arrangements should speedily be made'.[3] Goyder's idea was basically that, with the knowledge he already had and information from the other valuators, it was not necessary for him to leave the office, at least when runs in the north were being valued.

At about the same time, a petition was presented to the parliament claiming that the surveyor general's frequent absences from the office were costing people time and money, and requesting that something be done. By the end of November the situation had become critical. Strangways insisted that Goyder needed to be in the office more often because squatters were taking advantage of his prolonged absences to subvert the intent of the land laws with clever schemes which Land Office staff, lacking Goyder's knowledge and experience, could not recognise. J.B. Neales agreed. Mineral claims were in arrears in the Crown Lands Office, which was beset with 'red-tapeism', according to him. If it was necessary for him to go out, then an assistant should be appointed.[4] When a motion was moved that Goyder be relieved of the need for such frequent absences, Milne read Goyder's warning about his health to the parliament and his request that he be excused from going north again, pointing out that his language was not of the kind that could be disregarded. He assured the parliament that the government agreed with the motion and that arrangements had been made to allow the surveyor general to remain in Adelaide and address the work of his department for the following 12 months.[5] But as 1866 ended there was concern about the northern runs that were still afflicted by drought, and in the height of summer, either to inspect the country again or to continue valuing, Goyder was out again. He travelled up to Wonoka in the Flinders Ranges, returning 'overland', as he put it, via Coonatto before heading down to Kooringa (Burra).[6]

Taking on the South-East

When valuing runs in the South-East had been completed, Goyder began to urge again for the draining of the South-East. On the basis of 'extensive observation and sketches', he argued that drainage could double the available grazing country and rid the area of coast disease, which he continued to believe was caused by noxious vapours from the swamps.[7] When threats of secession were heard again, a select committee was formed, in early September 1866, to give the region's difficulties further consideration. Once again the focus was upon transport and communications, and once again a tour was planned.

Goyder urged, as he had in 1863, that a survey be undertaken to provide the basis of information on which to drain the whole region. Drained land was the most important benefit, he told them, and the process of creating it would produce good natural roads as a by-product.[8] Moreover, the major drains themselves could become part of the region's transport and communications network by functioning as canals. He suggested that the mail could be routed over the Coorong by steamer to Salt Creek, where it would connect up with a navigable canal – the main drain – that would lead to the tablelands of Mount Gambier. This drain would also possess navigable branches.[9] Such schemes must have been in the air, because a few years later, in 1871, the Grand Victorian North-Western Canal Company issued a prospectus for an irrigation and navigation scheme for the Mallee-Wimmera region of Victoria, using water from the Murray and its tributaries and pumped ground water.[10]

It was characteristic of Goyder to produce a vision of a new landscape traversed by canals while the politicians were fretting about boggy, rock-strewn roads. What is surprising is his enthusiasm for taking on the work himself, even allowing for his general belief in his own abilities. Many years later, he would explain that he was confident that his education had enabled him 'to undertake all that kind of work'. Because of his experience in the construction of railway bridges over water, he had 'no doubt' that he was 'thoroughly competent' to control the construction of any drainage system, although the only evidence of his ever having been anywhere near an entire drained region was a short period in his childhood when he lived in Hull, to the north of the East Anglian Fens and the Washes.[11] In 1866, that enthusiasm and confidence led him to tell the select committee that he would be happy to do everything he could 'to aid the works being speedily carried out to connect Mount Gambier with Adelaide, and to make the people of the South-East more South Australian than they are, by giving greater facilities for their commerce with us'.[12] Since it had only been four days since he had written divesting himself of the valuations on the grounds of ill health through overwork, it seems unlikely – but not impossible since the project was obviously dear to him – that Goyder had volunteered to take control of this work.

Whatever he had intended, responsibility for the drainage project was transferred from the Public Works Department to the Surveyor General's Office at the end of November, and Goyder was made officially responsible for the project in January 1867.[13] The first thing he did was take the levels he had been recommending since 1863. These demonstrated that there was a watershed south of Penola near Dismal Swamp. While most of the South-East drained along the parallel ridges to the north-west, as already understood, the comparatively small area to the east of the watershed drained towards the Glenelg River in Victoria.[14] Milne persuaded Goyder to divide his original grand unified scheme for the drainage of the entire region into two. The area south of a line drawn between Penola and the north end of Rivoli Bay would empty into the coastal lakes, Lake Bonney and Lake Frome. This work would be done first, since the soil there was recognised as suitable for agriculture. (This area of the South-East is pocked with volcanic craters, as Goyder noted, and has areas of good volcanic soil.)

In Adelaide, men made unemployed by the drought-induced depression were beginning to assemble in noisy gatherings. In August 1867, in an attempt to relieve the pressure, Goyder was instructed to select a party of a hundred men from among the unemployed for work in the area.[15] On 27 August, two new outfalls were opened to drain the swamps behind them into Lake Bonney. This took place exactly at noon, and by four o'clock that afternoon the water level behind one had fallen by 60 centimetres. Behind the other it was over 15 centimetres lower, but still deep. In his report on these works Goyder appears as the early nineteenth-century engineer of the kind he had probably been entranced by in adolescence, moving from one place to another, getting the gangs working, tirelessly checking and supervising the work, and elated when it all went to plan and was successful. By the end of the year, 37 kilometres of new connecting drains had been excavated.[16] His evidence to a committee only a few years later bore witness to the work he had undertaken during this period – testing flows and depths and analysing soils. He had examined 'every bit' of the South-East, he assured the members: 'I have travelled over it in every possible direction'.[17]

This effort was observed and publicly hailed by Ebenezer Ward, a journalist and former member of Finniss's expedition to the Northern Territory, who would later became a parliamentarian and minister for agriculture. Ward wrote an entire volume, *The South-Eastern District of South Australia: Its resources and requirements*, dedicated to Goyder, with his permission, 'as an acknowledgment of his interest in the welfare of the South-Eastern district generally, and of his promotion of works calculated to reclaim a large area …' Published in 1869, the book contained a history of the drainage works and lauded Goyder's efforts in 'converting a dreary waste of water into a very paradise of profitable settlement and rational prosperity'.[18] Ward reported that W.A. Crouch, one of the earliest settlers at Rivoli Bay, had always urged the government to consider drainage, and noted that Milne's support had been of critical importance, but he presented Goyder as the source of the vision, the motivating energy, and the transforming genius who had effectively created the South-East as the colonists had come to know it. 'Sceptics and croakers' had 'pooh-poohed' the main channels to Narrow Neck (on Lake Frome) as an 'absurdity', he wrote, but 'Mr. Goyder has proved himself fully equal to the responsibility he undertook, and he can now point to actual results to refute adverse predictions and justify his own action'.[19]

Nevertheless, the soil behind the two new gaps remained dangerously boggy.

The ascendancy of wheat

Although the work of his early years had been driven by the needs of the pastoral industry, it was while he occupied the position of surveyor general that wheat gained ascendancy in Australia – and in South Australia especially. South Australia had the largest area under wheat of any of the colonies and provided more than half the Australian crop in a good season, and because its farmers had to struggle with lower yields per hectare, they generated major technical innovations. From 1866, with many squatters broken by drought, it was wheat farming rather than squatting that became 'the symbol and the reality of prosperity and settlement expansion', a result that was pleasing to most.[20] As one South Australian parliamentary committee later declared:

> The substitution of a numerous yeomanry owning agricultural holdings for a few pastoral lessees on large tracts of land let by the State, has been the object of land legislation in all the Australian Colonies.[21]

At the beginning, though, the first arrivals had been plunged into despond-
ency at the sight of the Adelaide plains in summer. Nevertheless, the colonists
soon discovered that the alluvial soil of the plains was easily ploughed, and by
the early 1840s, South Australia had begun to export wheat and to generate
new contrivances with which to produce it. The Ridley stripper, which mowed
and threshed at the same time, appeared in 1843. During the 1850s wheat
production spread east towards the Murray and north through the central
hill country to fill the market provided by the Victorian gold rushes, but the
movement north was halted at Gawler by what was known as the Gawler
Scrub – mallee country to the west of the highlands that began at that point
and extended north, across and beyond the top of the Yorke Peninsula. Farming
had also begun in the South-East, around Mount Gambier. By the end of the
decade, South Australia had more land under cultivation than any other colony,
and wheat and its products exceeded wool as a proportion of exports.[22] While
there was always the possibility of privately purchasing or renting farm land,
most small farmers obtained their land by attending auctions held in the Land
Office. Although there were regulations intended to ensure that the agricultural
lots were used for agricultural purposes, prospective farmers, apart from having
to bid against each other, might find themselves competing against squatters
using proxy purchasers, or 'dummies'. They might also find themselves having
to deal with scheming agents and land sharks (who threatened to run the price
out of their victim's reach unless paid off). Farmers were also disadvantaged by
the physical and social effects of isolation and, very often, by illiteracy. Goyder
attempted to get the government to make changes that would assist them, but
he was only partially successful. He persuaded the commissioner to abandon
small, weekly land sales in favour of less frequent sales, where large amounts
of land were offered, so that farmers could plan to attend with some hope of
success. He also managed to introduce and establish the principle that the
survey should respond to the needs of farmers, offering land they wanted once
it was suitable for them to take it up (after a road for transporting produce had
been constructed, for instance). But he failed to get maps distributed to post
offices and district council offices to advertise land sales more widely, and when
he produced a lengthy report, in which he suggested that the outrages of the
auction room could be circumvented by the introduction of a system of sale by
tender, he was ignored by both the commissioner and the Cabinet.[23]

One way to obtain more land for agriculture, without disrupting the steady
outward movement of the survey or sending the farmers into undesirable isolation beyond the band of grassland dominated by pastoral estates, was to begin
clearing the mallee lands. The problem for farmers was that it cost twice as

much to clear the land as to purchase it; furthermore, in the years that followed, the roots of the eucalypts would sucker and sprout, making cultivation difficult. A different scheme for making land available would be required for cultivation in this country. Goyder was asked to report on proposals raised in parliament that would enable cultivation of mallee country, and in August 1866 he advised that a scheme was possible, provided appropriate limiting conditions were in place. If carried out, he advised, it would 'change the face of a large area of country from hopeless scrub to smiling fertility'.[24] In this departure from his usual practice of avoiding the use of metaphor in describing land, Goyder was invoking a much-quoted line by a popular writer, which proclaimed that, in Australia: 'Earth is here so kind, that just tickle her with a hoe and she laughs with a harvest'.[25]

His report led to the passage in 1867 of the *Scrub Lands Act*, which introduced a new system of tenure, allowing scrubland unsold at auction to be held on long leases for a small rent, provided that clearing proceeded. As predicted, a complete and radical transformation of extensive areas of landscape followed. Soon afterwards, a technique was developed for rolling and dragging down the mallee – rather than chopping it tree by tree – thus preventing the remaining roots from successfully suckering. Known as 'mullenising', after its inventor, the method was widely practised through the 1870s and was later used in tandem with the stump-jump plough, specially created to pass over the mallee roots. When even more liberal arrangements for tenure were legislated in 1877 and 1880, the mallee on the Yorke Peninsula were eradicated, and from there areas on the Eyre Peninsula, the Murray Mallee and to the east were cleared, with clearing continuing well into the twentieth century. Opening up the mallee was not enough to satisfy the farmers. By the end of 1866, it had become apparent that settlers were leaving South Australia and heading east to Victoria, with its more accommodating laws, to take up land there.

In early November 1868, Strangways became the head of government, but because of his legal background, he occupied the position of attorney-general rather than that of chief secretary. Strangways, like others, recognised the need for reform of the land laws and immediately attempted to stabilise agriculture – in the previous year, despite having a record area planted out, the failure of the crop (through disease) had been so dramatic that the future of South Australian agriculture was questioned. After legislation aimed at ensuring fairness in the auction process had been passed, a Bill proposing amendments to the *Waste Lands Act* was considered. In response to a request from the new premier, Goyder provided, on 19 November 1868, a long confidential report in which he identified and described six areas suitable for opening up to small farmers who would be purchasing on credit. He commented on proposed legislation and set out pages of new Regulations.[26] The Bill was presented and read a first time on 25 November. The first part of the confidential report, identifying the six 'Agricultural Areas', became the basis of a further report submitted on 21 December. Goyder had selected the areas according to 'soil, climate, and proximity to a market or port'.[27] The first two areas listed were in the South-East; the following two occupied areas of big runs to the north of Adelaide – Bundaleer and Booyoolee, the Hummocks and Crystal Brook. The fifth covered portions of runs on the foot of the Yorke Peninsula, while the sixth claimed part of one of the runs of the prominent squatter Price Maurice in the Port Lincoln District.

The Bill was passed in early January 1869, and became known as the Strangways Act. The Act gave power for 'agricultural areas' to be declared as required. This land was to be officially classed and made available for sale at various prices, depending upon the classification. A buyer could acquire up to 640 acres, and pay 20 per cent of the purchase price as interest and the whole of the principal at the end of four years. Outside the agricultural areas, the auction system would continue as before, the exception being that purchasers could also make use of the deferred-payment option that the Act had introduced for the agricultural areas.[28] Following on from the modest beginnings of the *Scrub Lands Act*, the changes that Goyder had recommended, enshrined in the Strangways Act, were the first major departure from the approach to land legislation adopted when the colony was founded. By directing agricultural settlement to areas identified as suitable for agriculture, and rating the land within those areas differently, the Act constituted official acknowledgement that land was not a homogenous substance; that not all areas were equally suitable for all types of use. Goyder wanted to see farmers succeed. His plan was to see agriculture restricted to areas with reliable rainfall, which meant farmers would have successful crops and would therefore be able to pay off their debts.

To the enchanted shore
There were much grander plans afoot for expansion than clearing mallee. The explorer Stuart had finally succeeded in reaching the north coast from Adelaide, and in 1863 South Australia annexed the Northern Territory. To finance a new colony there, the government, following the same model by which South Australia itself had been created, began to sell large areas of land well before even a site for settlement in the territory had been selected. Land in the Northern Territory went up for sale simultaneously in London and Adelaide on 1 March 1864, with the provision that the land was to be available

to the purchasers within five years of the first sale – on 1 March 1869, in effect. The North Australian Company bought lots in London, while in Adelaide members of the leading clique of successful businessmen and pastoralists, who were also central figures in parliament, formed the Northern Territory Company. Goyder, who had not long left town to begin the valuations, was asked for advice about surveying the new territory, and from his camp on the Murray Flats, he issued a bald warning:

> In a new country – whose settlement is only attempted on the strength of lands applied for and sold prior to the blocks being marked on the ground – delay in effecting the survey should be carefully avoided, otherwise the purchasers become dissatisfied, possession is loudly called for, and the survey is hurried on without considering the requirements of the community – at the cost of years of future trouble and expense in remedying errors that, had proper precaution been used, might easily have been prevented.[29]

With five years in hand, it looked as if Goyder's common-sense advice constituted a purely hypothetical warning, but by 1868, not only had the survey not begun, the site of settlement had still not been selected, although not for lack of trying. The government had sent two parties, led by supposedly competent men, and both parties had retreated unsuccessfully and in disarray, leading the *Register* to bluster irritably that:

> Nature seems to have intended the Northern Territory for the scene of a monster harlequinade. No sooner does the foot of a white man touch its enchanted shore than his character is metamorphosed.[30]

At least part of the reason for the failure was that no government had clarified the purpose of the settlement – there were several possibilities beckoning – to enable an appropriate site to be selected. The possibilities included pastoralism, for which the Victoria River region was favoured; a port and trading centre; agriculture, exploiting imported labour; and a livestock industry, in particular one providing horses for India. There was also the prospect of the telegraphic connection through Timor.

The leader of the first and main expedition was B.T. Finniss, who chose Escape Cliffs as the site for the town, and the Adelaide River for the agricultural lots, sites that were both regularly flooded. Despite a martial approach, Finniss could not control his men, who occupied themselves drinking and conspiring against him and each other. Finniss was also terrified of the Aborigines, to whose presence he responded brutally. Camps were raided and destroyed and two people were killed. The next party was led by John McKinlay, an experienced explorer of sound reputation. McKinlay had been sent in search of a better site, but he had delayed his departure from Escape Cliffs until after the wet season had begun. His party then took six months to struggle as far as the East Alligator River and only barely survived their return to Escape Cliffs, bumping around the coast in a coracle made of horsehide.

As a result of these failures, the wealthy backers in parliament were keen to get their money back, although the majority of parliamentarians still wanted the establishment of the Northern Territory to proceed. The government recalled its staff and called for tenders to survey 300,000 acres, this time between 'the east arm of Port Darwin and the Daly Range, south to Anson Bay; between latitudes 12° and 13°40'S, longitudes 130° to 132° E'.[31] Goyder was given the task of assessing the tenders, but he advised the government that it was pointless to proceed until the best land and most accessible site had been chosen, and he was not convinced that this had been done. He suggested re-examining the Victoria River area as a possible site for the settlement, and, if that proved unsuitable, assessing Anson Bay, Port Darwin and Escape Cliffs. Not until the best site had been chosen could the government realistically call for tenders. 'Were I in the position of a successful contractor … this course is the one I should pursue before conveying men or material to the scene of action,' he warned: 'any other course will terminate, unless under a combination of extraordinary circumstances, in renewed failure'.[32] This alarmed the government sufficiently for them to send Francis Cadell, the steamboat captain and explorer who had volunteered to navigate the inland in 1857, to investigate. Cadell delivered reports in what was perceived to be comically flowery prose and complicated matters by favouring the Liverpool River, at the top of Arnhem Land.

On 18 August 1868, another debate on the Northern Territory began in parliament. With time running out, it was recognised that this debate would have to produce solutions that were effective and final. Over three days, the parliament as a whole, government and opposition, resolved to proceed and tenders were called for the survey of 420,000 acres. In private, the Cabinet had asked Goyder to respond. On the last day of the month, he wrote a confidential memorandum explaining why he had chosen to decline: he had no intention of giving up the security of his post as surveyor general to undertake work as a contractor, in which he might find himself 'at the termination of an arduous undertaking – without personal profit and with perhaps impaired health', and worse still, with someone else appointed to his position. In any case, attempting to act as contractor and as surveyor general, he explained, would create a conflict of interests. Despite these remarks, Goyder was not unequivocally refusing: he was closing off options before stating his own terms. To soften up the desperate politicians he tantalised them with the prospect of fulfilment. 'In declining this proposal ...' he assured them, 'I do not wish the Cabinet to suppose that I have any doubts as to the practicality of effecting the survey within the time mentioned by the cabinet – and for a sum which would be economical ...' What he wanted was an increase in the rates of pay for both officers and men, 'a considerable bonus to myself', a formal instruction from the government to him as surveyor general to undertake the work in that role, and a guarantee that his position would be maintained for him both during the work and on its completion.[33]

The call for tenders was made on 2 September. Next day, the government asked 'the Surveyor-General' to prepare his own list of requirements for a survey, which would take from 10 to 12 months after arrival. A month later he produced his considered plan. He wanted three vessels – a steamer to take the whole party, horses and rations included, to the site; a small steamer or steam launch to enable him to travel up the rivers to the various sites where surveying parties would be at work; and another ship to go back and forth between the probable site of Port Darwin and Batavia (Jakarta) with mail. He also wanted to select the party 'from officers and men personally known to me ... it is of the first consequence that I should have confidence in them and they in me', and the opportunity to inspect and select the provisions himself, and to see that they were properly packed: 'it is equally requisite that those upon whom ... I have to rely should run no risk as to the quality of the provisions supplied for their use'.[34] A £2000 bonus was to be provided to pay the men pro rata for extra work, and he was to receive a £3000 bonus, half of which would be paid to his attorney the day he sailed and half to be paid on his return.

There were cheers in parliament when it was announced that Goyder was making preliminary preparations, but his personal bonus and the terms of its payment attracted widespread comment. With his by now famous 'energy' (it was during this period that he seemed to have acquired the nickname 'Little Energy') and his undoubted competence, it was acknowledged that Goyder could ask for whatever he wanted, and the government could only agree. He certainly had years of experience – and years of observing the experience of others – by which to learn that South Australia's businessman–parliamentarians happily legislated for developments to enrich themselves, but people who undertook expeditions took on the risk alone.

Arabian nights
Although the Moonta controversy was several years in the past, Goyder had a

difficult time in the two years before he left for the Northern Territory. After Kingston's report in favour of Mills, the matter of the Moonta claims had gone to court. Mills eventually settled for a payment in compensation, but Mitford, now on his own, struggled to keep his claim alive. He put petitions to parliament and tried taking the matter to court, but with no hope of success. For Mitford, though, there was more at stake than money. Claiming the Moonta and campaigning against Goyder and Strangways had provided him with a new identity and a new way of life that at least partially redeemed years of failure and frustration.

Nearly a year after the Tipara inquiry had dismissed Mitford's claims, a letter had appeared in the open column of the *Advertiser*. It was unusual in that it took the form of an elaborate literary conceit and sported a peculiar style. In the form of a story from the Arabian Nights and in florid language, it presented a comic allegory of Mitford's version of the Moonta story, taking advantage of the form to make otherwise undisguised allegations that the learned legal 'Moollah' of the Royal Mud Department, Bulli Khan, and his chief surveyor, Chayne Ghangh (a name that neatly linked surveying with crime), had both demanded to be bribed. He also began to develop the figure of Goyder as king of his own realm: 'I am the monarch of all I survey,' the evil surveyor chants, 'my right there is none to dispute …'[35] A month later a follow-up, reflecting on the general process of land and settlement, was published.[36] In the middle of October another letter appeared, this time in the *Register*. Under the heading, 'New invention for killing squatters', it was written 'straight' – or as straight as Mitford's tendency to hyperbole, allegory and excess would permit – and signed openly. 'No department has been the cause of so much trouble and depression to the country as the Crown Lands Office', it began, and went on to use the valuations as a new weapon with which to vilify Goyder. The letter also invoked the 'Survey Office Inland Sea', and referred to the 'Deputy Assistant Rainmaker-General', who would also 'catch all the locusts and put out all bush fires'.[37]

After this, there seems to have been a lull in Mitford's literary activities until, at the beginning of 1867, he established his own small weekly journal. It was titled *Pasquin* – a name traditionally used by the authors of satirical verses – and sub-titled *The Pastoral, Mineral and Agricultural Advocate*. In doing so, he had at last found his proper ground. As the chief contributor to *Pasquin*, Mitford commented on current events and issues, while returning endlessly, issue after issue, to the Moonta claim and his obsession with Goyder. Envious and enraged that Goyder continued to prosper, Mitford set out to discredit everything Goyder did. As surveyor general, Goyder made an ideal target: he could not – and did not – hit back. Unlike Mitford's attempts at farming, carting and prospecting, *Pasquin* appears to have thrived. It was unfailingly lively and could be entertaining, regardless of whether his readers shared his view of his subject. His writing could also be ridiculous, repetitious and self-pitying, but his work was without parallel. In the 'the city of churches', *Pasquin* displayed a refreshing disdain for the conventions of nonconformist piety – hymn-singing or 'howling' Methodists and their alcohol-free 'tea-fights' were a familiar target.

In 1866, during his attempts to divest himself of the need to ride out and examine runs, Goyder had also suggested that the actual work of inspecting mines be delegated to a subordinate. When the position of inspector of mines was advertised, Mitford had responded, undoubtedly looking for an opportunity to harass and annoy Goyder. His writings at this time suggest that he had taken to hanging around King William Street, and in 1869 an 'old friend' wrote of remembering him 'more recently in King William-street, in the well-worn blue coat, and small cloth cap stuck on the corner of his head'.[38]

Goyder's reply to his letter was at once absolutely straightforward and something of a challenge. Goyder set out all the skills that were required in a lengthy paragraph, and concluded by inviting Mitford to apply, should he consider himself competent.[39] Six months later Mitford made sport of this letter in *Pasquin*, turning it against Goyder by complaining that:

> So many arts and sciences were expected to be thoroughly understood, that it was impossible any one man could have included so much varied information sufficiently early in life to render him fit for the billet. He was required to be a surveyor, a miner, an assayer, a mineralogist, a lapidary, a geologist, and to understand surface exploration, timbering, steam machinery, dialling, the principle of drainage over large tracts of country – to be a thorough bushman and to know how to draw correctly – give reports, including all these principle of scientific operations – and construct complicated maps and plans of all these particulars to make his written reports perfectly intelligible to the head of the department.[40]

In the following year, Mitford developed this line of argument – the extent of the qualifications required for the job – to cast doubts on the breadth of Goyder's involvement overall and the quality of his expertise. 'Do we believe,' *Pasquin* inquired:

> that Frome or Freeling, as men of honor and education, would have degraded themselves and their profession by meddling with sheep and grass, intermittent thunderstorms, problematic mineral [*sic*], idiotic water-basins, or destructive drainage of a country preternaturally supplied with water?

The Ancient Mariner, Eustace Mitford. The cartoon satirises John Hart, a prominent politician and former sea captain, known as the 'Ancient Mariner' (after Coleridge's poem, 'The Rime of the Ancient Mariner'). Goyder's first map of his line was presented to Hart who was chief secretary in December 1865. [SLSA]

Design for a trophy to be erected at Tipara in honour of the "Tax upon the grass" by grateful squatters, Eustace Mitford. Goyder is depicted as the corrupt head of the Lands Department, borne up by the commissioner, H.B.T. Strangways. A noose hangs over Goyder's head and he has the face of a pig. The coins held behind his back are marked 'Price of Moonta survey'. The skeletal sheep are described as 'Prime mutton ... fattened on Goyder's valuations'. [SLSA]

Mutilated statue of the constitution, Eustace Mitford. This imaginary monument expresses Mitford's contemptuous view of democracy and the South Australian constitution. The scene includes a portrait of his editorial alter ago, a hookah smoking 'fakir of the howling wilderness'. The Survey Office across the square advertises 'Plans burnt or tampered — no bribes received'— with the word 'no' struck out. Under 'Opera company', a poster for a 'Methodist tea fight' offers 'Hymns and kiss in the ring by moonlight'. [SLSA]

Do you believe that these gentlemen would have confessed themselves impostors and charlatans by accepting responsibilities for which they knew they were unfitted by previous education? And do we believe they would have accepted public money as a reward for proving themselves dishonest blockheads?

No, we cannot believe anything of the kind. We know that officers in Her Majesty's Service, and gentlemen in any service, do not wish to hold themselves up to public contempt, or make themselves the mere tools of incompetent employers; and, under this conviction, we ask naturally – what are the duties of a Surveyor-General, and what has or does the country gain by the office as it has been conducted?[41]

While the style of his expression was excessive, the doubts that Mitford raised were essentially reasonable, and it is possible that he had simply taken up thoughts already being expressed.

Between 1867 and 1869, Moonta provided Mitford's imagination with the material for endless conceits, including a change of imaginative scenery, prompted by the visit of a Japanese delegation in 1868, that saw the unspeakable surveyor general take the name 'Inlansee'. Mitford's obsession also expressed itself visually: he was his own cartoonist. But despite the wealth of opportunities furnished by his imagination, Mitford was not above straightforward abuse, referring to Goyder on one occasion as 'the *ci-devant* clerk of a broken-down speculator [J.B. Neales] – a man whose very appearance would make any respectable dog vomit'.[42] This was in addition to repeatedly characterising Goyder as a criminal deserving punishment:

> The valuator's progress on the waste lands should have been accelerated from the 'the people's grass' into the nearest waterhole. If that failed to convince the small offender that he had mistaken his mission, then he should have been put in a bag with tar and feathers, and returned to the Commissioner …[43]

Even more irritating to Mitford was that he had evidently failed to destroy Goyder as a person of repute in Adelaide society. A scattering of references in *Pasquin* suggest that sometime in the immediate past, Goyder had even given an autobiographical address to one (or more) of Adelaide's cultural organisations.

When Goyder was hailed as the hero who would save the Northern Territory project, Mitford's spite knew no bounds. His wit began to turn genuinely malevolent, and more than slightly deranged:

> Captain Hart proposed immediate survey, and felt great confidence in the Surveyor-General. So do I: but I strongly advise that officer to remain in

safety. The Moonta question is as far from settlement as ever. While that remains open the office of Surveyor-General is intangible. No one dares molest the tenant. There are secrets of importance hidden in the Land Office – two or three fearful skeletons concealed in the closets of that department. No Ministry dare disturb them … Mr. Goyder is King de facto of the Land-Office, and independent of every Ministry that may come into power. Whisper a complaint – make a remonstrance against the gross injustice and fraud which has become a settled principle – and 'Remember Moonta' stays all further objection or investigation. But I should not advise the Surveyor-General to venture into the remote obscurity of the Northern Territory. Injury is not always forgotten – laws are powerless – opportunity is constant – vermin are obnoxious – arms are always at hand – satisfaction can be demanded, and, if not willingly accorded, can be enforced. I have seen such things done more than once – and the fairest justice administered; but justice is not by any means the most agreeable incident in the lives of some men, and, therefore, I advise the Surveyor-General to regard Adelaide as a place of safety.[44]

But the departure of the 'incubus' for the northern coast did not bring relief. In 1869, Ebenezer Ward's very real song of praise to Goyder's work in the South-East was published. Mitford reviewed Ward's work, carping where he thought he could.[45]

Mitford may have felt some satisfaction when, in February 1869, the *Illustrated Australian News* published an article, accompanied by an etched portrait, that introduced Goyder as the figure of heroic energy and ability who had been chosen to lead the next expedition to the Northern Territory. Although the article marvelled at the way in which Goyder had been able to carry on unaffected by the 'unenviable notoriety' and 'amount of public criticism that very few men have been called upon to bear', it commented that he had not been left 'scatheless' by the *odium mineralogicum* surrounding the Moonta mine, a comment that, without further explanation, effectively functioned as a smear.[46] In 1870 the *Australian Journal* published an expanded version of the same article, on this occasion describing the *odium mineralogicum* as something 'which will not soon be forgotten by those opposed to Mr. Goyder's views'. But by then Mitford was dead.[47] He was spared seeing Goyder's eventual triumphant return. He died in October 1869, about a month before his fifty-ninth birthday, when a cold from which he was suffering turned to inflammation of the lungs. The *Register*, which testified (after his death) to his position as the colony's leading satirist and to his popularity, nevertheless was in no doubt that his unsuccessful Moonta claim had been 'the crowning disappointment ... of a chequered career', and that its influence on his later life 'was obvious to all who came in contact with an otherwise keen and well-balanced intellect'.[48] His 'old friend' also considered that the Moonta affair had 'soured his temper', and added that *Pasquin* would have been 'a much more useful paper' if he had reined in his propensity to attack, often unfairly.[49]

That he had made a place for himself is borne out by the naming of a new boat built at Langhorne's Creek in 1870 after the title of his newspaper.[50] Readership was strong enough for friends to publish a collection of all the editions of *Pasquin* to assist his wife, Eliza Mitford, who had been left without support. Much later, in 1883, when Loyau's *Representative Men of South Australia* was published, Mitford was given a place for having 'contributed largely to the South Australian press', aiming, according to Loyau, by wit and satire, 'to reform and improve society'.[51]

Dazzling splendour

Goyder was confident that he could establish a settlement in the Northern Territory, and he would have been as keen to see the colony expand its territory north as he was to ensure that it retained its hold on the South-East. But as always, he had immediate, personal reasons for undertaking a project which, like the valuations, enabled him to earn a considerable sum in addition to his salary. In November 1866, Frances had given birth to Norman Underwood, the couple's ninth surviving child. Goyder now had a substantial household, probably also including Frances's sister Ellen, to support in a manner considered appropriate to his position, but the possibility of a conflict of interest prevented him from using his knowledge to increase his income through investment in land, pastoralism or mining claims.

During 1866, the cottage in Medindie was converted into a 'handsome villa' at the cost of £900. The conversion to a villa was as much related to the position in society and social obligations of a surveyor general as the need to accommodate a large family, and it was 'public' space that was added. The new areas consisted of a large drawing room with an oriel window, dining rooms, and a spacious entrance hall, while a porch and substantial verandahs were added to the exterior of the house. The changes were impressive enough to be reported in the 'Private residences and general improvements' section of the *Register*'s annual report on building improvements, which also informed readers that the entrance hall was paved with 'Minton's tessellated tiles'.[52] The tiles make Goyder's predicament plain. Halls paved with Minton's tiles can still be seen in the surviving mansions of the squattocracy. As surveyor general, Goyder was part of Adelaide 'Society', and expected to entertain its predominantly very wealthy members, as well as visiting guests – but to do this on the income of a public servant. (His salary was eventually augmented with a 'house allowance' in recognition of this.) In choosing to establish a villa in Medindie, he was about a decade ahead of his peers. After 1875 the area was developed by William Wadham, and modest cottages were soon outnumbered by the grander residences of other senior civil servants and professionals, merchants and pastoralists.[53]

There was a general outbreak of magnificence in the colony the following year, during the visit of Prince Alfred. The levee held for the visiting prince was a gorgeous full-dress affair, at which consuls sported gold lace. The South Australian officials naturally attended in their own finery, but this time, according to the *Register*, it was only 'rude boys' who commented on a similarity to the town crier. Rather, 'language would fail to describe adequately the effect which [the heads of department and under-secretaries] produced. We have no doubt their respective families admired them highly, as we did ...' Some of the writer's 'old acquaintances' (almost certainly including Goyder) were so transfigured that he shrank from approaching. Clad as he was in humble black, the writer: 'could only gaze and admire at a distance. It was indeed a proud day for the "gentlemen of the Civil Service", as they like to be called'.[54] Goyder was not the only member of his family to shine in the royal presence. At a special show conducted by the Royal Agricultural and Horticultural Society, the Prince presented Edwin Smith, now the Mayor of Walkerville, with the gold medal he had won for his collections of roses and ornamental plants. Along with everything else the Prince did, this was reported in the press.

In mid-December 1867, Frances went back to England, accompanied by her sister Ellen, all the children, and a male servant.[55] Frances had been badly affected by the birth and immediate death of the twins in 1863, and had given birth to 12 children from 11 confinements in 15 years.[56] Her marriage, too, cannot have turned out as she had expected. Although she had had the companionship and help of her sister, there was no getting around the fact that the young draughtsman she had married, once he had become assistant surveyor general, was more absent than present, and his ascension to the position of head of the Survey Department had only made his absences more numerous and more prolonged. He continued to be away for months at a time, and in addition had become the focus of an 'unenviable notoriety' that extended beyond the colony. If these constant physical and mental pressures were not enough to undermine her, she had be forced to manage the household and the numerous small children during a year of major renovations. Not surprisingly, she was exhausted and unwell, reportedly afflicted by 'great weakness and sleepless-ness'.[57] It was believed that a visit home would enable her to recover. It would be an extended stay.

With the family away, Goyder removed himself to the Adelaide Club, where he was a foundation member, on North Terrace.[58] They were still in England when he left for the Northern Territory, and their absence meant that, as well as meeting the cost of renovations to Hillside, Goyder had to support a party of three adults and nine children on a prolonged visit to the other side of the world. To do all this he had to mortgage part of his property, and while he was away, lease his whole estate to William Hanson, the engineer with whom he accompanied William Milne on the tour of the South-East in 1863.

CHAPTER ELEVEN

Larrakia country: the founding of Darwin

Out spoke the hardy little wight
I'll go so don't be nervous
But 'tis not for your silver bright
'Tis for the public service

– satirical poem[1]

Goyder had effectively staked his reputation on completing the survey in the Northern Territory within the allotted time, taking up the role of defender of South Australia's honour into the bargain. He had earlier warned that the best site should be chosen before sending men and materials, otherwise further failure would result. However, the luxury of investigating various sites was denied him because of the approaching deadline. That he would lead an expedition had been announced in parliament on 20 October 1868 and, according to the contracts, the land would be available in just over four months. Given the little time left, all Goyder could do was consider the available options, including the accounts and opinions of those who had preceded him, before setting out. He settled on Port Darwin, where it was expected there would be fresh water and a deep inlet, along with timber, bamboo, and access to the inland.[2]

Affairs proceeded swiftly, and were reported eagerly. Confidence grew. Even so, the pace did not prevent Goyder from exercising to the full the powers he had insisted on – of personally choosing his own stores and his own men. The press commented favourably on his 'commonsense and anti-redtape' approach.[3] It was also typically detailed: in the case of flour, for instance, he arranged for 20 tons of the best quality to be ground without the husks being dampened, because experience had shown that it kept best when treated in this way.[4] This flour was packed in biscuit tins, their lids soldered to keep them airtight. The tins were then packed into large cases made of substantial planks of planed wood, suitably for carpentry, which could be used on arrival.[5]

In what was a first, the applicants who had applied to accompany the expedition were required to undergo 'a strict medical examination, army fashion' before being accepted.[6] To avoid the antipathy and disunity that had caused the previous expeditions to fail, Goyder did not select the men himself. Instead, he relied on a structure of existing relationships and he only selected the men who would fill positions of responsibility and who would answer directly to him. As a result, all the surveyors were from his staff – including his young brother-in-law Arthur Henry Smith and his sister Sarah's son George MacLachlan, both first-class surveyors, and his nephew Edwin Mitchell Smith, who was one of the second-class surveyors.The surveyors for their part were free to choose the men who would work in the parties under them, and mostly they selected the teams they normally worked with. Edwin Mitchell Smith, in recollections written after his retirement, remembered this as an important factor contributing to the good order and efficiency of the expedition.[7] The men were also an unusually literate band, as was noted at the time: only two were not able to sign their contracts with their name.[8]

While the preparations went ahead without problems, there were some difficulties obtaining the ships. Steamers had proved too expensive. Two old sailing vessels, the *Moonta* and the *Sea Ripple*, were the best the government could afford. The *Sea Ripple* was to take the one-horsepower steamer *Midge* and supplies and make haste to the survey site, arriving ahead of the ark-like *Moonta*, a refurbished collier. A jetty and the stores would therefore be in place when the expedition arrived.

All preparations seemed to run to schedule, and on Wednesday 23 December, Goyder was issued with the formal instructions he had insisted upon.[9] The *Moonta* began loading, and from 2.00 pm the 135 members of the expedition came on board.[10] The members of the expedition and their friends and families, crowds of onlookers, and Goyder himself, all arrived at the port by train – the railway station in Adelaide had been crowded with people anxious to get to the port. Once there, Goyder set about overseeing proceedings. At 6.00 pm, tea in mugs and pannikins was served in a large store and at 7.20 pm a bell on the *Moonta* was rung to summon the party. Goyder called the roll of officers and then men and, as each man's name was called, he came up the gangway and boarded the ship. It was a nice piece of theatre that gave every member his moment of glory, while providing a satisfying spectacle for the very large crowd that had waited all afternoon in the heat. [11]

But the *Moonta* did not go far. For the next four days she remained at anchor, waiting for the wind and tide that would enable her to get over the bar. On Christmas Eve, as they sat out on the river, Goyder mustered the men and organised them into watches for the voyage. The six survey parties of about a dozen men would have to take turns keeping watch for two hours every day and night – a wise precaution, according to one member, because there was a great deal of hay on deck, a large store of gunpowder in the hold, and many men who smoked. The teamsters and Goyder's personal staff were made responsible for providing a constant watch over the animals. The remainder of the men were assigned various duties, which included the task of seeing 'that the officers behaved properly and that the men did their duty'.[12] Goyder then counselled them briefly about the value of maintaining good and friendly relations as the best way to ensure the comfort and safety of each.[13] He was addressing a receptive audience. 'We have a lot of good men, and we came here with the intention of doing all in our power to assist Mr. Goyder to complete the work in a satisfactory manner', one of the surveyors wrote in a letter home.[14]

Conditions on board were crowded, uncomfortable and noisy. To help pass the time amicably, a round of activities was organised. A weekly newspaper was put together, there was a weekly public auction, and the men staged concerts and performances. A series of boxing matches culminated in a final combat between the Queensland Slasher and the Trigonometrical Swell. Divine Service on Sundays was accompanied by a harmonium and a choir. According to Goyder, they had 'a pleasant voyage, wholly free from differences or disputes'.[15] To the men's surprise, it was also free of bad language and blasphemy, which Goyder would not tolerate. One young officer, who persisted in being foul-mouthed after having been warned, was demoted to the rank of chainman. Good fortune and careful management resulted in no serious mishaps. Two draughthorses and several goats died, but this was judged to be not a bad result considering the conditions.

Port Darwin

For all the careful preparations, when the *Moonta* sailed into Port Darwin on 5 February 1869 it was apparent that something was amiss, since there was no jetty, no cache of stores, and no sign of the *Sea Ripple*. Although they could not have known it, the *Sea Ripple* had been condemned as unseaworthy after their departure. Goyder was undeterred. In mid-afternoon the ship anchored about a third of a kilometre offshore, a ration of grog was downed, and then Goyder, the naturalist Schultze, and a few others rowed ashore to begin their examination. They established that the saddle between Fort Point and the plateau beyond was cooler than the higher land to either side, and it was decided that this would be where they would set up camp. A few likely spots were located where the well-sinkers could begin their search for fresh water.

The *Moonta* had anchored off a beach in Larrakia country, and that evening, two young Larrakia men, Billiamook and Umballa, paddled out in a canoe. The pair called out a few English words and were invited on board, where they entertained the new arrivals with renditions of 'Old John Brown' and 'Ole Virginny', in what was judged to be very good English. They also reported that everything at Finniss's abandoned settlement at Escape Cliffs was in good order. They were able to do this because one member of Goyder's expedition, a young draughtsman named John Bennett, had been with Finniss and had learnt some of the language of the Woolna people, who lived to the east on the Adelaide River.[16] Just a few years before, in 1866, the crew of the survey schooner *Beatrice* had played music that had so delighted some of the Woolna people, who had been in the mangroves listening, that eventually a whole group had come forward. They had ended up on board, 'dancing jigs and polkas with the men and joining in chorus to negro songs; indeed everything said to them in English was repeated by them at once, without mistake'.[17] But for the Woolna, contact with the newcomers had not been polkas and jigs all the way. By the time Finniss and his men departed, two of their people were dead, and they had been attacked where they were settled and their camps destroyed.

The next morning Goyder was up early and with the assistance of the captain located fresh surface water in several places. ('Mr. Goyder has been here, there and everywhere, examining the coast and the country', one member of the expedition recorded.[18]) The men were divided into parties for the various tasks: establishing a road to the campsite, trig-piling, timber cutting (to build their own jetty and store), landing and supervising affairs on the *Moonta*. The rest of the first day was tremendously taxing for Goyder, and the cruel incompetence of some of the men unloading animals caused him anger and distress. Since there was no jetty, there was no alternative to having the horses and bullocks swim ashore. The bullocks jumped from the ship and swam ashore readily, encouraged by men used to handling them, and protected from sharks by men in boats. The horses were less keen. When Goyder saw a black mare being 'badly handled' by men in a dinghy and nearly drowned by the ship's carpenter, he raced out from shore in a lifeboat and took the mare safely back to land himself. Goyder and the captain returned to the ship and issued instructions that the dinghy was not to be used again to land animals. This order was disobeyed, and Young Bobby, 'one of our best horses' as Goyder described him, was drowned. The records of this incident provide an opportunity to compare one of Goyder's diary entries with an observation made by an onlooker. In his diary, having described the disobeyed order and the loss of Young Bobby, Goyder only concluded tersely that, 'as I consider this loss solely attributable to one of the ships [*sic*] officers, I hold the ship responsible for it'.[19] However, a witness to the incident, George Deane, an axeman, recorded in his diary that: 'One horse strangled himself in water. Mr. G. awful wild.'[20]

This was not the only incident involving animals that raised Goyder's ire. The day before, when they had entered Port Darwin, his diary recorded that a Newfoundland dog named Carlo had died of distemper. According to his grand-daughter, this dog was Goyder's own. He also had another Newfoundland with him named Brownie, who later gave birth to pups, and an English terrier (a forerunner of the bull-terrier) named Tiger, who was actually his wife's pet. (Newfoundlands and English terriers were reported prominently among the breeds exhibited at a show attended by the visiting Prince in 1867 and were obviously popular at the time. The Goyder family file in the Mortlock Library contains a posed portrait of a large mastiff beside a small, curly-coated terrier perched on a cloth-covered box. The family dogs were apparently adored enough to be the subjects of studio portraits and obedient enough to sit still through the slow exposure of a glass photographic plate.) After the death of Carlo, the mishap with the black mare, and the drowning of Young Bobby, whose body was torn to pieces by sharks, Goyder discovered that Tiger had been taken ashore by a member of his staff and 'carelessly left', presumably without water, in the extreme heat.[21] The dog was found dead on the beach. Despite these losses, by the end of that day the horses and bullocks had all been landed and were back in the care of the teamsters, who took them to water at the well that had been dug. In his diary, Goyder described himself as still suffering from a 'bilious attack'. Presumably, these attacks were brought on or aggravated by exertion and stress, which, apart from outbursts of temper, he does not seem to have expressed or allowed to affect him in any other way. It can hardly have helped his stress levels that they had been informed that people from the Adelaide River area – Woolna – intended to kill two of them, although Goyder was probably not surprised, knowing what Finniss's men had done.[22]

Investigation and selection

The *Moonta* remained with the party for a month before setting sail for Adelaide. Goyder had wisely chosen not to wait for the *Sea Ripple* and had gone ahead with his plans anyway, extemporising and making do. A store, stables, and a blacksmith's shop had been built, and a vegetable garden had been established. The first investigating party began to head inland on 10 February, only five days after the *Moonta* had dropped anchor. Without the little steamer he had wanted, by which he could have travelled well inland along the three arms of Port Darwin, Goyder was seriously set back but not deterred. He investigated the plateau beyond the saddle where Darwin now stands, then, making use of the *Moonta* while it was available, investigated the navigability of the three arms

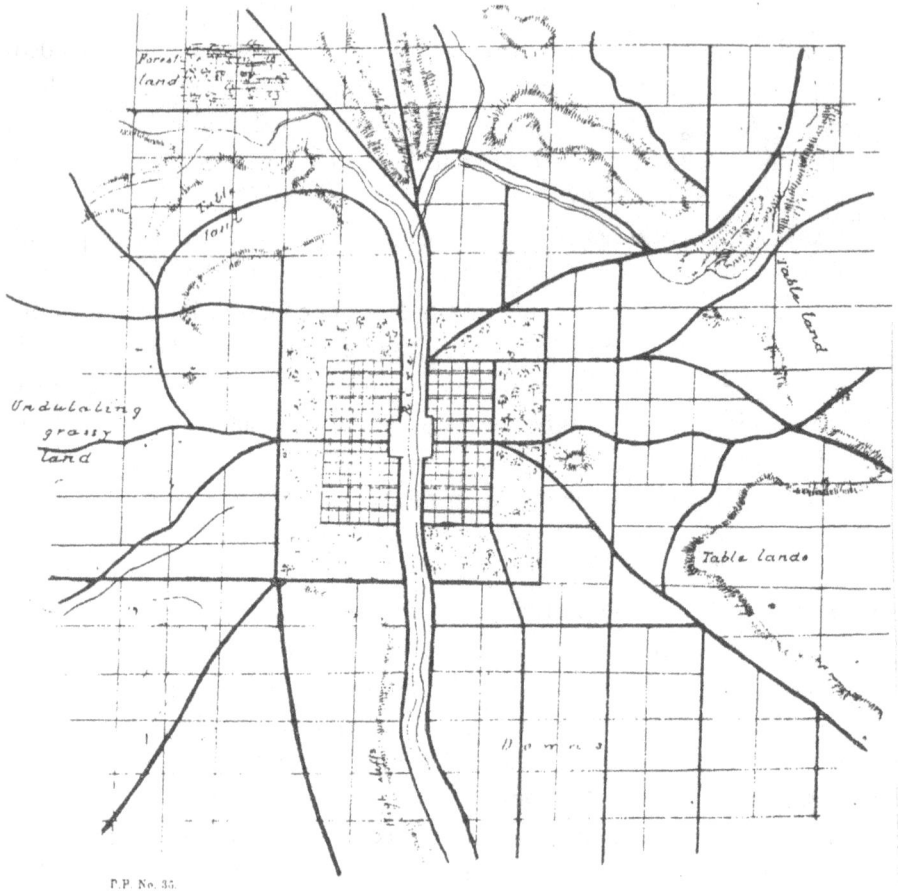

Rough sketch of the ideal town. [SRSA]

and made geological observations, looking for useful materials and gold. At the end of the East Arm, as arranged, he met up with the first overland party, which had set out on 10 February. Satisfied with what he had seen himself and heard from the surveyor in charge of the overland party, Goyder directed them to carry on with their examination inland as far as Fred's Pass.

The 'healthiness' of the site was a prime consideration and in consultation with the officers of the expedition, in particular the doctor, and after much thought and re-visiting of the country, Goyder decided upon three sites for townships: one on the plateau near the first and main camp at Fort Point and one at the end of each of the East and South arms. They were to become known as Palmerston, Virginia and Southport. (It had been established earlier that the town would be named Palmerston, but in his despatches Goyder referred to these townships by their locations, at Fort Point, and on the Elizabeth and Blackmore rivers respectively.) An inland township, Daly, was added at Fred's Pass much later in the expedition.

Goyder had evidently been confident about the Port Darwin site. Early in January, while they were still on board the *Moonta*, the official photographer, a man named Brooks, had begun laying out a design for Palmerston.[23] The main street was designated Smith Street, and the names of other members of the party were given to others. Goyder must not have liked what he had produced, because the diary entry for 11 February, by which time the site at Fort Point had been examined, begins: 'Parties proceeded to shore as usual at 7 am. Designed plan of township and set draughtsmen to work preparing six copies for use of surveyors.' The result was an adaptation of the model plan he had produced in 1864 when asked to provide advice on surveying the Northern Territory for Finniss. The same model was used to plan townships as South Australia was surveyed.

In his first official dispatch, written on 2 March 1869, Goyder announced that the survey of the principal township would be completed in 10 to 12 days, news that was well received when it reached Adelaide in late April.[24] Not only was 'Little Energy' finally getting the job done, but his ever-confident and optimistic gaze was giving them a new vision for the future prosperity of South Australia. The country, he told them, was 'first class ... for large stock', parts of the tableland were 'well suited for cultivation, and mostly rich', and the timber was 'fine and suitable for nearly all purposes'.[25] As he had been when he marvelled at the Lake Blanche in flood, Goyder was too quick and enthusiastic in his judgment, but this time, he was not alone. The seemingly endless variety of shrubs and small plants that so impressed him was there to be seen, just as the floodwaters had been, but neither Goyder nor his men understood that tropical soils, leached by heavy rains, presented their own obstacles to cultivation. For the moment, however, the fate that had doomed past efforts in the north seemed to be smiling on Goyder. Their very capable surveyor general had declared himself 'very pleased with what I have seen'. 'South Australia has no reason to fear her connection with this place', he had assured them. 'Sooner or later it must turn out well.'[26]

Settling in

News that the expedition was proceeding efficiently in a sober atmosphere of peace and cooperation was also received with pleasure. The *Register* hailed the 'gratifying contrast to former experience ... seen in the temper of the party. Perfect discipline has been maintained – not through martinetism apparently, but through the moral confidence of the men in their leader ...'[27]

The thorn in the side of this new rule of universal harmony was the surgeon, Doctor Peel. There was no doubt about Peel's ability, but from the outset he was was found to be pompous and selfish and there was serious dissatisfaction with his fondness for going exploring and shooting for days at a time, rather than remaining at his post and attending to the men. There was even talk of getting up a petition to Goyder to have him removed.[28] Moreover, the officers enjoyed free access to wine and beer – which had been shipped in boxes labelled 'medical comforts'. Peel, however, only allowed the men, who were doing the hard work, one wine glass of rum a week, for fear that more would produce fever. When Goyder became aware of the issue, he 'put a stop to [the] idea' that what was good for the officers was not good for the men by instituting a twice-daily ration of rum.[29] Goyder later testified that experience had shown that a 'stimulant' was necessary before breakfast and the evening meal, to enable the men to eat. 'I worked as they did, and found out what was required when the fits of lassitude came on,' Goyder explained, 'and treated them accordingly'.[30] After an incident of drunkenness, the men were required to drink their ration on the spot to prevent hoarding. Edwin Mitchell Smith, in his recollections, judged that this tight control over the consumption of alcohol was an important factor in the safe and successful progress of the work. Special occasions, such as the dinner held to farewell the *Moonta*, were different, and here Goyder seems to have been a generous host. George Deane, who had been promoted from axeman to a member of Goyder's staff because of his competence as a bushman, recorded that there had been great fun 'amongst some of us, all tight as a bottle, I fared well with the liqueurs ...'[31]

Without the *Sea Ripple* bringing fresh fruit and more rations, scurvy was beginning to appear alongside the boils, ulcers and prickly heat with which some men were already afflicted. The garden had begun to yield a little cress, but not enough to save them. Goyder anxiously checked the supply of tasteless tinned meat, to see how long it would last. There had been reports of a ship off the coast, but nothing materialised. The longboat and lifeboat were sent around to Escape Cliffs to see what could be salvaged from the remains of Finniss's camp. Carrying on with the work, Goyder set off to determine a township site on the East Arm. He returned in late March.

On the morning of his return, a few hours before he reached camp, the *Sea Ripple*'s replacement, the *Gulnare*, arrived, bringing all the goods they had been waiting for. Within hours Goyder was issuing instructions for a party to prepare to travel down the South Arm on the *Midge*, while a horse party was to travel overland. Once Goyder had determined the site for the town, they could begin surveying. Another party had already left to begin work at the town site on the East Arm.

After returning from a trip up the South Arm in late April, Goyder became ill and was obliged to rest for a day or so. He was weak with dysentery, he had injured his foot, and all his joints were painful. His hands were so stiff that he could not write and the accountant had to stand in as his secretary. Although the *Gulnare* had brought fruit and lime juice, with Goyder later reporting that the beginnings of scurvy among the men had quickly disappeared, it seems that in his own case the conditions had perhaps aggravated damage already done. Even so, after three days in camp he was off again. By the end of April, the three townships had been surveyed, and the survey of the rural sections had begun, with 40,000 acres already completed.[32]

Protector of Aborigines

Not everything went well. Although Goyder took great care to ensure he did not repeat Finniss's mistakes, the legacy of his predecessor's relations with the Aboriginal people haunted him. Unlike Finniss, Goyder had taken on the role of protector of Aborigines himself rather than delegating it to the expedition's doctor, no doubt because he anticipated situations involving the Indigenous people where great care would be required, and in relation to his own party, a very firm hand.

The post of protector of Aborigines had been established by the South Australian Colonization Commissioners, according to whom the rights of the Aboriginal people were to be a matter of first importance in the founding of South Australia, in contrast to the situation in the other colonies of Australia. They were to be 'placed under the protection of British laws and invested with the rights of British subjects'.[33] In keeping with these pronouncements, Light had been instructed to treat wild animals as the property of the inhabitants, and the resident commissioner directed to see that 'no lands which the natives may possess in occupation or enjoyment may be offered for sale until previously ceded by the natives to yourself'. If they chose not to cede their land, the role of the protector included ensuring that they were 'not disturbed in the enjoyment of the lands over which they may possess proprietary rights, and of which they are not disposed to make a voluntary transfer'.[34] But the reality was that these declarations were an afterthought tacked on to a very different set of assumptions. The preamble to the *South Australian Colonization Act* of 1834 declared that South Australia consisted of 'waste and unoccupied lands'. As George Fife Angas, one of the founders of the colony and one of the early Colonization Commissioners (who had also been involved with Wilberforce in the fight against the slave trade), explained to a select committee of the House of Commons, these words meant that the Aborigines were clearly excluded:

> from any advantage whatever arising from the land; it does not even recognise their existence. They have no existence in a legal point of view, therefore no provision could be made for them by the commissioners. The natives cannot purchase or hold land.[35]

Despite early noble sentiments from its founding fathers, in the end South Australia was no different from the other colonies: dispossession, disease and disruption had the same impact on the people there as elsewhere.

Goyder's prolonged expedition to triangulate the region south of Lake Eyre in 1859–60 had placed him at the limits of the European incursion into South Australia. This ensured encounters with people for whom Europeans were an unfamiliar presence, but if this caused him any anxiety, it is not recorded in his field notebooks, which contain very few entries about the Indigenous inhabitants. At one waterhole he recorded casually: 'blacks around us all about + camped to rest horses + get a wash …'[36] But the paucity of comment must have been a result of his note-keeping style, since he found himself compelled to elaborate a code of behaviour to govern the conduct of the members of the party. This was in the light of what he later described as the 'painful prominence' of the sexual advances by the women, which led to 'that objectionable familiarity which almost invariably terminates in resentment and attacks on the part of the Natives, and ultimately retaliations on the part of the whites'.[37] Goyder insisted that men of his party conduct themselves with complete restraint – this was an 'inviolable condition' for them, applying 'invariably, under every circumstance', and he was certain of the value of this course of behaviour: it meant freedom, security and peace of mind for the surveyors and Aboriginal people alike.[38] To this alone, he stressed:

> I attribute the immunity with which we were allowed to pass in every direction through the country, and to camp without molestation at the same waters, and in the midst of numbers of the Natives, who appeared to have thoroughly realised that – whilst we occupied ourselves exploring, and surveying in a method past their comprehension, they were held sacred with their property, and as free from danger as they appeared free from fear.[39]

The principle of reciprocity was employed in a systematic and deliberate way. According to Goyder, the surveyors returned a gift of food for any assistance they were given. The respect was mutual: the people they encountered 'never deceived us as to the direction of localities, or waters we wished to find, and they respected the records we left on the ground'.[40]

1. Abandoned farmhouse [Janis Sheldrick]

2. *Land of the Salt Bush*, John White, 1898.

The earliest known oil painting of the Flinders Ranges, *Land of the Salt Bush* was purchased by the Art Gallery of South Australia in the year it was painted. The saltbush country, in the light of dawn (or perhaps evening), is presented as a pastoral paradise of subtle colour and delicate form, the 'beautiful saltbush plains' the ploughing up of which was lamented as a crime against future generations. A rough brush fence can be seen on the right and two sheep stand on the track on the edge of the shadows. [AGSA]

3. David Goyder

4. Frances Goyder

5. The young George Goyder

The photographs of Frances and George Goyder appear to have been taken in the same studio, possibly to commemorate their marriage. Photographs courtesy of Vaughn Smith.

6. George Goyder [SLSA]

7. Sarah Anna MacLachlan, *née* Goyder.

This portrait was taken while Sarah was in South Australia. Photograph courtesy of Vaughn Smith.

8. Sarah Goyder, *née* Etherington.

Photograph courtesy of Vaughn Smith.

9. From the Razor Back Hill, looking south over Mt Bryan, Edward Charles Frome, 1843

Mount Bryan was the eastern point mentioned by Hawker when defining his northern grazing line. It is the little shoulder on the eastern side of the peak of Goyder's Line. Frome's watercolour therefore shows a section of the line long before there had been any significant European impact on the landscape. 'Inside' country is to the right of the hills (west); 'outside' country to the left (east). A difference in vegetation is evident. [AGSA]

10. First View of the Salt Desert—called Lake Torrens, Edward Charles Frome, 1843

Frome would have understood himself to be confronting Eyre's impassable horseshoe lake, and his watercolour presents his predicament. [AGSA]

11. *The sandy ridges of Central Australia*, Samuel Thomas Gill, ca. 1846.

Gill's watercolour of men chaining over sand dunes to 'Lake Torrens' was copied from a sketch by Charles Sturt. [NLA]

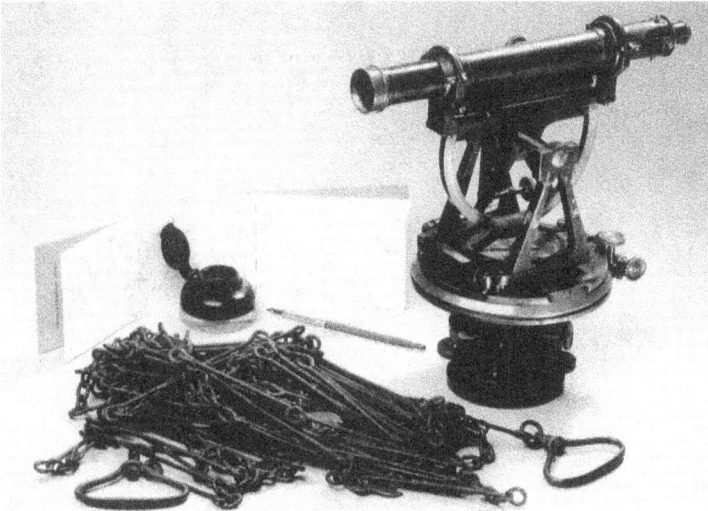

12. A theodolite, chain, and field notebook from the South Australian Survey Department.

Working chains had to be compared constantly to a base chain (or rod) because of distortion due to temperature and wear, so that necessary corrections could be applied to the calculations.

13. *Sketch of the district south east of Lake Eyre.*

A hand-drawn map, signed by Goyder and dated 1 March 1864, shows the detail that his system of observation and recording could achieve without a detail survey. The map covers the whole area examined by the expedition. [SLSA]

14. The nurseryman – Edwin Smith.

THE ROSERY—CLIFTON NURSERY.

NURSERYMAN AND FLORIST, WALKERVILLE.

EDWIN SMITH.

15. The rosery at Clifton Nursery, 1889. [SLSA]

16. **William Strawbridge**, as surveyor general, after Goyder's retirement. Unlike Goyder, Strawbridge is tall enough to wear his ceremonial sword with aplomb. [SLSA]

17. **Lionel Gee**, who accompanied Goyder on trips, and provided the only account of Goyder's presence as a person in his reminiscences of the South-East in the 1870s. Gee later joined the Mines Department and occupied a number of positions, from Warden of Goldfields to Chief Registrar of Mines and Recorder. [SLSA]

18. Goyder's Line, as first drawn two days after Goyder returned from the north.

This map was never presented to parliament or published in the parliamentary papers, and the original no longer exists. (It was evidently disintegrating when this image of it was made.) [Government of South Australia]

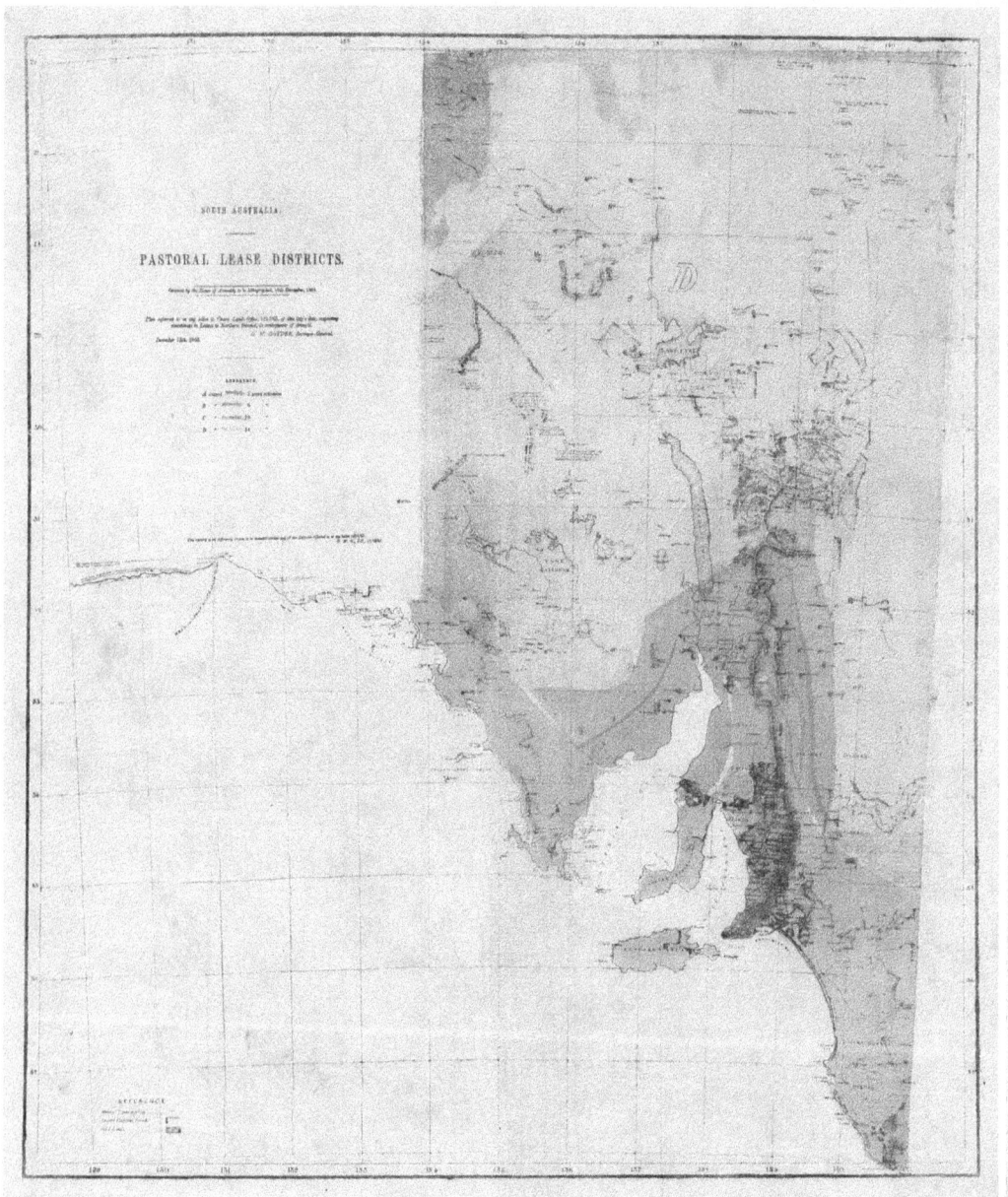

19. *South Australia, Pastoral lease districts,*

the map that accompanied Goyder's report on the northern runs (SAPP 1865-66, no. 82). The Line made its first public appearance in this map showing the areas covered by the different schedules of pastoral drought relief. It is the southernmost boundary of the scheduled areas – on this map it appears as a white line, but this is the accidental outcome of poor registration of the print. The map is now very faded. The colour below the Line was originally green, and the colours about it are described in the report as purple (A), red (B), blue (C), and yellow (D), the outermost area). [SLSA]

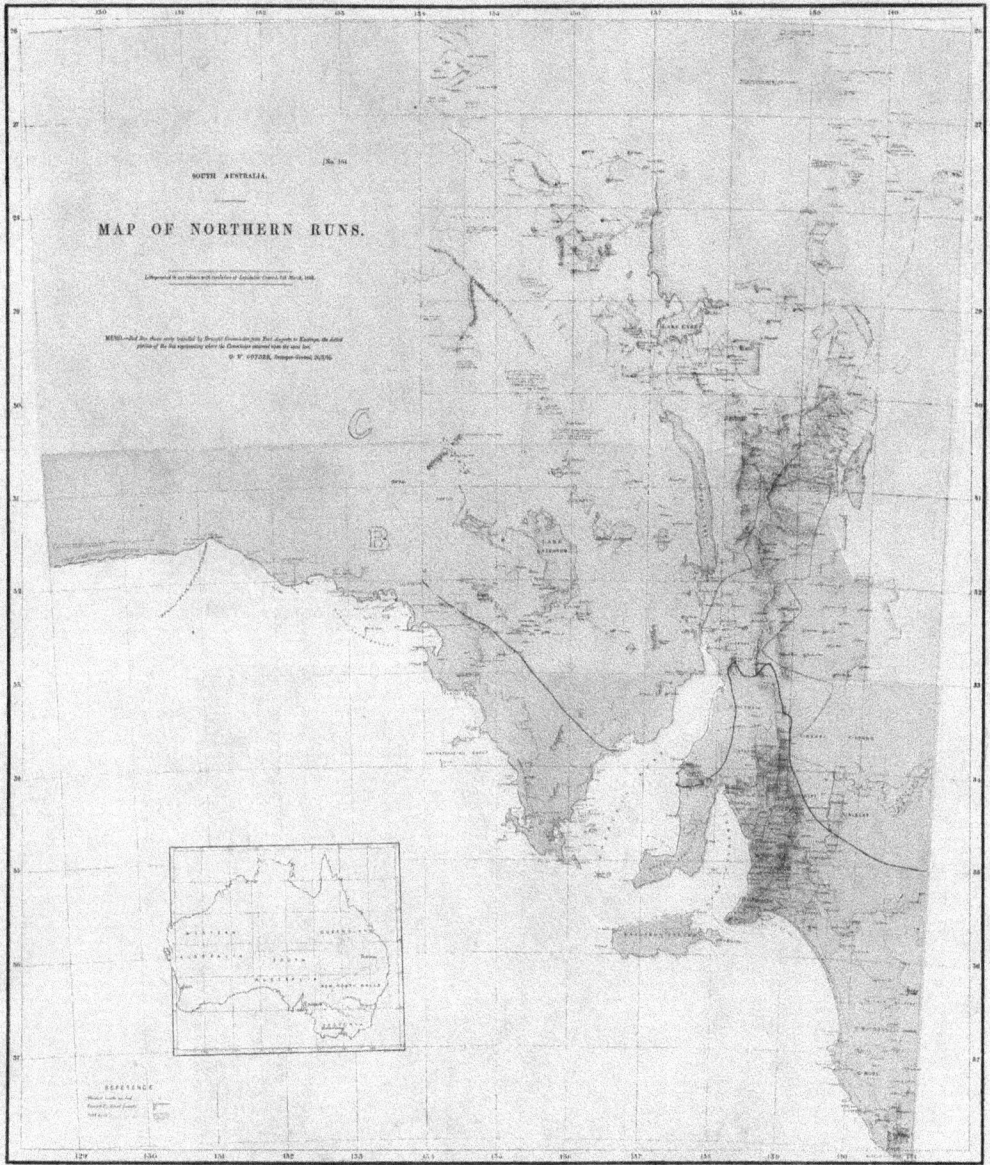

20. *South Australia, Map of northern runs.*

The Line first appeared as a separate entity in this map from early 1866. Prepared for the parliament (it accompanies SAPP 1865-66, no. 154), the map also shows the route taken by the commissioners sent to observe the impact of the drought on pastoral runs in the north. [SLSA]

21. Henry Bull Templar Strangways [SLSA]

22. Thomas Playford [SLSA]

23. Eustace Reveley Mitford [SLSA]

24. Members of the expedition to survey the Northern Territory.

Clockwise from top left: Arthur Henry Smith, George MacLachlan, Edwin Mitchell Smith (Goyder's brother-in-law and nephews) and J.W.O. Bennett, the young draughtsman who was speared and later died. [SLSA; photograph of George MacLachlan courtesy of Vaughn Smith.]

25. *Above*: Ellen,
sister of Frances and Goyder's second wife, and
(right) Ellen with the twins.

26. G.W. Goyder.

A photograph (*left*) probably taken after his return from the Northern Territory and a later portrait (*below*). [SLSA]

27. *Left*: Goyder in retirement [SLSA]

28. Goyder's long-time friends.

Above: Charles Todd and his own large family photographed at the Observatory. Todd is seated at the far right of the middle row and his wife, Alice, namesake of Alice Springs, at the far left.

Left: Dr Allan Campbell. [SLSA]

But even while Goyder insisted that his surveyors treat the people with respect, he must have been fully cognisant of the ultimate impact of the work in which they were engaged. It was plain enough to others. At around the same time, Bishop Short had written to the *Register* in response to reading that the pastoralist explorers, Hack and Swinden, had been helped in their search for good pastoral country to the west by two guides, Warrio and Pinegulta:

> Poor Warrio! Poor Pinegulta! Little do you think that ere five years have elapsed you will be deemed inconvenient intruders on the runs which you have laid open; that your emus and wallaby will disappear; that when the cattle run from you, or you are tempted by ravening hunger to rush a few sheep, you will be shot down, it may be, like the wild dogs; and when the [Budget] Estimates are moved the rents of that 'good pastoral country' may not be devoted to your protection and food and clothing and instruction, but diverted to other purposes.[41]

A few years later, J.B. Hughes, another of the leading pastoralists and proprietor of Bundaleer, also wrote to the *Register* from Yudnamutana, in the far north of the Flinders Ranges, protesting that:

> The government ruthlessly lets the entire surface of the country to its pastoral or mineral tenants. There is not a hill, a creek, or an area of any kind which the aborigines may retire to or look upon as their home ... If the country is let as a cattle station the native owners of the soil are flogged, frightened and driven away from the springs or water-holes that their presence may not scare away the cattle from drinking at them.[42]

In his proposal to the government, Goyder had explained the policy he intended to pursue in the Northern Territory. He had insisted that the expedition be provided with arms and ammunition as 'conscious security was necessary for the vigorous prosecution of the work'. Security would be maintained by the surveyors keeping close together, to provide support for each other and to 'avoid any appearance of isolation or weakness'. Adopting this tactic would discourage the Indigenous people who encountered them into following 'the promptings of their savage instincts, causing reprisals, which, though they may secure the object sought, only [do] so at a great sacrifice, and one that cannot be too much regretted'. The surveyors would take care not to harm or interfere with anyone, and only 'strong provocation' and 'persistent aggression' would provoke 'effective measures' to ensure that the survey continued.[43]

But as events unfolded, the task of establishing another colony would force Goyder to confront the contradictions inherent in holding the Indigenous people and their property 'sacred' – a term he could not use lightly as one who abhorred blasphemy – and his role as the chief instrument in dispossessing them of their land. Late in February, Goyder had issued a memorandum to all members of the expedition outlining rules to ensure that a real distance was maintained between the survey parties and the Aboriginal people. Except in special circumstances authorised by a senior officer, Aboriginal people were not to be allowed to approach within about 18 metres of the camp, they were not to be used as labour to fetch and carry wood and water, they were not to be given food unless infirm, or as part of an authorised barter, and no one was authorised to communicate with them after dark, unless circumstances demanded it.[44] Early in March, suspicious of activity during the night which suggested imminent attack to those with previous experience, Goyder rostered the cadets and draughtsmen to guard the camp after sunset.[45] On the morning of his return to Fort Point from the East Arm, he ordered the camp to be fenced off.[46] Only a pair of botanists and naturalists, the Schultzes, a father and son, remained outside, camped some distance up the hill, where they carried on working prodigiously, preparing box after box of dried and preserved specimens to send back to Adelaide. (In October 1868, Ferdinand von Mueller, the director of the Botanic Gardens in Melbourne, who was preparing a comprehensive work on the flora of Australia, had written to Goyder explaining his eagerness to obtain plants from the region of the north 'both geographically and phytologically as yet imperfectly known', adding, in English so splendidly unidiomatic that his German voice can be clearly heard: 'I trust to your well known scientific inclination, to bring with you as large collections back as you can'.[47] Together, the Schultzes collected and prepared over 8000 plant and animal specimens.[48])

In the first part of April Fort Point was visited by someone the old hands from Finniss's expedition had been expecting for some time: Mira, an old man they judged to be a chief of the people from the Adelaide River area. Mira was friendly and peaceful and had often protected Europeans from attack by speaking up in their defence. On this occasion he and some other Woolna people had brought two Malays, survivors from a trading vessel that had been shipwrecked two years before. Arrangements were made with the captain of the *Gulnare* to provide a passage for the two men back to Kupang, where one had a wife and family, and Goyder gave Mira food, to reward his 'humanity' in getting help for them.[49]

Despite the presence of Mira, and although Hoare, the doctor's assistant and scientific illustrator, continued to barter to obtain a collection of string bags and items such as necklaces and bracelets, relations between the surveyors and the Woolna deteriorated steadily. Security at the camp was increased, with the situation finally becoming so tense that Peel had to fire a warning shot to bring help. Hoare's diary entry for 8 May 1869 recorded the incident. Headed '200 Natives on the War Path', the entry noted that: 'A number of armed Blacks came and wanted to fight us. Old Mira tried to stop them. All the camp took up firearms.'[50] Goyder returned from visiting the camp of two surveyors and all was quiet again, but sentries were posted and a magnesium light was kept burning at night. But while pacifying the Fort Point camp, Goyder took care to see that the men at Fred's Pass, where Finniss's party had also been involved in a shooting incident, were put on alert.

One of the men camped at Fred's Pass was Bennett, the young draughtsman and former member of the Finniss expedition. On his first visit to the Northern Territory Bennett had been prepared to deny, as he put it, 'that these miserable degraded (animals more than men) are my brothers and sisters', and to reject the idea of calling 'a half-cannibal a brother, one that would spear you at the first opportunity'.[51] Nevertheless, Bennett had made friends with some of the people he encountered, Mira in particular, and undertook to learn the Woolna language. With Goyder's approval, he had begun to compile a vocabulary.[52] During May, Goyder and his party visited the depot camp at Fred's Pass. The survey party was working elsewhere, and the camp was deserted except for Bennett, who was plotting the survey, and a labourer named Guy who was also the cook. Goyder noted that the men in this area had ignored his instructions to ensure that they were always armed. He 'remonstrated' with Bennett, who nevertheless 'continued to trust the natives implicitly'.[53] Goyder also noticed that not only did Bennett consider himself perfectly safe, but others of the party considered his presence a protection against attack.

Goyder's warnings and instructions that the parties should stay together had little effect. Some days after his departure, three surveying parties moved on from the camp beside the lagoon at Fred's Pass, leaving only one party of five men, including Bennett, to finish a small amount of work that still needed to be done. When a group of 15 Aboriginal people turned up, Bennett instructed them to remain outside the camp, and they did so. Later, he went hunting with them, visited their camp and was painted by them.

The next morning, 24 May, Bennett overheard members of the group discussing an attack. He told those involved that it was no use trying this. The situation prompted him to load the weapons in his tent, but he did not take the threats very seriously. Late in the morning, the only other person in the camp at that time, Guy, the cook, made a temporary departure to look for firewood, leaving his revolver on a water keg. While Bennett was alone working in his tent, he was speared twice from behind in the chest. When Guy returned he was speared as well.[54] Despite being badly wounded – one spear had gone deeply into his chest – Bennett had nevertheless managed to obtain a rifle, and Guy, who had been speared while attempting to reach another man's revolver from under his mosquito net – he could not get near his own – was able to crawl to the tent. The two defended the tent until they were found in the early afternoon, exhausted from loss of blood, by a surveyor returning to camp. It was nearly two days before the doctor was able to reach them. After treatment, the two were taken back to Fort Point. There, a long spear tip was found and removed from the smaller of Bennett's wounds. Bennett died the next morning.

Goyder, still out exploring, did not get the news of the attack until he reached an out-camp of the Fred's Pass depot. He immediately ordered the consolidation of these camps.[55] From there he pushed on to Arthur Smith's camp on the South Arm, and to an out-camp at Tumbling Waters, where he learnt that Bennett had died. These camps, too, were immediately consolidated. On the way back to Fort Point, on 12 June, Goyder himself was threatened. As the party rode through long grass, they were surrounded and the grass around them set on fire. Presumably the plan was that, blinded and suffocated by smoke, the intruders could be picked off easily as they tried to break away in a panic. If so, it did not anticipate what Goyder described as 'the coolness of my men and our knowledge of the country'.[56] Able to avoid being overwhelmed by smoke, the party rode on, although when the flames around them increased in height, they were only just able to keep their frightened, exhausted horses moving; the crackling, roaring heat and smoke being enough to 'confuse the poor brutes', as Goyder put it, but not Goyder and his party – they escaped unscathed and avoided a violent confrontation. Goyder later reported, with an air of sanguine brutality wholly atypical of his writing: 'we could easily have shot one or two of them, as three were visible at one time, within twenty yards of me'.[57] But no shots were fired. Although 'ever ready for the worst', Goyder went on, 'we abstained from injuring them so long as there was a possibility to avoid bloodshed'.[58] Deane, the former axeman, was one of the party, and his diary confirms that they were ready to respond if necessary. 'We had all redy [sic] for them', he noted grimly, in one of the two sentences he lavished on the incident.[59]

Both sides showed remarkable persistence. According to Goyder, as they continued on the journey home their harassers 'kept firing the grass near and in advance of us for a distance of eighteen miles', although, to their 'yelling disappointment', he and his men kept evading the flames successfully, finally arriving safely at Fort Point at about five in the evening. The 'old native hangers-on', as Goyder described them, were about the camp as usual when they arrived.[60] On the basis of past experience, Goyder seems to have assumed that they already knew what had taken place. They were ordered away immediately and not allowed to return. That night Goyder heard the sounds of human presence in the mangroves. This time he fired in the direction of the sound. 'All quiet after', he noted.[61]

In early July, the junior officers and cadets were formed into a regular guard, armed with carbines and revolvers, and arrangements were made to ensure that all firearms were kept clean and in working order. Rifles were to be discharged and cleaned twice a week.[62] A few days later a dozen people turned up, but they were not allowed to come near the camp. The same day, Guy provided a written statement recounting what he could recall of the attack. In noting the incident in his diary, Goyder was about to write, 'the attack in which W. Bennett was speared', but wrote only the first two letters of 'speared' before changing his mind and writing 'murdered' instead.[63]

For the remainder of the month, an atmosphere of peace and security seems to have settled over the camp again, until on 1 August a disturbing event took place. In the early evening Goyder noted two empty canoes floating past, drifting up the harbour with the tide. On learning that several parties of his own men had been out, Goyder became suspicious and began asking around to see if any of his men were responsible for these canoes being in the water. No one knew anything. It was a troubling evening for him: immediately after this he had found the armoury left vacant for a quarter of an hour.

Later that evening, at eight o'clock, one of the first-class surveyors, Harvey, and a cadet named Greene visited Goyder. They had heard that he had been enquiring about the canoes and had come to inform him that they had found a group of canoes on the beach, and, in retaliation for the death of Bennett and the wounding of Guy, had smashed them all, except for two which proved too sturdy. These they had set adrift. Goyder was furious:

> I was grieved and annoyed at this act of malicious folly + pointed out the possible consequences in strong language to them as well as the injustice of the act – stating that we had now given cause of aggression on their parts + that this would probably follow – our boats might be cut away – wells filled in and all possible damage done by them + that now I could no longer blame them – set a watch to look out for the two canoes adrift – with orders to have them taken back if they could be discovered –
>
> Whatever aggression results from this act it is due to the want of proper feeling and great lack of judgement of Mr. Hardy + Cadet Greene.[64]

The next day, men rowed for three hours looking for the canoes, without success. As it turned out, their owners sought no retribution.

Managing the survey

From 1 May, in order to survey the huge area required, Goyder had initiated a system known as 'mileage', by which a set amount per day was required of the parties. Any work in excess of that would be paid at a higher rate. Eager to increase their pay, the men worked at a terrific pace, with only one small rebellion. Four surveyors rode in to complain that too much was expected of them in the tropical conditions. Goyder 'went into figures with them', made some corrections, and the dispute ended in an amicable chat.[65] Unlike the leaders of many exploring expeditions, Goyder had extensive experience in supervising teams of men and in settling disputes of all kinds and between very different types of people, and his swift resolution of the surveyors' concerns was almost certainly due to his negotiating skills, which were based on a real willingness to listen and a recognition that concerns must be genuinely addressed. A significant element of Goyder's success and effectiveness throughout his career was attributable to his ability to work with other people. One member of the expedition observed that, despite his 'very hasty temper', Goyder's good qualities far outweighed his bad, and that he was 'strictly honourable in his conduct, and very thoughtful of his men'.[66] He expected his men to work hard, but he worked equally hard and he understood that there were limits. When first-class surveyor Woods sacked two men for 'shirking work', and sent them back to Goyder, Goyder respected the decision of a senior officer and dismissed the men, but he wrote to Woods expressing concern that 'the men are not being pushed beyond their powers of endurance'.[67]

Goyder's firmness, resolution and the grand scope of his vision also served to maintain the self-confidence and self-esteem of his men: no one working under him was ever in any doubt about what was being done and why; and what, in particular, was required of each of them. Lionel Gee recalled that Goyder had:

> a way of imparting to those under him a desire to do their best, and it is notable that when workers are so employed very little time is wasted in complaining about the conditions or reward.[68]

At the same time, his combination of firmness and fairness provided a sense of security to those reliant on his leadership. It was understood that he would not permit dangerous negligence. Claims that the doctor had been inattentive in his duties were refuted by one member of the expedition, who, in a personal letter to a friend, observed that the doctor *could not* have been negligent: 'the Surveyor-General is not the man to have allowed it'.[69] It is also clear that they trusted his concern for their welfare. When, early in August, there was a problem with tainted pickled meat, one of the men brought the problem straight to Goyder, who listened, but also suggested that the man ought to have taken the matter to his senior officer. Goyder then issued a memo directing that the meat be withdrawn, and that in future his senior officers were to bring such problems to him immediately – 'their juniors and men of their respective parties had a right to expect that such be done'.[70]

Although Goyder seems never to have produced any written plan of his approach to the business of surveying over 200,000 hectares, accounts of the work indicate that, with modifications, he followed the plan he had devised in 1864, when he had been preparing advice for Finniss. In that document he had claimed that the task would take 'a dozen efficient surveyors from twelve to fourteen months', but since he had not anticipated the optimum number of assistants being available, he had based his schedule on the work of six surveyors. In 1869 he had his 12 surveyors; six first-class and six second-class, as well as a dozen cadets. The basic strategy described in 1864 was to decide upon the lands to be surveyed and the sites of the port and the inland or river townships and then to divide the surveyors into three parties, one party laying out allotments at the port and another at the inland or river townships. To enable the necessary main lines of road to be determined, the third party was to make a detailed survey of the country adjacent to the port and surrounding the inland town, showing the major topographical features and the position of the best land. By that time the survey of town lots was to have been completed, which meant that those surveyors could then be employed marking out the lines of roads. These were to radiate in the appropriate directions from the centres of population to the limit of the land proposed for sale, and so to form a basis for the survey, with the sectional boundaries becoming checklines. When this was done, the less experienced surveyors could mark out the sectional lines, defining the blocks.

Accounts of what was done during 1869 indicate that a similar strategy was followed. The plateau behind Fort Point was investigated and while the survey there was in progress investigations were made to determine sites for the river port towns, beyond the East and South arms, and an inland town at Fred's Pass. Once the key town site had been surveyed, the men were divided into three parties, or 'double parties', each responsible for surveying 90,000 acres and a township.[71] In the late 1960s, the then surveyor general of the Northern Territory, Peter Wells, guessed that individual blocks must have been pegged by chainmen or cadets for the work to have been done in the time. He also pointed out that the work had been expedited by the boundaries staying on or close to the cardinal bearings, with the roads (with a few exceptions) following the same principle.[72]

On 21 August Goyder recorded that the survey of the town allotments and sectional blocks was complete. In fewer than seven months the survey parties had surveyed and pegged out over 500,000 acres. While this was taking place, another party carried out a rough geological exploration, essentially aimed at finding gold. And at the end of all this, the expeditioners turned their hands to building a larger set of stables while they waited for the *Gulnare* to reappear. They also replaced the first floating jetty with a stone and timber structure.

When the *Gulnare* did arrive, it brought with it much-needed supplies of fresh fruit and vegetables – scurvy had returned again – and preserved mutton from Booyoolee. Captain Sweet also brought a dispatch that ordered the surveying of an additional 150,000 acres. (Sweet was a photographer as well as a sea captain, and many of the photographs of the expedition are his.) No sooner had this news been received than preparations were under way to begin. By this time Goyder was nearly exhausted. He had been sick again in early August, suffering from nausea, diarrhoea and stomach pains, complaints that plagued him all his working life. Many years later he would tell the Northern Territory Commission that he considered he had aged 'fully ten years', had lost a great deal of weight, and, since he had seldom taken 'tonics', had only just escaped with his life.[73] Fortunately for Goyder, the four first-class surveyors who rallied to the task surveyed 74,000 acres in 18 days.[74] The additional survey task was complete in time for him to leave on 27 September, as originally planned.

Along with the extra work, the government had also asked for a continued presence at Port Darwin, and Goyder called for volunteers to remain behind. George Deane's diary offers testimony to Goyder's powers of persuasion. Deane had a beloved in Adelaide, and he was eager to return and be married, but Goyder talked him into staying – to do nothing grander than lead a party through the surveyed areas looking for lost horses. 'I suppose I must stop', Deane wrote. 'SG had a conversation with me. I think him a trump.'[75] Deane's waiting future wife may not have agreed, especially since the search was understood to involve risk.

Early in August a makeshift theatre was established. The 'Theatre Royal', which could seat 150 people and housed a series of performances over successive Saturday nights, was the work of Doctor Peel and earned him the gratitude of the men. Goyder appears to have enjoyed the entertainment as well, although his abhorrence of blasphemy prompted him to walk out of one of the performances. The performance itself he judged to be 'pretty good', but one of the performers had uttered an 'impious remark', which was 'totally uncalled for'.[76] His diary always contained a general note about each Sunday's attendance at Divine Service.

On Monday 27 September, the last day of Goyder's stay in the Territory, a full holiday was declared. In the afternoon there was a farewell performance at the Theatre Royal and in the evening, a gathering in the mess tent. Goyder was presented with a signed address. In reply he expressed his pleasure at how well they had all worked together, pronounced them a fine body of men, and pledged either to employ them himself, or to use all his influence to find a job for any man who wanted one.[77] The next day the *Gulnare* set sail for Adelaide via Kupang.

Scarcely to be wondered at

Bennett was buried on top of Fort Hill, his grave eventually marked by a monument designed by Goyder: a little sketch of a cross on a mound sits among the calculations at the end of his field notebook. (Richard Hazard, a cook who became ill and died and the only other person to die on the expedition, was also later buried in this grave.) On the day before his departure while at Fort Point, Goyder wrote his report on the Northern Territory survey. He reported Bennett's death, and discussed the whole matter of the expedition's relations with the Aboriginal people at some length.

Goyder saw himself as having to account for two things: how the spearings that resulted in Bennett's death had occurred in the first place; and why he had not retaliated. Retaliation was what the men had wanted and was what they, and many in Adelaide, saw as the victim's due. Unlike Finniss, who had been condemned as too brutal, Goyder's position was at the other extreme: it might be he, and not the men who had smashed the canoes, who would be perceived as displaying a 'want of proper feeling'. (Since there had been no retaliation, Goyder chose not to mention this incident in his report. To do so would only have highlighted the difference between his own attitude and that of many of the men.) To avoid being seen as having failed the victims and the possibility of a retaliatory response from Adelaide that might ensue, his account of the events and his response had to be presented with a great deal of care. Explaining Bennett's death was a therefore a subject Goyder entered upon 'with unusual hesitation'.[78] Goyder presented the spearing of Bennett and Guy as a typical example of what was then seen as 'native treachery', referring to the 'apparent friendliness' which had deceived his officers, lamenting that 'we are liable to the consequences of their sudden and treacherous attacks', and warning that: 'Nearly all are alike treacherous'.[79] But having framed the events in these terms, he went on to present disobedience by his own men as the factor that had allowed the spearings to take place. The danger of attack was something he had foreseen, he explained, and they had been warned that there were people seeking revenge. Not only had he 'personally remonstrated' with Bennett for 'his persistent familiarity with the natives', he had 'written a special memorandum upon the subject', but Bennett had taken no notice.[80] Bennett's disregard of his orders and the friendliness of the Aboriginal people in general had led other officers and men to depart from his instructions by going out unarmed and by not remaining in one camp, thereby making themselves vulnerable.

The matter of retaliation was less straightforward. 'Both officers and men were naturally indignant at Bennett's murder,' he wrote, 'and, had the order been given for retaliation the punishment of the natives would have been a simple matter with a party so armed ...' But although Goyder had framed his account of events in terms of 'native treachery', when he came to present his reasons for not retaliating, these included explaining the behaviour of the attackers as both understandable and law-governed. Goyder argued that the attack upon Bennett and Guy was the consequence of 'feelings of revenge on the part of those who had probably lost relatives in some previous contest with the whites'.[81] Moreover, to retaliate, he explained:

> even could we have identified the murderers, would have been to secure to our successors, less able to defend themselves, a debt of lives to be paid for our act of reprisal, unless we annihilated the tribe, which was not to be thought of.[82]

Goyder could have safely left it at that – that he was fearful of further revenge from the local inhabitants – but having already described the prolonged and very threatening harassment of the small party he had led back to Fort Point, he launched into an account of what he clearly considered to be the fundamental issue:

> that we were in what to them appeared unauthorized and unwarrantable occupation of their country, and where territorial rights are so strictly observed by the natives, that even the chief of one tribe will neither hunt upon nor remove anything from the territory of another without first obtaining permission, it is scarcely to be wondered at if, when opportunity is allowed them, they should resent such acts by violence upon its perpetrators.[83]

Even if he and his men were attacked:

> retaliation on our part would, by many – and, I do not say without some show of justice – be looked upon as little short of murder, as we have no right to take the lives of these men without such be done in actual defence, or by the laws of our country.[84]

British legal procedures offered no solution, he added, since nothing said by either party would have been intelligible to the other.

What Goyder really believed was that the attack on Bennett and Guy was understandable and to be expected, and that resistance to the presence of the surveyors was not only just as understandable, but determined by Indigenous law. The rhetoric of savage instincts and treachery, while it may have given him an opportunity to express feelings of anger and distress, was used because it was expected. Goyder could not afford to distance himself from the views and assumptions of his contemporaries: for his voice – and his views – to be heard, he had to appear to be angry. Similarly, the words 'we could easily have shot one or two of them', even though they begin a sentence explaining why this did not happen, by their sheer casualness manage to suggest that this was something he might actually have considered.[85] His plight was clear enough to the leader writer of the *Register*. The day after the expedition's return, the paper observed that the fate of Bennett was so important to the future relations between settlers and Aborigines – so 'ominous' – that Goyder had had to exercise 'special caution' in presenting it. The paper went on:

> To some this passage of his dispatch may present a cold contrast to the mingled indignation and sympathy with which private letters [from other members of the expedition] allude to the same subject; but it must be remembered that Mr. Goyder had to *subordinate his personal feelings* to very grave considerations of policy [my emphasis]. [86]

Having subtly implied that Goyder might personally have preferred to wreak revenge, the article concluded:

> Mr. Goyder's policy towards the blacks, as illustrated on that occasion, will of course be judged by two very opposite standards. In South Australia, as at Port Darwin, there are nigger-haters, and also adherents of the Aborigines' Protection Society. The former will taunt Mr. Goyder with having preferred the philanthropist to the man, but his philanthropy has had a view to ultimate justice. It may prove the noblest manliness in the long run.[87]

The opposite assessment of Goyder's response was put in *Pasquin*, which supporters had kept alive for a few issues after Mitford's death. There it was argued that Goyder had 'palliated' a cruel murder that was the bloodthirsty act of an individual. Worse still, he had had the 'effrontery to talk of the matter as though the unfortunate victim had been killed in a fight between the natives and the settlers'.[88]

The ship carrying the returning party reached South Australia on

15 November 1869. Although the men were seen to be in excellent health, Goyder was observed to be looking 'careworn'.[89] A week later a lithographed plan of the survey was ready for presentation to parliament, along with a bird's-eye view of the Port Darwin district. The report demonstrated that arriving as the head of a small, isolated party attempting to settle land belonging to others – a fact not recognised at the time – had forced Goyder to confront the realities of colonisation in a way that his role as surveyor general in an established colony had not. Over 600,000 acres (more than 243,000 hectares) had been surveyed, and for the first time the surveyed land was identified by naming its traditional owners. It was defined as comprising portions of 'four native districts, viz., the 'Woolner', 'Woolner-Larakeeyah', 'Larakeeyah' and 'Warnunger'.[90].

The governor, Sir James Fergusson, requested that Goyder provide a separate report on the Aborigines of the Northern Territory. In this report he compared the people he had encountered there with those he had met in the interior, concluding that they were governed by similar long-held customs (although he reported other differences). He made plain his attitude towards future relations:

> I cannot but think that the first essential element in the successful treatment of the Natives ... is to inspire confidence by convincing them that no real harm or injury is in any way intended them. This gives them confidence, after which their language is easily acquired and they can gradually be made aware of and taught to appreciate the conduct and motives of those seeking to develop the resources of the country.[91]

He considered that the colonisers should first impart their own language and training in the 'discipline' of their way of life before any attempt was made to share the Christian religion. Goyder also believed that Indigenous people should be taught the use of European tools and gradually weaned from their nomadic way of life, but, typically, the reason he gave arose out of realistic foresight and a concern for their situation. Learning to use European tools was absolutely necessary because settlement resulted in the disappearance of game and made their traditional way of life impossible. Having presented this report, it was not until the end of his career that the issue of land and the Indigenous people occupied him once again.

Before he became ill on his return from the trip up the South Arm in late April, Goyder had planned to continue to expand his knowledge of the country by returning to Adelaide overland. This intention had become rumour in Adelaide, making his arrival by sea something of a surprise. By land or by sea, he returned a conquering hero. The author of the *Register*'s weekly literary column (the editor, J.H. Clark, under a pseudonym), who satirised affairs about town, penned a welcome to the tune of a popular air, which began:

> Oh, Goyder is it you, Sir,
> > Come safe to hand?
> They said what was not true, Sir,
> > That you'd come overland.
> We'd heard how hard you worked,
> > And it made us hope you'd come,
> For we know of many holders
> > Of land-orders here at home,
> Making game of your attempt
> > To survey their money's worth;
> So, Goyder, shake a fist; you
> > Are welcome from the North.

The final verse, however, hinted at Goyder's real anxieties:

> Your duties here, without you,
> > Have fallen in arrear;
> But with lots of tin [money] about you,
> > It's little you can fear.
> Last week the Upper House
> > Talked of knocking off your screw [salary],
> Though the Caucus doesn't sit there,
> > And their tricks you needn't rue.
> But you'd better not again
> > From your office thus go forth;
> For Goyder, we have missed you;—
> > Welcome from the North.[92]

CHAPTER TWELVE

Going home

Only a week after returning from the Northern Territory, Goyder left town to make a 'descent' on the drainage works in the South-East. He had finished the necessary reporting and was returning to his usual concerns. Portraying him as the hero of the hour whose every move was cool and expeditious, the *Register* marvelled (in a manner that to Goyder may have seemed distastefully close to blasphemy):

> Seven days to create a settlement, and seven days for its historiography – then the Surveyor-General calmly washes his hands of it, and returns to his ordinary duties as if they had never been interrupted.[1]

The call to the Northern Territory had come just as it had become apparent that further drainage of the Mount Muirhead flats and an area around Naracoorte, designated Mayurra, would be required. The flats round Mount Muirhead had been drained by the newly cut gaps, and the drainage scheme there had been constructed such that, by closing sluice gates, all the land behind could be flooded for irrigation, even in summer, a system that Goyder had already tested to his satisfaction.[2] This drainage was intended to prepare the land for grazing, but because the Mount Muirhead flats and Mayurra had been selected as agricultural areas under the Strangways Act, the government was committed to draining them sufficiently to allow agriculture to go ahead.[3] However, the credit selection provisions were not working as anticipated in the South-East: buyers were letting the land pass in several times before eventually purchasing, to keep the prices low, and the squatters were having no trouble obtaining the best land by using so-called 'dummies' to circumvent the law. (The dummies used their names to purchase land for the squatters.) The resulting revenue from land sales did not cover the costs of the pledged drainage.

After Goyder's visit, in order to get an appropriate price, the land was withdrawn from sale until fully drained, and the broad scheme for drainage was abandoned in favour of drying out particular flats for sale. Goyder was also concerned that, in a scheme combining public and private responsibilities, the holders of blocks several times removed from the publicly constructed main drain might find themselves in need of a drain that ran past all the intervening blocks, but they could not be certain all their costs would be covered. From this point onward the work proceeded in a reactive, piecemeal manner.

In Adelaide disquiet over unemployment was provoking public unrest, with protesters in 'occupation' of King William Street. In response, the government revived the previous solution of selecting unemployed men for work in the South-East.[4] Goyder was sympathetic to the realities of poverty, but would not tolerate idleness for any reason, and later recalled that some of the 'loafers' from Adelaide were 'among the most useful men I had, when they got used to the work'.[5] Another 108 kilometres of drains began to function and the earth dug from them was heaped up in embankments to form a network of roads.

Around this time Goyder spent two months in Victoria investigating the effects of the land laws there. The parliament there had passed yet another land act, taking the radical step of permitting selection before survey. Beyond the urban areas, and away from the goldfields, intending settlers could peg out blocks of up to half a square mile of crown land anywhere they chose, including on pastoral leases, and have their block surveyed by a contract surveyor. Free selection was the complete opposite of the South Australian approach and the South Australians were deeply curious to see the results. With Goyder on hand again, their curiosity could be satisfied by sending him to investigate and report.

'A lady accidentally poisoned'

Goyder's field notebook shows that, despite the hard work, the frenetic pace, and the tropical conditions – or, perhaps, because of them – while he was in the Northern Territory, his house and his family had been on his mind. In the back of his field notebook he had drawn and redrawn the plan of a house.[6] In addition to Frances and himself and their nine children, his plans show that he was also trying to accommodate Ellen ('Nelly' in the plans) and Marion, another member of the Smith family, who was to come to South Australia with the returning family. This was to be accomplished by pairing everyone into six rooms, except for Florence, Gertrude and Isabella, who shared one room. Mary Ellen, whose age placed her between Gertrude and Isabella, was placed with Ellen, suggesting perhaps that she had a problem with her health that required the presence of an adult, or some other issue which separated her in some way from her sisters. Marion Smith was to share her room with the baby, Normy. This left a small spare bedroom.

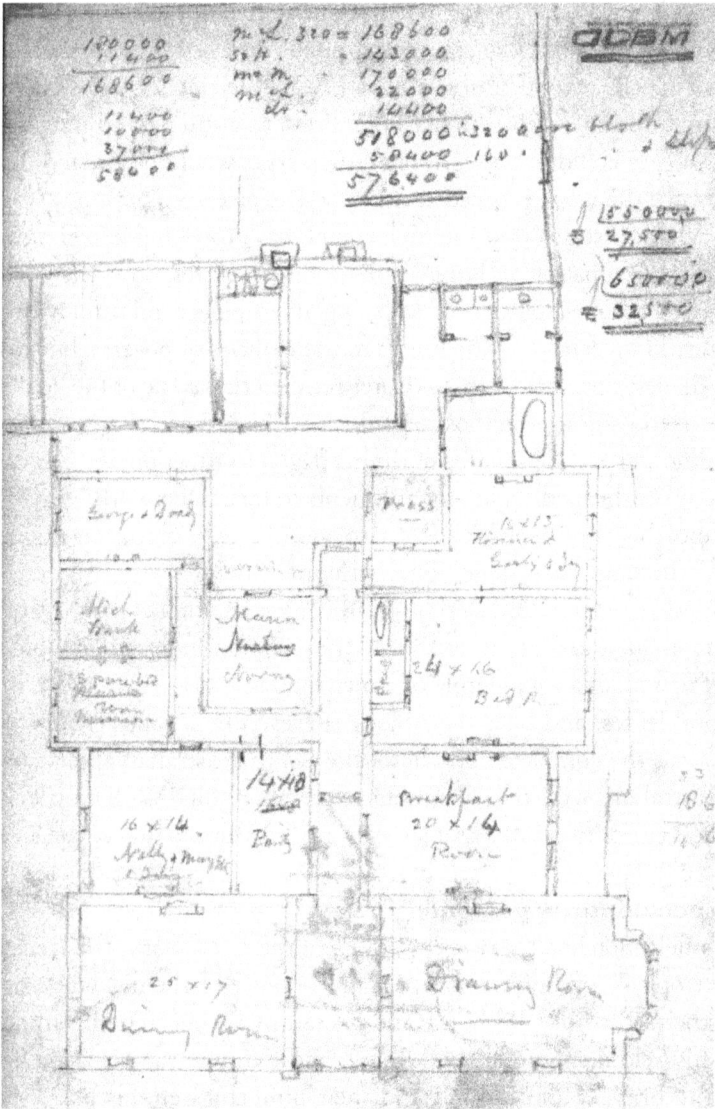

Dreams of home

The fifth, final and most elaborate of the house plans in the back of Goyder's field notebook of the Northern Territory survey. On the right, below what is obviously a bath, is a large linen press, a bedroom to be shared by 'Florence + Gerty + Isy', a breakfast room and a drawing room. On the other side, one bedroom is to be shared by the older boys, George and David, and another by the two younger boys, 'Alick' and Frank. Marion Smith was to mind the young 'Normy' (the word 'nursery' has been crossed out), leaving a spare bedroom. Another room, tucked in behind the dining room and the pantry, was to be shared by Nelly (Ellen) and Mary Ellen. At the top of the page are Goyder's calculations, apparently for the area of the survey. [SRSA]

In England, Frances and Ellen and the children had been living at No. 5, The Polygon, an irregularly shaped Georgian terrace of wide three-storied houses still standing in the Bristol suburb of Clifton. Clifton is situated on limestone cliffs overlooking the city and in the nineteenth century was home to wealthy merchants. Frances was supposed to have benefited greatly from the stay, but was still suffering from sleeplessness. Towards the end of the visit, she had been advised by her husband's youngest brother David, a doctor in Bradford, to take a few drops of Batley's sedative, a brand of laudanum. This was usually given to her by Ellen, mixed into a wineglass of water, and it had been found to be effective.

One morning, after passing a sleepless night, Frances took a dose of the mixture and slept for two or three hours. Later, Frances requested the key to the medicine chest from Ellen. Ellen subsequently went out and while she was away, Frances helped herself to more of the sedative. However, instead of adding a few drops to a glass of water, she downed a whole glassful. Finding Frances prostrate on her return, Ellen called a doctor and, although Frances's stomach was pumped, it was too late to reverse the effects of such a dose. Frances died on 8 April 1870. The next day, a Saturday, an inquest was held at a nearby tavern, the Adam and Eve, with both Ellen and the doctor giving evidence. The coroner found that there was no suggestion of suicide and that Frances had died from an accidental overdose. The coroner's records for this period were destroyed in a fire so it is impossible to determine how the conclusion was reached, but the brief reports seem to imply that she had acted in desperation and ignorance, knowingly taking far more than the recommended dose, but not with the intent of killing herself.[7] There seems no real reason to doubt the finding. At the time opium was freely available and the only reliable painkiller and sedative. Inadvertent addiction was not uncommon and disposed those it had ensnared to exceed their dose. To make matters worse, effective dosages were not yet standardised, and strength varied from batch to batch. It was a situation which invited misfortune.

The family had been due to return on the *South Australian* on its sailing for Adelaide on 10 May, but after the death, Ellen seems to have gone to seek comfort with the Goyders, perhaps especially with Sarah, whom she and the children knew from Adelaide. In Ellen's autograph book (which contained inspirational and uplifting poetry entered by relatives and friends, as well by herself) there is an entry by David Goyder, which shows that by 9 May she was at the small fishing village of Wivenhoe, on the east coast, near Colchester in Essex. This was where Goyder's sister Margaret lived – Margaret Harvey, as she had become – and where older sister Sarah Anna, and her oldest daughter, also Sarah Anna, were staying. Margaret Goyder had met her husband, John Harvey, while the family was living in Ipswich. Harvey's father, a prosperous shipbuilder whose family and business were long established in Wivenhoe, was a keen believer in the teachings of Swedenborg. The pair had married and moved to the village in 1857, and David Goyder had then attempted to found a small society of followers there, with John Harvey and his father as core members. Sarah and her daughter seem to have been staying to help with the family, because Margaret had contracted tuberculosis.

Although Frances's death was reported in the *Times*, there was no way the news could reach her husband, who in any case was out valuing runs. At about the time Goyder must have imagined that his whole family was setting sail for home, he was compelled to return suddenly to Adelaide and to send his deputy to value runs in his place. In a letter to the commissioner he explained that this was the consequence of ill health, and that it would 'take some months of quiet before my health is restored'.[8] The choice of words suggests that, as in 1866, he was feeling the strain as much mentally as physically.

The news of Frances's death reached Adelaide on 3 June. The terrible thing for Goyder was that a portion of the mail from England – the portion with the letter bearing the news to him – had been sent on to Melbourne. The English newspapers, however, arrived in Adelaide safely and the story was published in a rush edition of a Saturday paper– which is how Goyder and the Smiths learnt of the death of Frances.[9] On Monday 6 June, the daily papers dealt with the matter formally. The *Advertiser* reported that the 'bereaved husband' had been 'looking anxiously for their return and preparing their reception' and assured him of heartfelt sympathy. The *Register* reproduced the *Times* article, which contained the blunt remark that 'the evidence entirely negatived the idea of suicide'. After receiving the news, Goyder, facing further months on his own, went on working.

Knocked up

Around the beginning of August, he made a flying visit to the South-East and produced a report containing a history of the drainage projects.[10] His literary skills earned modest praise from a journalist whose spirits had evidently sunk at the prospect of having to read the thing. Goyder, noted the journalist with relief, was 'not at all a dull annalist. He handles his facts very firmly, and has a terse style of narration, while his practical acquaintance with what he describes is the most obvious factor of his description.'[11] As well as recounting the history of the project, Goyder continued to press for an idea he had presented earlier. Although he had initially recommended rail transport as appropriate for the region since it was difficult to obtain the durable stone necessary for roads, in this report he suggested that the 'discovery and successful application' of the 'road steamer' rendered laying track unnecessary. What he envisaged was one continuous road running from Rivoli Bay across Lake Frome to Penola, connecting a large area to the coast, with steam cars running on the embankments constructed of soil excavated from the drains.[12]

The report was Goyder's final summing up; at the same time he resigned from his role of supervising the drainage works. From September 1870, the drainage of the South-East was one responsibility he no longer had to bear.[13]

A necessary break

During August, Ellen and the children, and the newcomer, Marion, arrived home.[14] Goyder, of course, had worked on. It was not until late in January 1871 that he penned a letter to the commissioner requesting, on two weeks notice, nine months leave of absence on full pay. The letter itself was evidence that something was wrong. Written in a large, looping scrawl, the six short paragraphs roll across five foolscap pages. Goyder normally used one of two styles of writing – a small, squarish, flat hand, present in the field notebooks, and a larger, more rounded style, used in letters and memoranda. The handwriting in this letter is like neither. Becoming larger and untidier with every page it suggests that the condition that had prevented him writing in the Northern Territory had recurred. He intended to visit England, Goyder wrote, 'to recruit my health which, in consequence of continued application to official duties during the last eighteen months, begins to cause me considerable uneasiness – if not alarm'.[15] The letter was accompanied by a medical certificate from a prominent North Terrace doctor, Allan Campbell. Campbell, who had arrived from Glasgow in 1867, was a friend of Goyder's.[16]

Campbell certified that Goyder was 'suffering very considerably from nervous and muscular debility, arising from prolonged application to business, and that under conditions of a tropical climate of an exhausting character'. In Campbell's opinion, residence for some time in a colder climate was 'absolutely necessary to the restoration of his health'.[17] The appearance and urgency of this letter seem to show that Goyder had suffered a crisis or collapse, but other evidence suggests that this was not the case. Goyder proposed to catch a ship, *Queen of the Thames*, sailing from Melbourne on 18 February to return to England. The *Queen of the Thames* was a luxurious steamship pioneering a fast passage around the Cape of Good Hope from England to Australia. She had reached Melbourne in 58 days, a feat that seemed 'almost incredible' to the colonists and one that augured well for communications between the colonies and the mother country.[18] In his letter Goyder stated that he planned to visit Italy, Holland and America to increase his professional experience, and he appears to have viewed his passage on the *Queen of the Thames* in something of a similar light: he had an agreement with the two major daily papers, the *Advertiser* and the *Register*, to provide them with an account of his journey.[19] Since Goyder's health had been doubtful for some time, it is likely he had been giving the matter of taking leave and returning to England some consideration, but the arrival of the exciting new steamship and the real benefit of a shorter period languishing at sea seem to have contributed as much to the exact timing of the decision as physiological or psychological considerations. Goyder, who never took holidays, had no difficulty obtaining leave, but the promised account of the journey that he eventually provided was not one that he or the editors of the two dailies had anticipated.

The Queen of the Thames

After crossing the Indian Ocean on the way back to England the new marvel of comfort and speed came into sight of land on 16 March. The ship was making such good progress that it looked as if they would round the tip of Africa the following day, but at noon on 17 March a strong head wind set in, and their progress was slowed. That evening, not long before 8.00 pm, a fellow passenger pointed out to Goyder a faint light to their right. It did not seem to him to be a lighthouse, nor was there sufficient glare for a bushfire. While several passengers remained looking at the light, Goyder and others went downstairs for the musical entertainment held every Friday night. As usual, this was attended by the handsome captain and his charming officers. (It 'would have done you good to hear him sing "The Fascinating Fellow"', Goyder later jeered bitterly.[20]) There had already been grumblings among the passengers that some of these might have been better occupied on the deck, attending to the ship. Early in the program Captain Macdonald announced that the Cape Agulhas light was now abeam, and excused himself to go on deck – but lingered instead to sing a song.

When the concert was over, Goyder went on deck and noticed from the binnacle (a compass accessible to the passengers) that the course had changed:

I glanced up at the stars, and saw that the ship was going on a true bearing north-west, and knowing the distance that the ship had run from noon, and the direction in which the land trended, as well as the course steered from noon up to 8 o'clock, the time of my going below, I became exceedingly anxious, and remarked that unless we steered a south-south-westerly course for a time we should be on shore before morning ...[21]

Goyder was concerned that the ship was not where it was believed to be: that the light he had seen earlier was not the Cape Agulhas light and that they had turned before reaching the tip of Africa and were now headed towards land. He told a companion that he wished the night were over and went to his cabin to sleep. At 2.00 am, his fears were proved correct. The ship was aground. Goyder went up on deck and put lifebuoys out in case the ship broke up, then returned below to avoid the confusion on deck. In the main salon passengers were sitting quietly while a clergyman read prayers. Goyder returned to his cabin, gathered together his compass and a few other items, and put on his cork jacket under his coat. Thus prepared, he wished a couple of friends goodbye 'in the event of the worst happening', and returned to his cabin to sleep until daybreak.

Dawn revealed that they were close to the shore, just beyond the surf. The weather was calm, and the wind was blowing onto the sea, but attempts to take advantage of this to free the ship by raising her sails failed. Although land was in sight, but with no coherent response from the captain or organisation by the crew, it was soon clear to the passengers that they were not yet safe. It was not until one of the passengers demanded to be put ashore with his family that Macdonald agreed that the process of landing the passengers could begin. Goyder was in the third boat ashore, landing at about 7.00 am. True to his experience and concerns, once on shore he immediately set off with a companion to find fresh water, which he soon located in an old well.

All the passengers were landed safely, the captain coming ashore with the last of them at about nine o'clock. During the afternoon several passengers demanded to know what was to be done to find shelter for the women and children away from the beach. Macdonald merely informed them that they would have to look after themselves, and that he was not in any way liable. Fortunately, they were close to the Cape of Good Hope and Cape Town and the agent for the insurance company Lloyd's was already present: men landed in the first boat ashore had make contact with the company and the agent arranged for them to be taken to the nearest village. Supplies, valuables and documents were also brought ashore. However, for the passengers, success in being reunited with their possessions required bribing the by-now drunken crew. As evening approached, the wind increased and the last boat to leave the ship was overturned in the breakers. Three crew members, including the purser, were drowned.

The *Times*, reflecting later on the maiden voyage of the *Queen of the Thames*, judged that it was the 'old story' of a magnificent steamship raced heedlessly to her doom 'retold in every particular of the narrative'.[22] Goyder's account, published in London and South Australia, was angry rather than moralising.

In the *Times* and the South Australian dailies he attributed the loss of the ship in good weather to Captain Macdonald's navigational incompetence and inattention to duty before the grounding; and after it, to his drunken abandonment of all responsibility, leaving the crew and the passengers to organise themselves, and the ship, which could have been saved, to break up. In his letter to the South Australian papers his outrage was tinged with a sharp satirical tone. After the attempt to free the ship by sail had failed, Goyder claimed to have seen the captain on the gangway:

> I had read of the extraordinary coolness, energy, pluck, and indomitable courage of the commanders of our navy and marine, and prepared myself for the thrilling sensation of pleasure which I experienced when prompt and decisive action is taken in emergencies by those in power. I fully expected to hear the voice of our noble commander demanding silence on the part of all – boats out, guess-warp passed on shore, and kedge anchors astern, with a view to hauling the vessel off after she had been lightened of passengers, luggage, and wool; but somehow the captain took a different view, and stated from the top of his house that we must help ourselves.

Goyder vented his frustration and helplessness as one of 'so many undeserving miserables … with our large responsibilities and numerous belongings, at the mercy of the stupendous intellect and untold experiences [inadequate training] of this George Macdonald'.[23] His view was not in any way specific to him. Voicing her feelings in her diary, another passenger was even more scathing. She saw the captain's abandonment of his command as an unmitigated act of selfishness and indifference, and curtly rated the view that he was 'stupefied with drink or shame' as 'most charitable'.[24] Goyder agreed:

> I was not surprised to see him sway backwards and forwards, and then lay down and sleep before 10.00 a.m. I was not surprised to see him turn from looking at the poor fellows struggling after the boat had upset, and before the last of the three that were drowned had disappeared knock the neck off a bottle of champagne and sit down to partake of its contents.[25]

The ship's surgeon was also drunk, but in his case it was no surprise. Watching him as he tried to resuscitate the drowned purser, Goyder anxiously recalled the decisions he had made concerning his own family. 'Men with large families are often induced to decide in favour of a … particular vessel, if assured that proper medical advice can be had for their children', he later lamented, 'and it is but poor consolation when the necessity arises to be informed that [the doctor] is a very clever fellow "when he is all right" '.[26]

The stranding was investigated in a court of inquiry in Cape Town. While on board, Goyder had been collecting and tabling daily information about the ship's location and course from the captain for publication in the ship's newspaper. According to the captain the ship's log had been lost in the capsizing in which the purser had drowned. Therefore it was principally on Goyder's evidence – especially as someone experienced in taking bearings – corroborated by that of another passenger who had also been noting down the official statements of the ship's position, that Macdonald and his senior officers were found to have made errors of judgment that could only suggest incompetence. The court drew attention to the facts that the navigational equipment the captain had claimed to be inaccurate had not prevented the ship from being sailed safely to the point where it struck disaster and that the missing log was last seen when the captain had called for it immediately after the ship went aground.[27] After giving evidence Goyder continued on his way to England on board the RMS *Briton*.[28]

Goyder's account of the stranding was published in the *Times* on 22 May 1871, with the result that he was threatened with legal action by solicitors acting on behalf of Macdonald, who argued, understandably, that his remarks were calculated 'to injure very seriously the captain's future prospects in his profession'.[29] Lloyds offered to pay Goyder's expenses and obtain additional leave for him if he stayed in London and gave evidence for them. Goyder did not take up their offer, but, on the advice of the solicitor to the Customs Office, he made a deposition and was examined on oath by assessors before leaving England. He did so believing that he was acting 'in the interest of the shipping community and the whole travelling public'.[30] Macdonald and the officers attempted to evade any responsibility for the grounding, but a Board of Trade inquiry in May 1872 accepted all the accusations against Macdonald – in essence, the whole story as Goyder and others had told it, including the merrymaking and abandonment of the ship and the passengers – and suspended the captain's certificate for a year from the date of the finding. Goyder later attempted to claim 'damages and costs' from Lloyds for expenses he claimed never would have been incurred had he not given evidence.[31] There is no indication of the nature of the 'damages', but Goyder evidently regretted having chosen to give evidence, and his tone was angry and threatening.

The wreck of the *Queen of the Thames* and the legal struggles which followed were not the only things to destroy whatever opportunities for recuperation a visit home had held out to Goyder. Tuberculosis, the disease which had threatened and consumed members of his own and his wife's family, was continuing its grim work. At the beginning of May 1871, while he was still in

England, his sister Margaret finally died. Three months later, he left England on the *South Australian*, taking with him his 21-year-old niece, the younger Sarah Anna. The young woman had also contracted the disease and was probably being sent back to the warm South Australian climate to seek recovery, but she did not survive the voyage.

The *South Australian* arrived at Port Adelaide at midnight, firing rockets to announce her arrival. Goyder was landed as soon as possible, sometime in the morning of 17 October, and returned by express to Medindie with the dismal news of the deaths of his sister and his niece.[32] His sister Sarah returned to Adelaide in 1873 and for the next couple of years lived in a seven-roomed house belonging to Goyder on the edge of the parklands in Medindie. Goyder appears to have purchased this house so that she could live nearby, as the house only appears in the council records as belonging to him (but occupied by his sister) during 1873–75. Like her brother's, Sarah's considerable strength and robust outlook must have been tested to the limit during this time. Her daughter Sarah's death on the voyage to Adelaide meant that, as well as losing her husband, she had lost three of her four children. Less than 18 months later, tuberculosis would claim her last child, her firstborn, George, who died in the Northern Territory in 1873. Sarah herself was long-lived, like her mother, and died at the age of 86 in Massachusetts in 1909.

A decent marriage

By surviving this nightmare of a recuperative break, Goyder seems to have bluffed the malign fates into retreat. While he was in England, Queen Victoria had assented to the passage of a Deceased Wife's Sisters' Bill, a piece of legislation the South Australian Parliament had passed and presented to her five times since the inception of independent government. Although they violated the directions of the Church of England, marriages between widowers and the sisters of their deceased spouses had been customary, for very practical reasons perfectly illustrated by the Goyder family, in which Ellen had long been an honoured part. However, in 1835 changes to the marriage laws prohibited such marriages in the future, and stamped them socially unacceptable. Lavington Glyde, who as commissioner had sent Goyder out to value the runs, had married his sister-in-law as a widower in July 1870 at the nonconformist Clayton Chapel, only to have the governor declare this arrangement 'indecent' and refuse to invite the couple to a ball at Government House.[33]

Nevertheless, less than a year later, in June 1871, the change to the law to make such marriages legal was gazetted in the colony. Margaret Goyder Kerr has reported that family tradition claims that Queen Victoria relented in her opposition when her loyal surveyor general's case was put to her, but this may have been a coincidence of timing. South Australians had been petitioning their Queen for repeal from the moment they had a legislative voice and there had been debate about the matter in the colony and in England. Even while Frances was still alive, it had been rumoured that the Queen would relent if the legislation were presented again, even though she was personally opposed to such marriages.

On 20 November George Goyder and Ellen Smith were married at St Luke's, a seemingly unusual choice since the church is located in the south-west of central Adelaide and served a predominantly working-class area, at least later. As many Anglicans would not countenance a divergence from the church's practice in England and continued to view the marriage of a man to his deceased wife's sister as virtually incestuous, St Luke's may have been chosen because its incumbent, the Reverend James Pollitt, was one Anglican clergyman who was happy to perform the ceremony. Immediately after the ceremony, the couple were married again in the New Church, as their announcement stated clearly.[34] This ceremony was performed by E.G. Day, the New Church minister who had earlier been employed as a storekeeper in the Survey Department.

While the decision to marry Ellen was an eminently practical one from Goyder's point of view, the question of what this meant, personally and emotionally, and to Ellen especially, inescapably presents itself. Ellen's life up to this point seems to have been spent in caring for her sisters and their families (even if for Mahala only briefly), a role that seems to have precluded any prospect of her marrying and having children of her own. But the evidence, in particular that gleaned from her autograph book, suggests that she did not see herself as a drudge, and her marriage, when it came, was real to her and not just a convenient arrangement for the rearing of children. Her grand-daughter's testimony to her fondness for the Goyder family suggests that she had found satisfaction as a member of this family. That she was regarded as a family member is borne out by Goyder's quite striking naming practices in his early explorations in 1857 and 1859–60, when he named springs and watercourses in pairs, after his wife and her sister, very much as if he had two wives. In most respects other than the sexual, this seems to have been the case, and responsibility for the care of Ellen, as with John and Susannah Smith, almost certainly fell to him. Goyder seems to have taken up the role of husband in the same spirit that he took up the role of head of a government department – accepting responsibility readily and creating his own borders.

Ellen's autograph book also reveals that she experienced her marriage as a liaison founded on a true emotional bond between partners, very much in accord with Swedenborg's teaching. She was clearly a believer on her own account: her autograph book contains a picture of Swedenborg and his house, as well as a brief guide to life and conduct for his followers. One of her own entries is a poem titled 'Mizpah', a term which signifies the biblical passage from which it is derived: 'The Lord watch between me and thee when we are absent one from another', Genesis 31:49, as it is quoted by Ellen under the title. The first verse is unequivocal:

> A broad gold band engraven
>
> With word of Holy Writ –
>
> A ring, the bond and token
>
> Which love and prayer have knit.
>
> When absent from each other
>
> O'er mountain vale and sea,
>
> The Lord who guarded Israel,
>
> Keep watch 'tween me and thee.

There were three verses, and after the third Ellen had written the destinations and dates of two of Goyder's later journeys away from Australia.[35] Interestingly, Ellen's autograph book begins and ends with a dedication, obviously written by the giver of the volume, 'To Miss Ellen P. Smith. 1st January 1857'. The title 'Miss' is spelt in the archaic way which characterised Goyder's writing at the time, the 'ss' written as 'fs', and the writing could possibly be his.

With his marriage to Ellen, Goyder's personal life moved out of its dark and dramatic interregnum and entered an era of stability. The couple's first child, John Harvey, was born at the end of September in the following year, and twin girls, Ethelwynne and Margaret, were born in December 1873. They were the last children born, and their arrival brought the number of surviving children from both marriages to 12. Despite his new happy circumstances the combination of exhaustion and the eruption of tragedy in the years from 1869 through to 1871 had left a permanent mark on Goyder, evident in photographs. The portrait of him in evening clothes, probably taken not too long before the departure of the expedition to the Northern Territory Expedition, shows an alert and confident man with an obvious capacity for firmness and the exercise of authority. Another photograph, probably taken in the 1870s, shows a different person.[36] Although physically worn, his expression is still alert, but animated by a different spirit, softer and more benign, and, unexpectedly, with a faint air of something like innocence. Despite Goyder's capacity for anger, his pain and grief seem to have been transmuted not into bitterness, but into greater humanity – or perhaps the camera has just caught something of the friendly sociability that was held to characterise his off-duty personality.

Part three:
UNIVERSAL GENIUS AND CLERK OF THE WEATHER

CHAPTER THIRTEEN
Nature's Line

I like to see men occupying the country doing well, and it is a source of misery to me, as it is to others and the men themselves, if they do not do well on the land. I like to see farmers and pastoralists alike doing well, and if they are not succeeding there must be something requiring alteration somewhere.

– George Goyder, 1891[1]

Where is the great medicine man of the farinaceous village – he who 'delineated the limits' of our happy hunting ground – Rain-Maker-General and Clerk of the Weather ...

– *Pasquin*, 1868[2]

With his marriage to Ellen, the very dark phase of Goyder's life came to an end, but the period that followed was not an easy one. Members of his family were severely affected by illness, and his financial anxieties continued. Goyder, too, was repeatedly debilitated by poor health, but through the 1870s up to 1883, he continued to be extensively involved in shaping the colony, shouldering his usual extraordinary workload. It was during this period that he had to fight his major battles over the climate.

In 1865–66, when the Line was drawn, Goyder had not acknowledged the gap between the simple instructions he had been given when sent into the drought and his own goal to map the end of the reliable rainfall. The various aspects of the Line – its role as an administrative instrument, its connection to a massive drought, its potential function as a major land-use division, and its nature as the border of a climatic zone – had never been teased out and considered in relation to each other, with the result that the phrases 'Goyder's Line' or 'Goyder's rainfall' might invoke any of these meanings. This lack of clarity about the multiple meanings and purposes of the Line meant that, despite the prodigious effort Goyder had put into examining and assessing pastoral runs and the determination and resilience he had shown in asserting the real value of land in the Mid-North, in people's minds the Line was associated with pastoralism – and to the suspicious, with the political manoeuvrings of the squatters.

Although Goyder was disappointed that there had been no move – not even the suggestion of one – to have the country divided by law into agricultural and pastoral zones, he had not given up hope that this would occur, since the idea

at least was still in the air.[3] In 1867, giving evidence before the Commission to
Inquire into the State of the Runs Suffering from Drought, Goyder was asked
by the chairman, John Bagot, a former lands commissioner, if he thought it
would be possible 'to divide the country into two districts, one pastoral and one
agricultural', and, if it was, would his line perform the task. It was this that
prompted the answer, since repeatedly quoted, that the Line did perform this
role 'to a certain extent'. But Goyder's reservation was about country *inside* the
Line. There was country, he explained:

> where, although the soil is eminently adapted for tillage, and will grow
> anything, the peculiar position of it, and its openness to hot winds, render it
> such as can only be safely continued as pastoral land. That is inside the Line –
> and outside it, the whole of the land is fit only for pastoral purposes; that is as
> far as we know of growing cereals at present.[4]

Bagot had then asked if there was any value in dividing the northern country
into hundreds, giving Goyder the opportunity to explain that it would be best:

> to subdivide it into blocks of thirty miles each, and by offering them at
> auction (or say from ten to thirty miles) give farmers in the inside the means
> of becoming middlemen by combining grazing with their tillage, and the
> extreme outsiders the chance of getting paddocks, in which they may get
> their stock fat ... and so enable them to enter the market in a better position.[5]

This answer was much the same as a reply he had given in 1865, when he
had first publicly suggested a line of reliable rainfall.[6] Together with the map
which defined the various schedules for pastoralists, the reply makes clear that,
from the beginning, Goyder had seen his line as the southern edge of a zone
of intermediate country, and had understood that a new and flexible response
was in order.

While responding to this commission, Goyder also had the foresight to
warn that, although the drought had been severe enough to alarm people about
the real nature of the country, if the present lessees abandoned it and good
seasons occurred again, 'the sad experience of past years would be forgotten
by new comers', the runs would be overstocked again, and the same results
would very likely follow. 'No class of persons has been more deceived than the
stockholders themselves', he advised.[7] But that, of course, was assuming that the
north remained grazing country.

The commissioners eventually arrived at 22 conclusions, the seventh expressing
their acceptance of the Line in the role Goyder had actually created it for. Their
report declared that, according to the evidence:

the rainfall boundary, as fixed by Mr. Goyder in his map of December, 1865, is very accurate, and fairly defines the line which at present divides the purely pastoral from the rest of the colony.[8]

They did not recommend that rain gauges be kept throughout the colony, as he had hoped they would, but on the whole, Goyder must have been pleased. The line of reliable rainfall seemed to be meeting with understanding and favour.

Report on Victoria

Goyder's opportunity to present his ideas to the colony as a whole had come at the beginning of 1870, when he was asked to report on the effects of legislation permitting selection before survey in Victoria. As well as acting as a lure for South Australians wanting to escape the depression in their colony or in search of more accessible land, Victoria was also a place where the squatters had acquired greater wealth and power – and where corruption in the administration of lands had been a serious issue.

Goyder left Adelaide by ship on 17 February 1870 with his letters of introduction and arrived in Melbourne two days later. After meeting politicians and officials, he spent the next two weeks in the company of Alexander Skene, the surveyor general of Victoria. Skene conducted him on a tour which took them from Ballarat and the gold-rich areas to the north-west, down to Hamilton and Warrnambool in the south-west, then east to Geelong, and finally back to Melbourne. After a few days' rest the pair started off again, taking the Sydney Road as far as a German settlement north of Albury, and returning via Beechworth and Yackandandah, visiting towns on either side of the route as they went. On arriving back in Melbourne, Goyder learnt that he was urgently required in Adelaide. He returned immediately, taking the opportunity to observe 'another line of country' from Ballarat to the western border as he did so.[9]

In Adelaide, his report was being anticipated with interest, but because of pressure from the work he had been called back to address, it was nearly a month before he completed it. The report was presented on 23 April 1870, and it included his regrets that he'd had to complete the project in such haste. Much of the report was devoted to tracing the history of land legislation in Victoria and the impact of the various acts, but it concluded with a definite rejection of the practice of selection before survey, which Goyder damned as unjust and wasteful. It had unfortunate effects on the lives of individuals, on society, and, ultimately, on the land itself, he claimed: settlement became dispersed, children grew up without the benefit of education, and destructive farming practices were transmitted from generation to generation. Those in the know could 'pick the eyes' out of the land, enriching themselves 'at the expense of the entire country', while reducing the squatter's leased run to the 'shadow of a real property'. Not only, as he later put it, could a pastoralist 'wake up in the morning to find his woolshed pegged out', but he might also be harassed by a malicious selector in the hope of being bribed to go away.[10] 'To me,' he wrote, 'it appears infinitely preferable to divide the whole country, viz., that which is at present only fitted for pasture from the remainder, and to exclude it from any occupation save that for which it is really adapted ...' What was recognised as suitable agricultural land should not be wasted on grazing, but opened to the selector 'not all at once, but in districts – the selections, in fact, radiating from the centres, and the unselected portions remaining as common lands until the whole is absorbed'.[11]

Having set the context for his vision for South Australia, Goyder was then able to present his felicitious, if local, solution to the problem of conflict between farmers and pastoralists that had been a source of tension in the eastern colonies as well as across the Pacific in America. 'In South Australia,' he explained:

> nature has clearly established a line of demarcation beyond which permanency of tenure may be given to the pastoral tenant, without detriment to the agriculturalist; and a certain portion of Victoria, viz, the north-west, partakes of the same character, and, as to the soil and climate, is very similar to the north-east part of South Australia; and, it appeared to me that this, at the present time, purely pastoral country might fairly be reserved for this occupation with great advantage to all concerned.[12]

Unfortunately, even with his usual penetrating foresight, Goyder did not seem to realise the opportunity this report represented for disseminating his ideas widely, and did not explain either the nature of this natural demarcation, or how it could be recognised. He must have anticipated that this report, like all his important reports, would be reprinted in the daily papers and that his observations and reflections would be greeted with the usual widespread interest. Farming and agricultural settlement were the concerns of the day, and now that monopolising pastoralists were no longer seen as a threatening force, the ideal situation for reintroducing and explaining the Line to the broadest possible audience now presented itself. As it turned out, the report was of such interest that the *Advertiser* reprinted it as a pamphlet and sold it for a penny. But, because the report had been prepared in a rush at a time when Goyder's health was deteriorating, the opportunity to promote and explain the Line was missed. Instead, he seems to have addressed himself only to the government members and parliamentarians, who were his primary audience, and to have assumed that, by this time, his line was already well understood by them and that little needed to be said. Perhaps, too, he had some idea of how close he was to seeing his hopes fulfilled.

A novelist reports

In mid-1870 the agricultural area of Belalie – up to this point a station on the Canowie run – was drawn up. Including the country around what is now Jamestown, this was the area Goyder had long expressed concern about, particularly because of the withering impact of hot winds on maturing crops.[13] Although the squatters did what they could to buy up land, by December 1871 the Belalie area had been transformed from a run dotted with isolated shepherd's huts to a place in which 'the busy hum of reaping machines may be heard the livelong day on scores of farms'.[14] The effects of the Strangways Act and a good season in 1870–71 led to what was termed an 'earth hunger', manifested as the steady passage of farmers pressing on towards the Booyoolee area.[15]

As the golden tide gathered force, South Australia was visited by the English novelist Anthony Trollope, who was on a journalistic tour of inspection of the Australian colonies. The result was his book *Australia*, in which the Line had the honour of being presented in a chapter on agriculture in South Australia. Addressing his English readers, Trollope wrote:

I must first explain that South Australia is a country peculiarly subject to drought,–more so than are the other colonies,–and is especially so in the interior. This is a fact so well acknowledged, that all who know the colony are aware that wheat can only be grown in certain parts of it. In order that the government might have some guide to tell it what portions of the land it would be expedient to throw open to the agriculturalists, and from what portions it would be expedient to exclude them as being unfit for agricultural purposes, a line has been drawn. The surveyor-general, Mr. Goyder, has drawn an arbitrary line across the map of South Australia, which is now known as Goyder's line of rainfall. It is anything but a straight line. It runs from a point on the eastern confines of the colony somewhat south of the city of Adelaide, in a direction north-west nearly as high as the top of Spencer's Gulf. Then with irregular curves it comes south half way down the Gulf, which it crosses below Moonta and Wallaroo, and then runs north by east [*sic*] till it loses itself in unknown deserts. North of the line, or rather beyond it, no farmer should locate himself. South of this, or within it, he may expect sufficient rain to produce wheat. Of course, Mr. Goyder gives no guarantee as to precise accuracy, but I found it to be admitted in the colony that the line had been drawn with skill and truth.[16]

Trollope had evidently gained his information from someone who was in no doubt about the variable nature of the inland climate and who was confident that the need to confine wheat growing was widely accepted. While he was in South Australia, Trollope had been a guest of Thomas Elder, so he would have gathered his information from among the influential elite and probably directly from Goyder himself.[17] The phrase 'Mr. Goyder gives no guarantee' suggests a personal conversation, and Trollope's account of the Line is distinguished both for its omission of any reference to pastoral drought relief, and for being true, overall, to Goyder's intent. (Nevertheless, despite emphasising the problems of drought, Trollope, like others at the time, was not able to maintain Goyder's almost unwavering focus on variability as a critical measure of the rainfall, referring in the end only to 'sufficient rain to produce wheat'.)

If Goyder was indeed one of the sources for Trollope's claims, then Goyder had every reason to be not merely confident, but ebullient, about the future of his line. The Line had been included in Trollope's account of the state of agriculture in the colony because a Bill proposing to enshrine it in law was before the parliament, and, as Trollope advised his readers, it stood every chance of being passed. Trollope was right. Produced in response to problems with the agricultural areas and containing proposals which had been gaining support since 1870, the Bill became the *Waste Lands Alienation Act*, which came into force on 15 August 1872. Under its provisions, all the land south of the Line was to be surveyed and offered for selection. Incorporated as the First Schedule of the Act, the Line was taken to the northern bounds of the hundreds through which it passed – hundreds that Goyder, with this day in mind, had already drawn up, so that the northern bounds of the relevant hundreds expressed the Line as closely as a series of large rectangles could (with the exception of two hundreds on the western tip that had been drawn up as special surveys in the early years of settlement). During the passage of the Bill, not one voice had been raised in parliament against the limitation of the agricultural lands, or against the Line as the expression of that limit. Goyder's vision was at last being realised.

An 'arbitrary' line

Trollope had tried to convey to his readers that the Line was the product of human intelligence, rather than a naturally formed border like a river, by describing it as 'arbitrary', a term that is so heavily imbued with connotations of bureaucratic insensitivity, and even capriciousness, that his intention is not at first apparent. His choice of terminology was unusual. At the time, the word most frequently used to indicate that the Line was the product of human appraisal – even though it expressed a natural reality – was not 'arbitrary' but 'imaginary', a term that instead tends to suggest the made-up or fanciful. Both terms could be used derogatorily, and both eventually were.

Probably because he had described the Line as 'arbitrary', Trollope went on to make the seemingly curious comment that the Line was 'anything but … straight'. In Australia at the time, any course marked on a largely empty map was likely to take its creator's name and be termed a 'line'. The route taken by an explorer was his 'line', and in South Australia even a road laid out by W.S. Chauncy was known as 'Chauncy's line'.[18] But Trollope's audience was in England, and expectations there of what might constitute someone's 'line' were likely to have been influenced by such things as the navigational grid of latitude and longitude, which had enabled the expansion of empire and mass migration to the colonies; the often straight borders that were imposed on the colonised lands; and the growing network of railway and telegraph lines. Against this background, Trollope must have guessed that many of his readers would suppose that an 'arbitrary', humanly created line would inevitably be straight, unless told otherwise; perhaps he had expected a straight line himself.

At the same time that Goyder had been confronting Australia's inland climate, another line, a major natural boundary, was being identified to the north of Australia, and this, too, would be named after its creator. As a result of his travels in Borneo and Sulawesi during the 1850s, the naturalist Alfred Russel Wallace had realised that species came into existence as a result of a process of evolution, shaped by natural selection (prompting Darwin to speed up his own work and publish *On the Origin of Species*). But Wallace did not stop at that remarkable insight. In 1858 he wrote that he had discerned a boundary in the islands of the Malay Archipelago between two distinct fauna. At the western end of the archipelago the Asian fauna included large mammals such as tigers, but no marsupials, and few parrot species among the birds, while the Australian-type fauna of the eastern end consisted almost wholly of marsupials, with a rich array of large parrots. The line dividing the two regions – Wallace's Line – made its first public appearance in 1863 when its author presented a paper to the Royal Geographical Society. From this point Wallace took yet another major step: he realised that the obvious sharp change between the species of Bali and nearby Lombok, where there was no impassable natural barrier to cause the species to develop separately, indicated that there had once been an impassable boundary, but a very long time ago. In fact, Wallace's

Line identifies the area where the Australian plate, having separated from Gondwanaland and set sail for the north, collides with the Eurasian plate. In this region two separately developed fauna mix.

Wallace's Line and Goyder's Line made their appearance within a space of three years in the early 1860s, and although very different, the lines did have something in common: both interpreted borders discerned in the living world as pointers to physical realities that were not directly observable. However, while Wallace was operating in a scientific context in which the bases and implications of his line-drawing were a necessary part of the debate, Goyder was not. As surveyor general, Goyder was principally responsible for the division of land for sale, and that role blocked any real recognition that he was making a major contribution to the understanding of the Australian climate.

The great question

Trollope's opinion of the South Australian wheat farmer, or 'cockatoo' is also likely to have been the product of a conversation with Goyder. On this subject he announced confidently that there could be:

> no doubt that the cockatoo of South Australia is a very bad farmer, – and that he is so because he has hitherto been able to make a living by bad farming ... The ordinary cockatoo knows nothing of the word fallow, and attempts to produce nothing but wheat. Year after year he puts in his seed upon the same acreage, and year after year he takes off his crop ... He does his work without any attempt to collect manure, or to give back to the land anything in return for that which he takes from it. He even burns his stubble ...[19]

The result was soil that Goyder described as 'cropped out'.[20]

While surface mining was booming, a more considered approach to agriculture was also coming into its own. On 9 February 1874, for instance, the leading article in the *Advertiser* concerned lucerne as a 'green manure' and was the sixth instalment of a series, 'Science in Agriculture'. It was accompanied by a report on the harvest in the north. The article began by remarking on the wonderful progress in the north and the conversion of sheep runs to fields laden with golden grain. It reviewed the history of false prophecy that had attempted to limit the wheat lands. The paper then boldly suggested that in future, South Australia would not merely lie to the north of Adelaide, as was commonly considered, but to the north of Clare. The local member was reported as having told a crowd at Port Pirie that the rainfall was as great at Clare as at Mount Barker (in the hills south-east of Adelaide), and as great at Mount Remarkable as at Clare. The 'great question', the paper acknowledged, was whether or not dependence could be placed on that 'succession of genial showers' – the kind that fall on beaming cultivated land – needed to bring the crops to perfection. 'Should it turn out that the North is no more subject to drought than the South,' the paper proclaimed, 'the brightest dreams of the most sanguine may be realised; but only after the experience of a series of years can anything be confidently affirmed. With high hope for the future we await the result.' If the public voices in Adelaide were humming with measured but impatient prudence as they waited to see what the future would bring, further north, in Clare, the wait was considered to be over. There, each day was seen to constitute additional proof that South Australia was destined to become 'one of the finest agricultural colonies in the world'.[21] At Port Pirie, where carts stacked high with bags of wheat were being hauled to the ships, it must have seemed self-evident.

Urged on by the editors of the local papers, the farmers held meetings to protest against the restriction imposed by the Line. Their actions and attitudes, and the consequent changes, show how shallow and restricted had been the acceptance accorded to what Goyder had considered a 'clearly established' natural boundary, how few people there were with any real experience or insight into the inland country and climate, and how quickly and completely memories of the massive drought of 1864–67 had been forgotten or dismissed. Within a very short space of time, the idea of a limit to the agricultural land suddenly seemed nonsensical and doubts about the rainfall a trifling concern. In May 1874, only two years after the passing of the *Waste Lands Alienation Act*, a petition reached the parliament requesting that further land be opened adjacent to the hundreds forming the eastern flank of the central section of the Line. This is the waterless area of poor soil beyond the road which now neatly divides the wheat lands to the west from the saltbush to the east, and even the *Farmers' Weekly Messenger* departed from its usual promotion of expansion to oppose this idea.[22] Significantly, the petitioners did not rely simply on pointing to the steady rainfall and rich harvests that had been experienced in recent years. In addition they argued that:

> the question of the suitability or otherwise of this land for agriculture should not rest on the dictum of one individual; it being notorious that much of the land now under culture and growing good crops, has been formally pronounced totally unsuitable for that purpose by people apparently well-qualified to form an opinion.[23]

Claims about limits to the agricultural land had been discredited in the past by the efforts of 'practical' farmers, the argument ran, and once again the

surveyor general's *ipse dixit* was proving a stumbling block. Just as it had during
the valuations, Goyder's judgment prompted public meetings throughout the
rural areas, but this time the meetings invariably expressed opposition. The
notion of a rainfall line was ridiculed, and, since the Line secured the tenure
of the leases to the north, claims were made that it was not merely 'imaginary',
but a corrupt fabrication: 'the most absurd thing known in squatter ingenuity'.[24]

A 'supposititious' line

It did not take the government long to respond. The chief secretary and leader
of the government at the time was a thoroughly urban figure, Arthur Blyth, an
ironmonger (like his brother and fellow parliamentarian Neville Blyth), and
an influential figure in business. In mid-1865 he had been on the committee
which had heard Goyder outline his notion of a line of reliable rainfall to define
the limit of the agricultural land – not that he appeared to understand the
concept. In mid-June he introduced a Bill to amend the 1872 Act. One of its
three goals was to dispense with Goyder's 'arbitrary' line. Land would then be
surveyed beyond the Line as it was required. Despite having been profoundly
impressed by Goyder's work in the South-East, Ebenezer Ward offered his
support. Now the member for Gumeracha, and editor and publisher of the
Farmers' Weekly Messenger, which set out to rival the *Chronicle* and the *Observer*,
the weeklies published by the *Advertiser* and the *Register*, Ward assured Blyth
that he was 'perfectly right' in seeking to abolish 'that supposititious line known
as "Goyder's rainfall"'.[25] In his second reading speech, Blyth presented the Line
strictly in terms of its first official appearance, as an administrative tool associated with pastoralism – as if the climatic events that affected pastoralists were
of no relevance to farmers – and implied that all the attempts at agricultural
containment were the work of squatters attempting to preserve their realms.
The Line, he said, had been:

> laid down on the map entirely in connection with the pastoral lessees of the
> Crown. It was a little singular that the question of what was pastoral and
> what was agricultural land had been gradually extending year after year.
> He was old enough a colonist to know the time when it was asserted that
> the land north of the Para River was not fit for cultivation, and could not
> be ploughed. Subsequently agricultural settlement was extended to Gawler,
> and they had gone on step by step until a feeling was entertained two years
> ago that the line of rainfall laid down by the Surveyor-General for defining
> the classes amongst pastoral lessees might be fairly taken as the limit of the
> agricultural land ...[26]

This misunderstanding was the consequence of Goyder's failure to openly clarify the Line's real nature and purpose when it was first drawn in 1865 and again at the time when he explained the Line in the context of his report on the Victorian land regulations. The very idea of a line seemed ludicrous and any ideas about limits on wheat cultivation were 'mere theoretical baubles'.[27] In the *Farmers' Weekly Messenger*, a man who had never before been north of Clare pontificated on the excellence of the soil and the abundance of water north of the Line, and sneered:

> I came into the store at Pekina, on the other side of the 'rainfall', on Tuesday, May 4, about 9 o'clock in the forenoon, wet to the skin, and it rained steadily all that day and night and part of the next day, and I defy Mr. Goyder or any other man to say at which side of the hedge the most rain fell.[28]

Clerk of the Weather by the inland sea

Interpreting the Line in such a superficial way and then dismissing it for being absurd was nothing new: this was an approach pioneered and polished by Eustace Mitford when the Line had first appeared, although Mitford's target, of course, had been Goyder – the Line had simply been material associated with Goyder that could be easily exploited, as in this passage:

> Have you seen Lyon's new map of the colony? – it's the best going. There you will observe a broad red line drawn in a zig-zag from the east to the west boundary of South Australia. 'What's that Mr. Lyons?' says I. 'Goyder's rain-line,' says he. – 'On what authority?' says I. 'Chief Secretary,' says he. – 'On which side does it rain?' says I. 'On whichever side you like,' says he; 'sometimes on one, and sometimes on the other, according to circumstances.' – 'Is it of any use?' says I. 'Yes,' says he; 'about as much as the fifth wheel of a coach, or the mile-posts on the Clarendon-road.'[29]

Earlier, he had dismissed the Line by asserting that 'rain had effaced every vestige of that barrier', a joke that was repeated in different guises.[30] Mitford's attacks also demonstrated that the misconception that the Line primarily expressed annual rainfall was present from the beginning.

Mitford's other main approach in using the Line to discredit Goyder had been to cast him as 'Clerk of the Weather', a self-inflated bureaucrat who 'absolutely defined the limits of God's power – virtually told the Almighty to put his head in a bag ...'[31] The inspiration for this title seems to have come from another source. In October 1865, as a late contribution to the debate over the valuations, John Chewings of Woorkoongoree had written to the *Observer* protesting about the impact of the valuations on the northern squatters; very little rain had fallen after Goyder's visit. As well as sole valuator of runs, he declared, the government should have proclaimed Goyder 'Clerk of the Weather', with the power to grant the rain needed to match the valuation.[32] It was just the sort of thing to appeal to Mitford, and in the issue of *Pasquin* of 6 July 1867, under the sub-heading 'Clerk of the Weather!', Mitford played with the idea for over a column. Goyder was likened to King Canute, delineating the limits of the ocean, and Mitford suggested that the 'Summary' for England prepared by another paper (the dailies prepared regular summary issues to send home) should advise that they had left off praying for rain, or for anything: rather 'we "delineate the limits", which is more convenient for all purposes – law and lessees, religion and the rainfall.' This was followed by a hymn – 'an original air, found at the fag-end of a hot wind' – composed in honour of the Clerk of the Weather:

> The sea, the sea! The Inland Sea!
> I am where I ought never to be,
> Counting dead sheep in the North Countree;
> With heat above, and dust below,
> And ruined stations wherever I go,
> And nothing left but a half-starved crow,
> And the limits of Rain laid down by me!
> If the drought has come and killed the sheep,
> What matter? my billet I shall keep;
> I can gammon the whole of the Ministree.
> There's Grenfell-street [the *Advertiser*] and the Clerk of the Weather,
> Did you ever see two such rogue[s] together?
> There's the Clerk of the Weather and Grenfell-street,
> And the devil to pay when these two meet.
> We're both afloat on my Inland Sea,
> And we paddle our own canoes, for a spree,
> In the limits of rain laid down by me!
> Amidst oceans of dust on the Inland Sea!

Evidence that Mitford's satires did influence people's perceptions – or at least gave some of them words to attach to their irritation with Goyder – is provided by John Baker, a leader of the attack on the valuations. In parliament in 1870, Baker remarked with satisfaction that the days when 'Goyder was King' were over.[33] It was in *Pasquin*, of course, that Goyder had been derisively crowned.

The steady lapping of the inland sea must also have undermined Goyder's position as an interpreter of climate, and this was raised in papers apart from *Pasquin*. In 1868 the *Observer* and the *Register* published a series of original stories. One, by an anonymous author, titled 'The Inland Sea', began with the statement that nothing could convince 'Jones, of the Survey Department', that he was 'wrong in his belief that a vast inland sea was waiting to be discovered in the northern part of South Australia'.[34] The 'clues' to the writer's source of inspiration were more like billboards: apart from the conjunction of the 'inland sea' with the Survey Department, there was also the matter of names. Jones is the stereotypical Welsh name, and the Welsh origins of the name 'Goyder' would have been quite evident to the colonists. Jones, however, was not Goyder in disguise. Rather, the author seems to have drawn on Goyder for a pool of characteristics. Accompanying Jones was a Captain Kinlock, who constantly made notes and wrote reports about the landscape and vegetation and who was devoted to homoeopathic medicine (which had strong adherents in the New Church). In the story – a heavy-handed satire on the events of 1857 – Jones wakes up in the country around Mount Bryan to discover that it has undergone an overnight transformation into a sort of Lake District.

The evidence provided by a voice such as Mitford's, irrespective of his personal animosity toward Goyder, also suggests that the attempt to deploy an unfamiliar climatic limit was doomed, not only because of simple ignorance and the lack of conceptual preparedness, but because of inchoate and hence largely unchallenged assumptions that were powerfully at work in the popular imagination. In September 1868, Mitford took it upon himself to review a textbook that had been prepared by Julian Tenison-Woods, *Geography for Australian Catholic Children*. 'We had no idea that Australia was such a miserable invention as Mr. Woods declares it to be', Mitford complained. 'He says Australia is nearly all sandy scrub, without a rainfall; that the country is fit for habitation, but will not support an extensive population ...' Allowing for his characteristic wild exaggeration, Mitford was probably speaking for a substantial proportion of the settlers when he repudiated this assessment. 'Our experience, so far,' Mitford asserted:

> serves to prove that Australia is capable of supporting a larger population per square mile than any country under the sun – only give the people a chance of turning the lands to advantage.[35]

What is expressed here is the underlying belief that human beings, not geography and climate, were the dominant factor and that, if people worked hard enough, the world could be modified to meet their needs. As the

1870s progressed, this widely held assumption crystallised around the idea, widespread at the time, that through agriculture and tree planting settlers could change the climate to suit themselves.

Thinking things through

From the current perspective of a world facing a whole raft of major environmental threats – global warming, ocean acidification, massive loss of biodiversity and instability of ecosystems, and the increasing scarcity of many resources, including water, oil, and important minerals – the debate about Goyder's Line has a familiar ring to it. To listen to the voices of the 1870s is to experience a curious kind of *déjà vu*. The dismissal of Goyder's Line finds a parallel in complaints about constant environmental doom-mongering. The man at Pekina who defied anyone to say 'on which side of the hedge the rain fell most' seems to prefigure the outbursts of scepticism about global warming that accompany a violent cold spell, despite predictions of increasingly extreme weather of all kinds. The unwillingness to allow a limit to be set by one man, even though he was uniquely experienced and uniquely and impartially committed to identifying the extent of the safe agricultural land, parallels the continuing unwillingness in our times to be guided by the advice of comparatively small numbers of expert scientists; while the suggestion that the Line had been devised to protect the interests of the northern squatters is matched by charges that scientists (climate scientists in particular) are in some way partial, and are really concerned with advancing their own interests.

But the real similarities lie deeper. Goyder was attempting to protect the settlers from an aspect of the climate, which, as global warming is to us, was new to their experience and unrecognised in their culture, and which could not be experienced in a simple, immediate way. His concern was climate, not weather. But, given the force of habitual thinking and established practice, people appear to have been unable to grasp the issue and either dismissed his concerns as unreal, or simply ignored them. Related to this is what seems to be a refusal, also deeply rooted in culture, to acknowledge and respect limits to growth. Mitford's claim that Australia was capable of supporting a larger population than anywhere else on earth 'only give the people a chance' seems almost comical now, but nevertheless, in its basic assumptions, it represents an element of how some, even today, envision the world and our place in it. The assumption of limitless economic growth on a planet with finite resources is the example that looms largest. A similar basic refusal to countenance the idea of a limit seems to have militated against the acceptance of the Line, especially in a community caught up in a land rush fuelled by the unpretentious needs and dreams of thousands of small farmers or would-be farmers.

The question of how a limit could be determined appeared to have been addressed when it was asserted that it should be up to practical farmers to determine what was, and what was not, farming land.[36] But in the context of a compulsive 'earth hunger', allowing the farmers to settle the matter themselves amounted to granting a licence to abandon all prudence. This, in a colony, it must not be forgotten, which had once believed itself contained in the north by a vast salt lake and in which, during the drought of the mid-1860s, there had been discussion of abandoning the north as unsuitable even for pastoralism. Allowing ordinary people to conduct a mass experiment was certainly a way of working things out – but a way that recalls the American poet Robert Frost's grim assertion that:

> Society can never think things out:
> It has to see them
> Acted out by actors
> Devoted actors at a sacrifice.[37]

Unfortunately, the sacrificial actors included not only the farmers, but the whole finely tuned ecology of the lands brought under the plough.

Caveat emptor

As the Line was swept away, the issue of government responsibility, which had nagged at Goyder from the beginning, was also brushed aside. As well as invoking the early claims about the limits of agricultural land that had been proved false, when Blyth developed his argument against the Line he also argued that when the 1872 Act was passed there had been little sense that the Line would be 'crippling' and would prevent the government from meeting the 'bona fide demand for land'. This was an argument that introduced a completely different implicit definition of agricultural land, one by which agricultural land was simply land demanded from the government by actual or would-be agriculturalists. As Blyth had put it earlier, this made the 'whole colony north of Goyder's Line ... one agricultural area'.[38] He claimed that:

> it was no use for the Government to say, 'This land is suitable and that and is not,' but that the duty of the Government was to survey the land and let the people select the land which they considered would be suitable for their purposes.[39]

The sudden adoption of a laissez-faire approach was a complete reversal of the protective, ordered policies of the earlier governments, and later it would become clear that many small farmers had not grasped the change and did not understand that they were being set free to conduct an experiment on behalf of the community, but at their own expense. At the time, this profound change in approach passed without comment. The focus was on the immediate demand for land.

In the course of the debate, the government asked Goyder to comment on the amount of land left within the Line and the prospect of opening up new hundreds beyond it. He took the opportunity to reiterate his concerns to the commissioner and the chief secretary. 'Beyond the limit of the First Schedule,' he explained:

> from Melrose [Mount Remarkable] northward and north-easterly, the land, except in the ranges, is mostly good agricultural soil: its extent is very great, but the rainfall, hitherto, has not been reliable – the result of farming operations on such land is therefore doubtful.[40]

When his words were read aloud in the Upper House they provoked laughter and cries of, 'Hear, hear'. By November, Blyth was talking of 'Goyder's absurd line of rainfall', and repeatedly asserting that it had been prepared for 'another object', and had nothing to do with the present situation.[41] On 6 November 1874 the *Waste Lands Amendment Act* was passed and the Line was abolished. It was the end of Goyder's plan for a climatically rational pattern of land use in the colony, and his disappointment must have been profound. He could foresee the ultimate cost of the amendments and he knew that all that had been said in parliament about the Line being absurd and created for another purpose was politically expedient blather. Years later, he told a major inquiry that members of the government had thought it unlikely that the northern country could be farmed successfully, but had released it anyway because 'there was such a clamour ... that no Ministry could remain in office that declined to bring the land into the market'.[42] Just as he had been unwilling to berate the northern squatters for overstocking their runs before the 1865 drought because of the 'deceptive' nature of the country, Goyder seems to have been resigned to political reality and unwilling to blame the politicians. His deputy, William Strawbridge, who also gave evidence, put the matter more bluntly, stating that the government had known that the northern lands meant almost certain ruin to farmers. This being so, the government had not merely withdrawn from the responsibility of ensuring that land it sold to settlers was suitable for the purpose it was sold for, it had abrogated responsibility altogether. If the small farmers were ignorant of the dangers they faced and the meaning of Goyder's Line, it was not solely because Goyder had failed to exploit the opportunity that had been available in 1870 to tell them; it was chiefly because the government had cynically chosen not even to attempt to explain.

Three days after the Line was withdrawn, Goyder set off with the commissioner for what the *Advertiser* described as a journey 'through the pick of the farming country in the North' – out to World's End and as far north as Arkaba, Wilpena and Kanyaka – and then south to the foot of the Yorke Peninsula.[43] Viewing the Willochra Plain where, in 1859, Goyder's feelings had been touched by the plight of his exhausted, dehydrated horses struggling to remain on their feet in the fierce heat of a dust storm, the irony of his situation must have hit hard. In 1857 he had been mocked for having reported a mirage, and in 1864 he had been attacked for supposedly having been misled by the grass that had been 'waving on Boolcunda' (a run near the Mud Hut). Now, in effect, he was being mocked for trying to tell others that the water couldn't be relied on and that the growth wouldn't last, while the only future the government was concerned about was its own. Perhaps one of the most remarkable things about Goyder was that he never succumbed to bitterness, or became cynical himself.

In the middle of the year a poem appeared in the columns of the *Register*, 'The old shepherd's lament on the settlement of Belalie', apparently penned by a shepherd and not a conceit adopted by an urban poet. For the baffled old shepherd it truly was the loss of a pastoral idyll – 'How sweet it was in former days, / How pleasant on Belalie!' In his view the land did not delight with the touch of the plough; rather, 'Nature mourns her alter'd home / And weeps o'er cultivation'. In a string of verses he recorded the destruction of eucalypts and pines as the land was cleared, while the birds that had dwelt in them had vanished, along with the emus and kangaroos of the grassy plains. The disturbance was profound:

> The ridges in the creek are gone,
> The very springs are failing;
> Everything has lost its tone,
> And Nature's self seems ailing.

Like the farmers, the shepherd, too, had a complaint to make to Goyder:

> Ah, Goyder! were the southern plains
> So circumspect and narrow,

That – woe is me! 'twas in the North
That you let fly your arrow.[44]

Hot winds or no hot winds, north or south of the Line; by the end of 1874 it didn't matter.

CHAPTER FOURTEEN

Following the plough

'It was like a garden of Eden when we first saw it.'
– a pioneer description of the area around Amyton
[a town that no longer exists] on the Willowie Plain.[1]

For six years after the Line was withdrawn, from 1874 until 1880, Goyder had the task of overseeing what has been described as 'the most outstanding episode in the history of South Australia before the twentieth century'.[2] As people poured into the north, the numbers increasing every year, it was his responsibility to anticipate the demand, give pastoralists the necessary warning that their land would be required, plan out the new counties and hundreds, and send teams of surveyors out to ensure that a steady supply of surveyed agricultural lots was available for purchase. All this he did with characteristic efficiency, despite his own beliefs about the wisdom of the enterprise.

In the space of a few years vast areas were transformed into wheat lands, with frontier districts extending not only into the north, but across the central and lower Yorke Peninsula, into the scrublands from the Para River to Snowtown, and on land to the east of the Barossa Valley. However, the north-ward movement was the most dramatic, since it involved the greatest number of people and the largest expanse of country. Between 1871 and 1876, population in the districts from the Broughton River to Orroroo increased from 2000 to 12,000 and that was inside the Line. In the second half of the decade, the dramatic expansion continued beyond it – from 1876 to 1881 the population in the districts from Port Germein (outside the Line and close to its northernmost point) to Hawker (very far north of the Line) increased from 6000 to 21,000.[3]

Many of those who moved north were farmers whose 80-acre sections (the standard size in the early surveys) in the established farming districts were cropped out and no longer able to support their families. Not only did credit selection allow them access to new land, it enabled them to select much more land – up to 640 acres – and, later, a thousand acres. The procession formed by a farming family on the move was a familiar spectacle. Leading the way, on a wagon loaded with the furniture and the winnower, was the farmer, followed by the older children or hired hands driving a dray, behind which

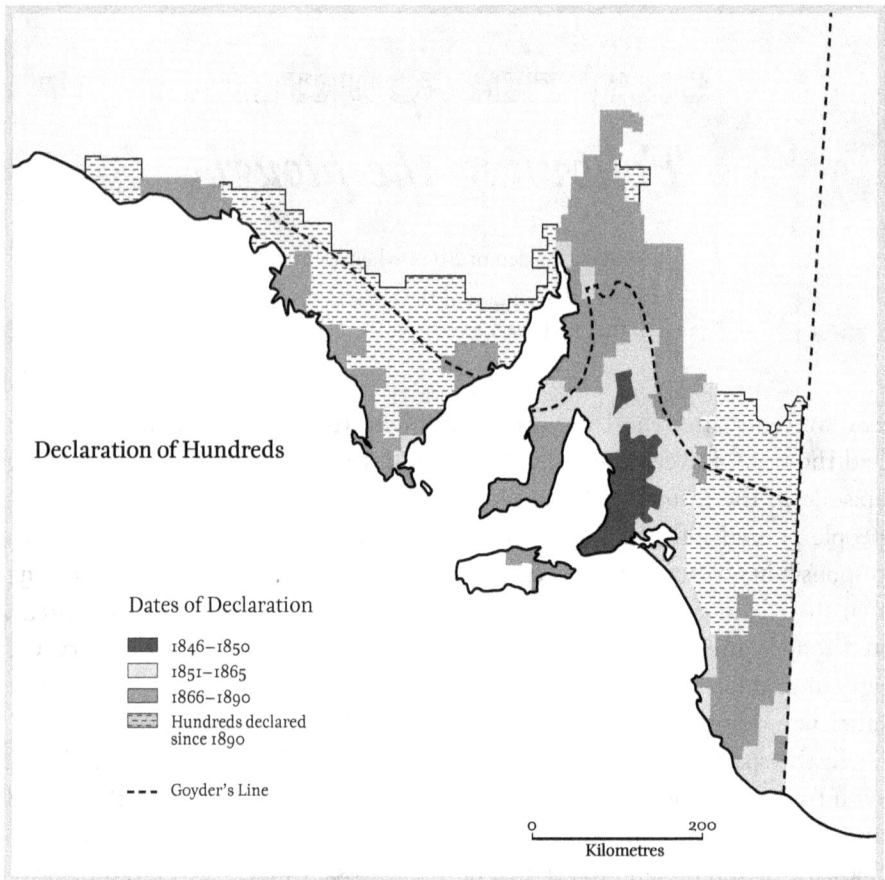

Declaration of Hundreds

Dates of Declaration

- ■ 1846–1850
- ☐ 1851–1865
- ▨ 1866–1890
- ▦ Hundreds declared
  since 1890

- - - Goyder's Line

0   200
Kilometres

The survey and occupation of South Australia:
Declaration of hundreds to 1890.

came the reaping machine. The farmer's wife and younger children brought up the rear in a spring cart. In some old farming areas, the loss was such that townships went into decline, and some of those who had provided services decided to follow the farmers north.[4] The sons of the 80-acre farmers were also compelled to take up land in the north if they wanted to continue with farming themselves.

These emigrants from the south were largely the 'practical farmers' – the farmers whose experience would decide how far the wheat lands extended. However, not all those pressing at the frontiers were farmers, and the proportion of non-farmers increased the further north settlement spread. In the hundred of Arkaba in the area around Hawker in the Flinders Ranges – Arkaba was a pastoral run which still survives – 40 per cent of the selectors were not farmers and had no significant farming experience in the south or anywhere else. Those without farming experience represented a wide variety of backgrounds, although labourers and teamsters made up significant proportions – 20% and 10% respectively. Fifteen per cent were single women (including widows).[5] Some would have been immigrants – during this period immigration expanded the population of the colony.

In 1873, in an attempt to lure immigrants to its shores, South Australia had reintroduced a subsidised migration scheme, which aided over 25,000 new arrivals over the following eight years. At the same time the government commissioned the journalist William Harcus to write a handbook for intending emigrants. Three years later, he was commissioned again, this time to produce a history of the province and its 'resources and productions' for an international exhibition. To do this, Harcus supplemented his original material with chapters by experts on various subjects. The result was a lengthy promotional brochure. In his chapter on agriculture – an undisguised invitation to come to South Australia and enjoy the good life as a wheat farmer – Harcus mentioned the history of claims that the colony's resources of agricultural land were severely limited and cheerfully admitted that the province's farmers were 'slovenly' by English standards. But all was well: their methods were adapted to the new conditions, the soil brought forth year after year regardless, and there were 'many farmers who have grown rich in this way'. The expression, 'tickling the soil with a hoe', was 'almost literally true here', Harcus claimed.[6] His book also provided prospective settlers with an introduction to the colony's land laws and a guide to the terms of credit selection. While Harcus had to admit that there was no selection before survey, he assured his readers that:

> an efficient staff of survey officers is always at work surveying the land as fast as it is required. Hundreds of thousands of acres are always open for selection, and the work of the surveyors is still going forward.[7]

What Harcus didn't report was that the surveyor general was consistently trying to warn farmers off the northern lands. The law required that, as each hundred became available, a description of the land had to be provided, and Goyder used these descriptions to post his warnings about the unreliable rainfall. The prospective immigrants heard nothing of this. What Harcus articulated was a glowing vision of the future of the north:

> I have heard said that the present law has worked with singular success. Immense acres of land in the North have been surveyed and offered for sale

on credit. Half-a-dozen years ago most of this land was used as sheep runs – supporting a dozen or a score of persons. Now it is covered with smiling homesteads and prosperous farms, on which many hundreds of families are settled, with every prospect of future success. In the course of a few years these farms will be freehold estates of a steady and intelligent class of farmers, farming their own land, who will constitute the pith and the strength of the Colony. A few thousand farmers, each farming his own freehold estate of a square mile, or a thousand acres, would form an independent and prosperous class, of which any country may feel proud.[8]

An engraved illustration was provided. Titled *Breaking new ground*, it showed over 20 ploughing teams on a flat plain, with hills in the background. To ensure that the scale of the landscape was understood, the rapidly diminishing sizes of the draughthorses indicated the extent of the plain – a plain where the plough encountered no obstruction. Unfortunately, the intended vigour and liveliness of the scene was not captured by the engraver, who had worked strictly from sketches of teams at rest. As a consequence, right across the plain men and horses alike appear to have paused simultaneously for a moment of reflection, like a partially mechanised version of Millet's peasants, heads bowed for the Angelus. Not a single foot is raised. Still, the result accorded with the overall message – land was cheap, terms were easy, and while there was real work to be done, there was no doubt or anxiety about the quality of the newly opened lands.

Into the saltbush

The collective drive to farm the north was inevitably hard work for the surveyors, and certainly imposed stress on the squatters. As Goyder himself explained, it had been the practice of the department to survey land only as it was requested by intending purchasers, but 'under ministerial and Legislative instruction the surveys [had] been gradually increased, irrespective of applications, from [a] rate of 380,000 acres per annum in 1870, to 1,000,000 acres per annum in 1876'. It was planned to increase this to 1,500,000 acres in the following financial year.[9] Twelve of the department's 26 survey parties were already in the country 'north and east of Melrose' – Goyder's code for the disputed country beyond the peak of his line – by July 1876.

Goyder was in no doubt that he was required to oversee an act of destructive madness, and when asked to provide a reply to a question raised in the House of Assembly about a deputation to the commissioner of crown lands and letter of protest from the squatters, he took the opportunity to put his views fluently and forcefully. The squatters were complaining of the 'severe manner' in which they were being pressed 'for the benefit of the settlers', he reported and had warned that:

> the resumptions include salt-bush lands, which, from the character of the vegetation, indicate a different climate, and coupled with the limited rainfall known to obtain, do not justify the attempt to grow wheat.

Some of the leases to be resumed had actually been extended as compensation for losses in the 1865 drought, and on one of the runs, so little grass was available that 50 pounds a week was being spent on fodder for horses. Where such is the case, he explained, 'the country is manifestly unfit for profitable agricultural settlement'.[10] His reply seems to have been the only occasion on which Goyder made plain that the dominating presence of saltbush indicated a different type of climate, not just low rainfall.

After stressing that the Survey Department was the faithful and effective executor of government policy, Goyder made his own beliefs unmistakably clear. He supported the squatters' request for a postponement of further resumptions until the agricultural viability of the lands already resumed was established, warned repeatedly about limited and unreliable rainfall, and asked the government to consider evidence of his own – since often quoted. 'During the last twenty years', he warned:

> I have crossed and recrossed the country in question during all seasons of the year, and have seen the surface in good seasons like a hayfield, teeming with rich, rank, and luxurious vegetation; and during the drought destitute of grass and herbage, the surface soil dried by the intense heat, in places broken and pulverised by the passage of stock, and formed by the action of the wind into miniature hummocks, surrounding the closely-cropped stumps of salt bushes, & c., and the soil blown away in places to a depth of several inches, the drift covering the fences of yards, troughs, & c., and so denuded of feed as to be altogether useless for stock of any description. Had the soil been ploughed at the time the whole of the depth of the furrow must inevitably have been blown away.[11]

Goyder also made the point that over one-half of the credit selectors who had completed their purchase to the end of 1875 had then sold their land – implying that what would be postponed, if the squatters' request was granted, was not permanent settlement, but a form of plough-and-run profit-making that would ruin good pastoral country. It was a confirmation of his 1870 prediction about the impact of very liberal terms of credit, although he did not say so.

In counselling delay, both Goyder and the squatters had good reason to believe that the climate was about to intervene to support their case. There was already talk of a 'present drought', with the season of 1876–77 yielding poor returns, prompting the commissioner, John Carr, a former pastoralist, to accompany Goyder as he visited selectors in the Terowie district, at the edge of settlement in the north-east. The farmers claimed that they were already over-extended and would be ruined by another bad season.[12] The journey was to little avail. Carr lost his position in October 1877, and early the following year six new counties were proclaimed. With their proclamation, two-thirds of the area surveyed since the introduction of credit selection lay outside the Line.[13] The government's only response to Goyder's warnings was to authorise the establishment of an experimental farm. Goyder chose a spot at Manna Hill, in the far north-east of Terowie, on the way to the Barrier Range, well beyond the existing agricultural frontier.

'Rain follows the plough'

While Goyder and the northern squatters hoped that delay would demonstrate that their warnings were not without foundation, the farmers were demanding the opportunity to surge north on precisely the opposite grounds: they believed that the good agricultural seasons of the 1870s were not manifestations of varia-bility, but a real change in the climate caused by the practice of agriculture itself. In the United States, a similar drama was unfolding as settlers moving west during the 1860s and 1870s encountered the arid and semi-arid expanses of the Great Plains. This monumental western march was given legitimacy around the 1880s when a speculative town builder and amateur scientist, Charles Dana Wilber, began to promote his view that 'rain followed the plough' – an expression that had already been used in South Australia in the late 1870s and was being abandoned there as Wilber was promoting it in the United States. Interpreted as a projection of the American notion of Manifest Destiny, Wilber claimed that it was a divinely imposed duty to keep moving west until the whole continent was settled, believing that God had created the deserts capable of transformation to farmland: 'so that in reality there is no desert anywhere except by man's permission or neglect'.[14] His claim makes explicit Mitford's similar, although secular, assertion that the land in Australia could support a massive population 'only give the people a chance'. Logically, the idea that the climate was changing did not support the belief that the Line could be abandoned because there was no natural limit to the wheat lands (or only a very remote one), since if the suitable country (land and climate together) was almost unlimited, the climate did not need to change. In practice, the two views worked in tandem to promote the movement north, and both were linked with another view, that establishing trees could significantly increase the rainfall. In late 1875, a year after the Line had been repealed and the first season after the lands beyond Melrose had been opened, heavy rains had threatened, and partly damaged, the harvest. On the last day of the year the *Farmers' Weekly Messenger* trumpeted that a change of season had been established and suggested that in future there might be more need to worry about flood than drought.[15] It was as if the words 'reliable' and 'unreliable' had either never been uttered or were incomprehensible items drawn from the lexicon of an unknown tongue.

After the dry year of 1876–77, there was heavy rain in early 1878, and throughout the north-east creeks were in flood, and lush waving grass was in abundance. Seduced by the promise of sufficient rain and with the evidence of rich pastures, the farmers settled further north, until another boundary appeared. It was the one the squatters had warned of and it was visible in the way that Goyder's imaginary line was not. The farmers had reached the saltbush. The editor of the *Port Augusta Despatch*, already excited by the pros-pect of the port handling a vast array of produce – wool, copper and wheat – from a bounteous hinterland, argued against this newly identified barrier, the 'unfounded prejudice against Salt Bush Country'.[16] By this time however even those who supported the agricultural expansion were describing the movement into the remote north as folly. The railway line that would terminate at Farina was under construction, although when Goyder had first recommended that such a line be built, it had been to enable minerals and wool to be brought to the port.

An invitation to recant

The last good northern harvest for many years to come was that of 1878–79. In mid-1879, a select committee of the House of Assembly began to invesigate the operation of credit selection, although its concerns were purely administra-tive. There appeared to be no alarm or anxiety about forthcoming seasons and the felicitous situation seemed to have proved Goyder wrong. Inevitably, he was called before the committee and asked if he thought the present agricul-tural settlement – now extended far beyond his line – would be permanent. Surprisingly, he agreed, explaining:

> I think it will be permanent, because I think the pressure of circumstances will be so great as to induce the farming population to consider the subject, and adopt a proper system of agriculture. There is no question as to the quality of land, it is simply a question of the method to deal with the land,

and after a time, when we have absorbed a large area, and there is no new ground to break, the farming population will begin to do what is really right. We need not, however, look for this new era until the greater portion of the land is separated from the Crown. It eventually will come, however, and, in my opinion, the agricultural settlement of the land will be permanent.[17]

In his answer, Goyder had shifted the emphasis of concern to the practice of 'cropping out' the land and moving on, which credit selection permitted, but the chairman, F.E.H.W. Krichauff, was looking for Goyder to admit that he had changed his opinion about the rainfall. Krichauff, a liberal parliamentarian of German origin, had a strong interest in land issues. He asked Goyder directly if he thought the farmers would be able to hold on beyond his 'former line of rainfall'. 'I think so,' Goyder replied, 'with the combination of agriculture and grazing ... it is essential in precarious districts that selectors should have sufficient grazing land to tide over a bad season – the loss of a crop'.[18] He continued to stress that selectors would need to overcome their ignorance of good agricultural practice and to begin to return to the earth what they had taken.

Goyder only appeared to recant; this was a screen erected for the purpose of warding off attack. While he had acquiesced in having his line tidied away into the past and had accepted the reality of settlement, his beliefs about the climate had not changed at all. He still expected disastrously bad seasons, and he evidently did not believe that farmers beyond the Line could survive by farming alone. As he explained to Krichauff, for a reliable income they would have to turn to grazing as well – and he had been talking about an intermediate zone of mixed grazing and grain-growing since 1865. To achieve this mixture, farmers would need to be assisted by more flexible arrangements. Goyder recommended adjusting the size of holdings to the quality of the land, and introducing perpetual leases.[19] As ever, he was utterly pragmatic in his approach. The farmers had invaded the pastoral regions, however unwisely, and destroyed the natural pasture. The question now was what could be done to secure them.

In 1880 – as in the past – the rains, about which most people had been so confident, began to fail, and once again Goyder was in a position to reiterate his beliefs about the rainfall. In 1881, when the good years of the previous decade were still a close and persuasive reality, he was asked if he had modified his opinions. Goyder replied:

I have always held that the land is suitable for growing wheat; but the only difficulty is as to the rainfall, and in that matter I have not modified my opinions at all. I am still of the opinion that beyond a certain line settlement is precarious, not on account of the soil, which is splendid, but simply on account of the rainfall.[20]

He added that he did not believe that cultivation increased the rainfall, and that settlement around Orroroo had proved a failure.[21] When it was pointed out to him that people were still holding their selections there, he answered:

I know that and I am only giving you my opinion. I have seen the nature of the improvements, and have had conversation with the people, and my opinion has undergone no change whatsoever. I have had but one feeling – that of regret – to see men labouring year after year and not making themselves independent. Instead of that, so far as my information goes, with one or two exceptions, it has been a struggle and gradual retrocession.[22]

Further east, Goyder reported, around Yalpara, people had told him they regretted having moved there. Far from expressing satisfaction that his predictions had proved to be correct, Goyder stressed that if the area had to be abandoned, no one 'will deplore it more than I should – not so much failure of settlement, as the failure of the people'. Stop-gap measures, such as reducing costs of cartage, would only protract that failure, he argued. What was needed was 'a new system'. He suggested an investigation of what was successfully grown in other parts of the world 'where the rainfall is precarious'.[23]

The 'former' line

By 1882 the situation was dire and the push into the north was clearly at an end. The beginning of a more mature understanding of the Line began to emerge. Addressing the House of Assembly nearly 20 years after Goyder had himself described the existence of a marginal zone, pastoralist George Hawker identified Goyder's Line as being a band of marginal country just beyond the original Line, claiming that it was 'absurd to say it was a fixed line', but rather it was a 'strip of debatable ground' 40 or 50 miles in width separating reliable farming country from the arid inland.[24] Only a few weeks before, R.W.E. Henning, a politician of German background, had read a letter he had received from a farmer, Friedrich Kuttler, to the House. After conjuring Goyder's 'ghost', the farmer concluded sadly: 'like so many others, I thought rain follows the plough, and I paid dearly for my folly'.[25] Henning identified others who, knowingly or unknowingly, would also pay. When he passed through 'those once beautiful saltbush plains,' he lamented:

and saw how they had been destroyed for the sake of trying to grow wheat, he felt that a crime had been committed against the rising generation in the colony.[26]

In July a petition was received from northern farmers asking for relief. It mentioned the 'former' line by name, stating that it was 'evident that special legislation is needed for the land lying outside of Goyder's rainfall'.[27] The same newspapers which had scorned the Line and had insisted that 'practical farmers' were the ones ideally placed to determine the extent of agricultural land now criticised the government for having allowed the situation to occur. Goyder had already produced a memorandum advising the government on the options available for relieving the plight of the northern farmers. Increasing the block size was dismissed as impractical, and allowing the farmers to go broke was unhelpful. His third option was to encourage the selectors to surrender their lands and to select scrubland or other lands 'which though they may be inferior are within the limits of reliable rainfall'. The amount already paid in interest would be credited to the account of the second holding, and 'fair value for improvements' would be allowed to enable the stricken selectors to pay their debts. They could also be employed for 12 months, constructing water reservoirs on or near the surrendered land, enabling them to stay in their homes while receiving a regular income. The land would then be let in large blocks of up to 10,000 acres on 21-year leases, with the option of cultivation, so that mixed farming and grazing could be pursued.[28] It was a compassionate response, entirely free of triumphalism or even the faintest sense of resentment at the contempt with which his warnings had been dismissed.

But the government could not countenance an admission of failure and a retreat from the north, and instead passed an Act which allowed settlers to surrender their sections and repurchase them at a lower price – meaning, in effect, that as long as the selectors did not concede defeat, the government was prepared to allow things to continue as they were.[29] Since the Line had no legal existence, selectors who had paid high prices for selections south of the Line in the early years of the rush were also allowed to surrender their selections, and many got them back at the basic price of a pound per acre, or not much more, because the farmers had a general agreement not to bid against one another in the repurchase of a block. The result was not the resolution of the difficulties of the northern farmers, but the consolidation of the good land in the Mid-North as freehold. The Act also made provision for selectors to surrender and reselect within a year, taking the value of the interest payments and improvements on account to the reselected land, a provision which bore a superficial resemblance to Goyder's suggestion but had a very different meaning, since the farmers were not required to reselect in an area of reliable rainfall. This provision treated a farmer's failure as if it were merely a matter of having selected land with poor soils, which could be remedied by the opportunity for a second try. In any case, this option of 'subsidized retreat' was taken up by only nine settlers.[30] Most selections were reselected, although a large number were simply abandoned unconditionally, as Goyder later testified.[31]

By 1882, the original map showing the much-contested Line was so worn that it was replaced by two new copies.[32] In the intervening years mapping of the colony had advanced, and on the new and detailed map the Line could now be seen in relation to many well-known points. Not surprisingly, the Line on the new map had been subject to a number of changes, with alterations at Franklin Harbour, Crystal Brook and Pekina.[33] One change in particular caused later confusion: on the eastern side of the map, the long curve out to the Victorian border had been lowered. This modification excluded what is now the wheat-growing area of Pinnaroo, perhaps because Goyder had come to hear of an arid belt lying along the original line, perhaps because he had always been in doubt about this section.[34] But when he visited Pinnaroo three years later, in 1885, he was favourably impressed, and this impression was confirmed when he visited the area again in 1892.[35] In his last year in office, 1893, these lands were proposed for opening to agriculture, a proposal which to others looking at maps showing the 1882 line implied that he had released land for agriculture in 'outside' country. But Pinnaroo was no longer outside the Line in Goyder's mind. In the privacy of his office and without discussing the matter with anyone, he seems to have modified one of the two copies of the 1882 map, raising the Line again to re-include the Pinnaroo area.[36] He also corrected the original map, returning the Line to its first position. The surviving image of the original map shows signs of erasure in this area, and traces of another erased line, falling further south, are apparent. Since there was only a single copy of the original 1865 map on which Goyder had drawn the line by hand, it cannot be deduced from that alone which line was put down first. However, the maps printed for parliament in early 1866 confirm the higher line as the original one, so it appears that Goyder had modified this map, presumably when he redrew the Line on the new maps in 1882, and then restored it to its original state. Pinnaroo was inside the original Line of 1865, outside the Line as it was first shown in the 1882 maps and the modified original map, but inside the Line again once Goyder restored the 1865 map to its original state and modified one of the 1882 maps to match.

Once again, Goyder's view was out of line with that of his contemporaries.

After the collapse of the 1880s, South Australians were now convinced that the Line really did express an agricultural limit, and the map then in use was the 1882 map on which the area around Pinnaroo was excluded. There was no interest in opening this country and the Bill proposing to do this lapsed. Perhaps if Goyder had not been driven out of office at the end of 1893, he might have argued for Pinnaroo's restored position and for his restoration of this section of the Line to its original position. As it was, these changes to the Line remained a mystery understood only by its author.

CHAPTER FIFTEEN

Fresh water and peculiar country

This, it must be remembered, is a nomadic country, and every opportunity
should be taken to conserve water.

– George Goyder, 1891[1]

Although Goyder's chief official concern was land, it was inevitable that in the
driest Australian colony his overwhelming preoccupation would actually be
water. His reputation as a public figure had been launched on a 'discovery' of
water, and his life had nearly come to an end for lack of it. Through the Line
his name has become associated with agriculture, but it was through activities
connected to the pastoral industry that he had gained much of the experience
that led to his insight into climate and his focus on water. Pastoralism, in that
sense, had shaped him, and he in turn influenced both the pastoral industry
and pastoral landscapes.

Wells, dams and artesian dreams

Goyder's first major contribution to the pastoral industry in the colony had been
the opening of country in the north while he was still deputy. Once he became
surveyor general, he assisted squatters by establishing lines of public pasture,
along which stock could be driven from remote areas to markets. Reserving
stock routes was an idea that had occurred to him early in his surveyor generalship, 'very likely ... in a conversation with someone else', as he later recalled, and,
as he would do with forest reserves, he began to put his ideas into practice well
before specific legislation was enacted.[2] Through his influence on the development of railways, especially as chairman of the Railways Commission during
1875, Goyder also attempted to facilitate the movement of stock to market.

But stock needed to be watered as well as fed. The South Australian
Government had first allocated money to the Survey Department in the early
1860s for the construction of wells and reservoirs along roads and stock routes
to enable stock to be moved to market, and away from drought if necessary,
and be watered along the way. Goyder selected the sites for these wells himself.
However, plans for strings of wells to facilitate the movement of stock went
back several years before his surveyor generalship.

In 1857, Goyder had encountered the Rocky and the Reedy springs, as he had named them, in the country near View Hill, just beyond the northern end of the Flinders Ranges. Although he was unaware of it at the time, it was here that he actually encountered the permanent water that does exist in the inland. These springs form part of the arc of springs that tracks the south-western edge of the Great Artesian Basin and their existence assured him that underground water could be located. At the end of the report in which he described his notorious freshwater lake he recommended that the government build a series of wells on the flat land to the west of the Flinders Ranges. This could be achieved by boring an initial well at Port Augusta, after which the equipment could be made available to the pastoralists occupying the plains to the north. Those travelling north were forced to rely on the surface water found in the ranges, and wells would mean that a road could be established on the plains and the difficult mountain terrain then avoided. It was, he wrote, 'well known that the large quantities of water flowing from the ranges to the plains, are not lost by evaporation, but by absorption, and that it would again find its way to the surface, if not prevented by intervening strata of rock or clay …'[3] Along with his theory relating to the formation of Wilpena Pound, his notion that the plains beyond Port Augusta were an ideal place to bore for water simply provide further evidence of his geological innocence at that time. It was another elaborate conjecture rectified by Selwyn, who reported after his visit in 1859 that the boring attempted at Port Augusta had been soon abandoned.[4]

It was because Warburton, in pursuit of the dawdling Babbage in 1858, and Stuart, scouting for pastoral land on behalf of the Chambers brothers, had already encountered groups of springs that Goyder had been sent to triangulate the country south of Lake Eyre South in 1859–60. The expedition had included experienced well-sinkers, but they had encountered nothing but salt water in the four attempts made.[5] These failures were offset when Goyder's party came across other springs that edge both Lake Eyre South and the Great Artesian Basin. Although his party had discovered a number of springs, Goyder was especially taken by the Blanche Cup group, discovered by Stuart, around Hamilton Hill: 'I need only say that the Cup alone contains a reservoir of 85,883 gallons, with an overflow without diminishing the quantity in the reservoir of 14,290 gallons per diem, and that the quality of the water is improved by the channels being kept open'.[6] A small drawing of the cup, and a diagram of its structure accompany the calculations of capacity and flow in his field notebook. Goyder's measure of the overflow translates into just over 45 litres a minute, but only a trickle usually runs down one side of the mound now.

During the early 1860s Julian Tenison-Woods had been mulling over the contradictory experiences of explorers in the inland. In the paper he published in 1864, Tenison-Woods used the evidence of the explorers to conclude that the inland and the north received a great deal more water than was generally believed and that on occasions these areas were substantially inundated for long periods. The inland was a desert, he explained, not because there was no water, but because the continent was flat and the water drained away, either to salt lakes where it evaporated, or into the earth below. The phenomenon of the drainage into the limestone in the area of Mount Gambier was understood by Tenison-Woods, and he was also aware of the existence of limestone around the Great Australian Bight. Working from the written record alone, Tenison-Woods had grasped the essence of the geography of the central region of the continent; he had also examined the evidence to reveal the same essential aspect of its climate that Goyder had recognised from his experience – extreme rainfall variability. Tenison-Woods's research also led him to suspect, as Goyder did, the existence of a reservoir of water under the centre of Australia. Some of the water that drained away, Tenison-Woods believed, came 'bubbling up in the wonderful springs of Central Australia'.[7] (The recharge for the Great Australian Basin is actually along the mountains of the east coast.)

By 1867 the Survey Department was using boring techniques to supplement well-sinking by the traditional method – the swinging of picks – and Goyder seems to have maintained a familiarity with the results of their work and with the progress of well-sinking in general.[8] When asked by a parliamentary select committee investigating the prospects for a railway linking Clare to Wallaroo about obtaining water in the area, he immediately described in great detail the salt and freshwater wells in the area, with indications of depth.[9]

Town water

As with the wells for the travelling livestock, Goyder's enthusiasm for building reservoirs also predated the government's allocation of funds to the Survey Department for carrying out this work. In 1861, Goyder had been dispatched to Wallaroo to identify what was required for the new copper town. While there he had decided on three sites for reservoirs, one of which had already been selected by his deputy Christie. Goyder was at pains to point out that, since all the mine shafts nearby and all the wells sunk in the northern end of Yorke Peninsula had encountered salt water, and since the rock strata was steeply inclined, it was unlikely that fresh water would be obtained by sinking wells. He chose to 'beg to direct particular attention' to the subject of fresh water:

feeling assured that, if care were taken to preserve a portion of the rain-water that annually runs to waste down the numerous water-courses, not only would little inconvenience from want of water be experienced during the summer months, but a large area of otherwise useless country would be made available for pastoral purposes.[10]

In particular Goyder wanted to dam a watercourse running to Wallaroo Bay and to wall in the sides to prevent contamination of the dammed water by drainage from the town. But he begged in vain for the attention of the politicians in Adelaide. No provision was made for water collection at either Wallaroo or at Moonta, and the situation became desperate. The demand was met by selling water distilled from that pumped from the mines.

According to the *Register*'s Wallaroo correspondent, by the beginning of 1863 the universal cry from the residents was 'Water, water', with someone complaining soon after in a letter to the editor, that if private enterprise would not provide a tank or reservoir, then the government must step in.[11] The writer added, with an air of resignation, that the: 'Surveyor General was to have inspected a site for such a work, but I suppose it has not been done …'[12] By the next month's end Goyder and the commissioner had paid their visit, but even so, still nothing happened.[13] Two years later the drought was beginning to bite hard and the situation was so dire that it merited an editorial from the *Register*, which opined that if 'the rains that fell were stored, instead of being allowed to run to waste', the annual danger of '8,000 souls' – the town's population – 'perishing from thirst' would be removed.[14] The severe conditions of the drought must have produced some sort of result, because by September 1867 Goyder was able to report that a two-year supply of water for Wallaroo had been collected.[15] But despite being involved with reservoirs from his first years as surveyor general, Goyder did not undertake the construction of a reservoir himself until the early 1870s. That reservoir, for public use, was located at Warnertown, near Port Pirie.[16]

Nomadic country
The big drought had also raised the question of whether or not the northern country was of any value at all, even for pastoralism. When Bonney and Valentine were appointed valuators along with Goyder, the trio had been required to submit three separate reports on the leases they were valuing rather than provide a joint one. A battle between Bonney and Goyder followed. (Valentine admitted to being sidelined whenever the giants clashed.) Bonney bluntly declared that valuing the northern country was putting a price on

something that had been proved worthless, but Goyder dissented.[17] He had already told a committee that the north was a 'very peculiar country', which was impossible to overstock in good seasons when the grass had seeded the year before, but which could be utterly destroyed if stocked during a bad season, and he argued that it was not just the drought, but the added factors of 'inexperience in the country and overstocking', that had led to disastrous results.[18] If the appropriate lessons were learnt, pastoralism could continue. 'Let this condition be borne in mind,' he warned, ' – let it be an indispensable condition in the future management of this part of the country – that portion must always be reserved in anticipation of bad seasons …'[19]

Over two decades later Goyder revealed more of what he had learnt, not just from the 1865 drought, but earlier during his long spell in the country south of Lake Eyre. Answering a question about rainfall variability he explained that:

> a pastoralist requires a greater extent of country; so that whenever there was a local rainfall he might follow it up and avail himself of what supplies he can. I have known in the outside country the blacks break up their camps to move in the direction of a storm that has occurred forty or fifty miles away. This, it must be remembered, is a nomadic country …[20]

In recognising what he called 'nomadism', Goyder had grasped a basic strategy for meeting the challenges posed by an extremely variable climate, and one in which variation occurred across both time and space. This strategy has since been recognised in species adapted to this climate. A spectacular example is the banded stilt which breeds only when brine shrimps emerge and multiply in flooded salt lakes. Breeding events for these birds may be many years apart and their location is entirely determined by the presence of the shrimps.

After the Line had been drawn and the drought-relief measures effected, peace was made with the pastoralists, and Goyder had no further complaints from them about valuations. In revaluing the runs when he did, Goyder had taken them on at the height of their power; after the big drought the squatters never regained the position they had occupied in the early 1860s. Although the rich squatters on the prime runs south of the Line had escaped the drought of the mid-1860s, they, too, were affected when the price for wool dropped later in the decade, and the market for sheep was non-existent. Even reputedly impregnably wealthy squatters were broken.

As the 1870s progressed, things had settled down and the squatters seemed to be headed for an easier (if far more modest) time, provided the weather held good. This scenario evaporated when the 'earth hunger' began in earnest. Squatters in the path of the farming surge faced resumptions on six months notice. Some attempted to delay giving up the land or tried to purchase at a pound an acre, but their resources were now depleted, and after the 1872 Act, it was not legally possible to purchase land for grazing – keeping pastoralists off agricultural country was as much a part of Goyder's strategy as imposing an agricultural limit. Inspectors of credit selections were on the lookout for dummying.

The overthrow of the Line in 1874 further disrupted the situation. Pastoralists in the south had always known they would eventually have to leave leased land, but those to the north had proceeded on the assumption of secure tenure until 1881 or 1888 at least, and by the time their land was resumed they had expended money on improvements such as fences, wells, reservoirs and homesteads. They received only partial compensation and watched as some of the land remained unsold. Their complaints led to the warning that Goyder delivered to Carr, in which, along with his concerns about misuse of the land, Goyder had also complained of the 'one-sidedness' of the land laws, in favour of the farmer. But squatters no longer dominated parliament the way they had in the past, and only Carr (briefly) and Hawker appeared in Cabinets in the late 1870s to speak on behalf of the pastoral interest.

The squatters' complaints about the hardships of having to leave their runs at short notice resulted in the period of warning being extended to three years – which led in turn to what became known as the 'Playford resumptions' of the late 1870s. Because of the government's willingness to allow agriculture to spread north without restriction, Goyder had to advise the commissioner, Thomas Playford, that notices of resumption for huge areas would have to be given first to ensure a supply of land to meet demand for the next three years. Logical and inevitable as this response was, these massive resumptions were not the sort of result the squatters had wanted and they were outraged. Although the land was not needed in the end and Playford waived the notices of resumption once it was clear that expansion had stopped, the squatters were not appeased and made what Goyder later described as a 'bogey' of the resumptions, continuing to complain about them whenever they felt hard done by.[21]

Practical, efficient boring

In 1867 Goyder had requested that the government place boring and well-sinking operations under the charge of someone with more 'leisure'; nevertheless, Goyder continued to be interested in boring for water. Soon after arriving in the Northern Territory he had attempted unsuccessfully to sink a tube well at Fort Point.[22] In 1875 he produced a map which showed the water drainage of the eastern colonies into South Australia, indicating that the possibility of exploiting an artesian reservoir was still on his mind.[23] In 1878 artesian water was at last located in New South Wales. It now seemed that the pastoral future that Goyder had envisaged for the arid country was real and required only determination and hard work to become reality. The dry early years of the 1880s forced the government and parliament to turn their attention to the issue of water, and in 1881 a quarter of a million pounds was placed on the loan estimates to develop water resources.

On 6 January 1882, Goyder sailed for England on the mail steamer *Carthage* to purchase more powerful machinery for constructing dams and sinking wells, a journey which his obituary writers later attributed to his having 'repeatedly represented to the Government the absolute necessity of deep boring and of partial irrigation'.[24] On the same day, the *Register*, while agreeing that fresh water was the vital issue for the future of the colony, expressed regret that Goyder would be absent, 'just now when the land question is assuming a very difficult shape'. If the writer perceived any irony in lamenting Goyder's being on leave while the 'land question' was taking on the difficult shape he had long been warning of, it was carefully concealed.

In England Goyder purchased six complete sets of steam scoop machinery and chain pumps for the construction of large surface reservoirs, as well as other equipment to assist in the drainage of the South-East, costing after alteration and improvement to match his specifications approximately £24,000. In April he crossed to America, where he claimed to have personally inspected nearly all of the boring appliances in New York, but deferred the purchase of more equipment, buying only about a kilometre of piping, the best he could identify for casing the bores. He also engaged the services of a man named Waddell, with whom he had 'a long chat' on the subject of boring and boring equipment, at the same time assessing him (without his knowing) for employment in South Australia operating that equipment.[25] Goyder considered Waddell 'one of the most efficient practical borers in the States'.[26]

He travelled across America to San Francisco, arriving there on 29 April and settling in the Palace Hotel. A correspondent who interviewed him there found him, 'looking remarkably well, though a trifle wearied from his rapid movements and the labour consequent upon the execution of his mission'. Goyder stayed a week in San Francisco, apparently having something a little closer to the usual notion of a holiday, busying himself, as the correspondent reported, in the inspection of various local places of interest: the Fire and Patrol Departments, Golden Gate Park and Conservatory, the Mint, the still-incomplete City Hall, the Art Gallery, Free Library, Safe Deposit Company, 'the Stock Exchanges, Cable-car depôts, & c'.[27] He sailed from San Francisco on 8

May and arrived back in Adelaide in mid-June.

The equipment he ordered for dealing with South Australia's needs comprised a machine that could bore rapidly by percussion – in America he had seen pipes driven about three metres through soft soil by a single blow – and which also used a diamond drill bit in very hard rock. The latter had the added attraction of producing an intact core, so that 'we should see the character of the rocks we were going through, and whether they contained valuable mineral deposits', as Goyder later explained.[28] But the problem with a diamond drill was that it required a constant supply of water to keep the bit from jamming. By using a machine that combined both of these features, Goyder planned to drill by percussion until sufficient water was obtained to enable them to continue with the diamond bit.[29] Bringing his own percussion drill, Waddell followed Goyder to Australia and set about sinking wells, although not long after the drilling machinery had arrived, Waddell's contract expired and he returned to America. When the new machinery did arrive, it was despatched to Albalakaroo on the Nullarbor Plain, near Eucla and the border with Western Australia, where an attempt was already being made to reach artesian water.[30] On Waddell's departure, supervision of the drill was transferred to a Mr Smith of the Survey Department (presumably Goyder's brother-in-law, A.H. Smith), who, urged to greater speed by Goyder (at his own admission), pushed the drilling on with too little water, causing the bit to seize.[31]

Goyder's attempts to provide water for pastoral uses made him liable to charges of partiality yet again. Finding water by sinking wells, often in a dry creek bed, was a chancy business that was usually responsible for about half of the capital expenditure on a run. So when artesian water was finally discovered, the squatters waited for a decade – until the 1890s – as the government experimented with the new procedures for extracting the water.[32] The construction of the bores at Eucla (several spots were tried) resulted in Goyder being defamed in 1887 in parliament by Harry Bartlett, who had entered the House of Assembly only in March of that year. Bartlett accused Goyder of having the Eucla wells sunk, not as experiments, but to benefit certain squatters.[33] Bartlett was a prospector and farmer who supported small farmers, selection before survey, and agricultural settlement on the eastern plains and the Nullarbor Plain. He wanted land at Fowlers Bay, Streaky Bay and Denial Bay divided into small blocks and was convinced that Goyder's attempts to contain agriculture were the fruits of a vicious partiality.

While the Survey Department was faced with increasing pressure for the provision of water in the pastoral country, demand was also increasing in the settled areas. Several areas of government were involved: the Public Works

Department employed a hydraulic engineer, also equipped with a diamond drill, who offered to undertake works to provide water for country towns, with local authorities bearing the cost of maintenance and interest. The engineer-in-chief was also involved, and the railways, too, had a vital interest in water, being wholly reliant on steam power. In Goyder's eyes, the government had done good work in constructing wells, dams, masonry tanks and scooped reservoirs, mostly along the main roads leading from ports and inland towns, but the results were desultory.

On 10 May 1883, on his own initiative and in an attempt to communicate the fruits of his years of single-minded dedication to the pursuit of ensuring sufficient water resources for the colony, he produced a seven-page report titled 'Water Conservation and Development'.[34]

The grand plan

The report began with the central conclusion of his whole career: that the chief obstacle to the development of European settlement in South Australia was the lack of 'sufficient and reliable rainfall'. As the population increased and the area of settlement extended, Goyder warned, the situation could only get worse. The solution was to conserve the rain that fell in times of flood – the phrase 'water conservation' in documents from this period really means 'water collection' – and to develop access to underground water supplies. From the evidence of widespread irrigation in ancient times and accounts of contemporary irrigation practices in China, India and the American states of Utah, Colorado, Nevada and California, he inferred that, in relation to 'deficient and irregular rainfall, South Australia is by no means exceptional', and drew the cheerful conclusion that South Australia could benefit from irrigation in the same way. The inference is not strictly logical – Goyder was presumably not in a position to understand how unusual the extent of rainfall variability in Australia is, in terms of a global comparison – but he took heart nonetheless.

The language and tone of the report indicate that Goyder identified with and had a comprehensive understanding of the business of government and perceived the government's role as facilitating the life and wellbeing of people of both present and future generations. His creative energies found their ultimate expression, as they had in the past, in establishing an appropriate administrative structure to undertake the necessary work. The problem being faced was that the administration was highly centralised, while the work was being undertaken over large areas and at great distances from Adelaide. Tendering the work out was no better, he considered, because capitalists only 'go in for a big thing', as he put it, and would not produce works of the standard the community required for the modest returns that could be expected. Employing men from the private sector on large salaries was also not worthwhile, Goyder argued, because experience had shown that once relieved of direct responsibility for the cost of the project, such temporary public servants very often ceased to care about costs at all.

The solution Goyder proposed followed the district-based approach, with elected district councils or boards responsible for the development and conservation of water for all purposes within their districts and capable of appointing officers to carry out the work required. Works would be funded by an assessment on properties in the district, the funds from which would be matched by the government. In his report he identified three distinct areas of the province – the hill country near Adelaide, the Far North and the South-East – and considered their needs separately.

The hill country, he explained, required irrigation to increase fruit, vegetable crops and fodder crops, although this need was not always understood. By this time Goyder himself owned a property in the Adelaide Hills, Warrakilla, and the report was actually written there. Positioned on a hillside on the road between Mylor and Echunga, Warrakilla included old cherry orchards and Goyder was possibly in the process of establishing his own orchards at that time. Certainly, by the end of the decade, he was about to begin irrigating there.[35]

Informed by his experience in 1857, his vision of the future for the Far North was on a typically grand scale. Referring to the floods he had seen, Goyder pronounced himself vindicated against the charge that he had been duped by mirage by later observations of flooding:

> subsequent experience and accurate geographical information has shown the region to abound in large lakes and lagoons of fresh water; and it is all but certain that the whole of the country referred to can be made of permanent value, and to give a large percentage upon any necessary expenditure by the construction of proper dams and ditches to retain and distribute the immense quantities of flood waters which it is known periodically inundate that country by the flooding of the Diamantina, Mulligan, and other rivers ...[36]

He admitted that there was no doubt that the works required would be extensive and costly, but he considered them to be essential to the continued prosperity and progress of the colony. To support his proposal, he presented evidence of extensive irrigation works in India. Goyder had to make clear that he was not talking about flood irrigation – the creation of water meadows – which was the practice in England, but about irrigating only to germinate the seed, and later, to mature the crop. He also stressed that irrigation should be applied to fodder crops and pastoral land, as well as to fruit and vegetables. To ensure that the works undertaken benefited all and were not restricted to a particular class, he advocated that the government continue to be responsible for all the works required along roads and on crown lands. The district boards would direct work necessary in the district and individual landholders were to carry out works on their own properties. Modifying the terms of leases would ensure that the activity at individual landholder level was undertaken.

The third region comprised the country Goyder had been struggling to drain over 20 years, but his report advised that to promote settlement in the South-East, it would be necessary at times to irrigate. Not surprisingly, he addressed the situation there in considerable technical detail.

When it came to the possibility of constructing dams, Goyder referred to a report by a British colonel which provided an account of the vast network of irrigation works, 'of purely native origin', established in Madras. (It is possible that his attention had been drawn to this by Colonel William Barber, who had been an officer of the Madras army for 25 years, and who served on the Forest Board and the Railways Commission after settling in South Australia.) What had impressed him was that, although the irrigation works formed a vast network, they had been created using simple technology by the irrigators themselves in response to the specific features of their location. But care would have to be taken, he warned, to avoid all water sources containing mineral salts.

The report also mentioned a scheme about which there had been a great deal of talk – tapping the waters of the River Murray by means of a canal from the north-west bend to Adelaide or to the plains to the north. Goyder warned that these proposals would prove impractical once levels were taken. Instead, for use in that area he suggested pumps driven by the force of the current to raise water into reservoirs near the bend. The report recorded that settlement and clearing already appeared to be having a marked impact on the Murray. At Moorundie, Goyder noted, soldiers' barracks had been built and had been occupied the year round. Two decades later, when the river was high, water reached the top of the walls of the barracks and was threatening to inundate the flats between the north-west bend and Lake Alexandrina. It had been suggested that this was the result of increased run-off following the compaction of the soil by stock around the tributaries in New South Wales and Queensland. Goyder's response was a characteristic piece of optimistic opportunism. Rather than lament the problems caused by stock-management practices that South Australians could not control, he chose to regard the additional water and silt load from the river as a gift. He suggested building banks around the flats and sluice gates across the river. Floodwaters could then be held back and slowed until they had time to deposit their load of silt on the flats, enriching the land with soil from upstream and gradually raising the level of the earth.

It is no surprise that Goyder recommended that it would be worth the expense to send geologists out with surveyors to record appropriate sites for the development of water resources. He urged the legislature to address the matter of 'water conservation' immediately, because the value to the state could not be exaggerated. The commissioner of the time, Alfred Catt, brought a Water Conservation Bill based on Goyder's report before the parliament, introducing it at the second reading speech as 'perhaps the most important that could come before our legislature'.[37] Catt used much of the argument presented in Goyder's report in his address to the parliament, even quoting one of his sources at length, but in the Legislative Council the Bill was rejected as impractical.

The report on water conservation was something of a swansong, because, while presenting his vision, Goyder had also been attempting to shed any further responsibility for the realisation of it. In this he succeeded. He had proposed that the task of managing water resources had grown large enough to establish a water conservation department, and the person he recommended to head this department was the deputy surveyor general, J.W. Jones, who had been responsible to him for this work. In essence he was launching Jones independently, and the Department of the Conservator of Water which resulted was 'merely a new label for the Surveyor-General's regular borers'.[38]

Ensuring an adequate urban water supply was also an issue during the early 1880s, and Goyder's opinion was sought about the location of dams for supplying drinking water to Adelaide. It appears to have been the one thing to which his attention had not already turned, but by 1889, when examined by the members of the Barossa Water Commission, he had become a mine of information.[39] He had considered all the potential sites for dams and was even able to add his voice to the recommendations for a new American form of dam construction, the earth and rock-filled, or, as it was then termed, 'rip-rap', dam. His replies indicate that he understood the principles of the construction of these dams and the reasons for their stability and strength. The commission also provided him with an opportunity to reassert his belief in the value of irrigation and to express his admiration for the projects established by the Chaffey brothers – who had gained their experience in California – at Mildura in Victoria and at Renmark in South Australia in 1887. In Renmark the water was pumped from the Murray on to vines, citrus and apricots. 'As soon as the Chaffey irrigation colonies are working', he informed the commission, 'anyone desirous of conducting irrigation on his own land can go to Renmark or

Mildura and see for themselves the most modern systems'.[40] His field notebooks show that in 1891 he visited Mildura himself.

CHAPTER SIXTEEN

Tree theories

But Mr. Goyder does not believe in being a lay figure and he is perhaps encouraged to an excess of assertiveness by the fact that the Government treat him as if he were a universal genius not simply a surveyor, but master of the sciences of hydraulic and marine engineering, mineralogy and geology and other subjects requiring knowledge, experience and patient toil to be thoroughly acquainted with. It is no wonder in what probably appears to him so simple a matter as forest culture Mr. Goyder should consider himself wiser than the conservator of forests … – *Advertiser*, 1882[1]

That South Australian settlement accommodated and adapted to its scarce and unreliable supply of fresh water was an issue that Goyder felt compelled to pursue at all costs. Because of the belief that trees could be used to control the climate (and therefore that they should be used for this purpose), he found himself in another struggle over unreliable rainfall in the north, even as he addressed the scarcity of another major resource: wood. Goyder was more interested in trees and plants than has generally been recognised. Ferdinand von Mueller had appealed to Goyder's commitment to science when seeking to obtain as large a collection of specimens as possible from the Northern Territory, and Adelaide Botanic Garden director Dr Richard Schomburgk had described him as 'animated by a zeal for botany' when contemplating the huge haul that Goyder had arranged to be collected.[2] After his return from Darwin, Goyder began to consider the fate of South Australia's forests and the reality of the colony's need for timber.

Although settlements on the east and south-east coasts and in Tasmania were surrounded by forests, in South Australia tall forest was confined to the Mount Lofty Ranges and the far South-East, covering a total of only 1.2 per cent of the area of the colony. Because the rainfall was barely adequate to support substantial trees and the soil was poor, the South Australian forests were open forests, consisting of stringybark with a dense understorey of hard-leaved plants adapted to dry conditions. Nevertheless, the situation had not been immediately apparent to the newcomers. In 1836 one writer had announced that there was enough timber for the colony to receive 20,000 emigrants every year for the next century.[3]

Clearing timber was regarded as an inevitable first step in settlement, accepted even by those who lamented the ugliness it created. Trees had to be felled to create cleared land and to provide materials for fencing and building, as well as fuel for domestic purposes and for powering steam-driven machinery. The stringybark forests on the hills around Adelaide became home to camps of shingle and paling cutters, while the large red and blue gums, dotted across the flatter land, were felled to provide construction and fencing material. Even so, it was necessary to import good, straight wood from the other colonies. Around the copper mines at Burra, Moonta and Wallaroo, great tracts of land were utterly denuded, cleared mostly of mallee, which was fed into the fires of boilers and smelters. In the mid-1860s, when prospective farmers joined the attack on the 'scrub', the mallee and mallee heath communities that had once covered nearly three-quarters of all of what are now South Australia's agricultural lands met the same fate as other, less dense and tenaciously rooted woodlands. Over the Yorke Peninsula smoke from the burning of cleared timber lingered in an obscuring haze.

As the population increased and settlement expanded, the demands on the forests increased. In 1869 it was anticipated that the extension of the railway line north to Burra would require between seven and eight thousand red gums for sleepers.[4] Earlier, in 1859, one observer had described a 'war of utter extermination' being waged by timber cutters and agriculturalists in Peachy Belt, a peppermint gum forest which stretched beyond the Para River to north of Gawler, and warned that 'unless peace be proclaimed in time, their descendants will look in vain for a tree in that which was once Peachy Belt'. By about 1880 the prophecy had been fulfilled.[5]

As early as 1851 a petition had been presented to the first Legislative Council which included the demand for the establishment of a 'proper forest culture'.[6] Significantly, it had come from a progressive section within the colony's community of German immigrants. The German settlers had brought with them a cultural heritage that included a concern for trees that was the result of bitter experience. The great forests of Germany had been devastated in the seventeenth century by the Thirty Years War and by the exportation of timber for shipbuilding. The loss of these mighty forests had given rise to a tradition of forestry which aimed to preserve the old oaks and the landscape integral to the German being; at the same time a scientific approach to the management of the fast-growing softwoods was adopted. The petition of 1851 was to no avail, but it was a significant forerunner of things to come: until the goal of creating a government structure to superintend and administer forestry was achieved, apart from Goyder and his brother-in-law, Edwin Smith, the leading figures of activity and influence would be German.

The first government action was taken in 1860, when, in an attempt to limit extensive wasteful despoliation of the forests, licences for timber cutting on crown land were introduced. The regulations were strengthened throughout the decade, but since adherence to the conditions of the licences was not policed, no amount of modification of the terms ensured stricter control, and the destruction continued. By the beginning of the 1870s there was real concern about the future of forests, not only in South Australia, but elsewhere in Australia and the colonised world. Growing public discussion resulted in stirrings in Australian legislatures. In South Australia, serious debate had begun in parliament in 1869. Around that time the *Register* published a series of articles on forest trees by Ernst Bernhard Heyne, who had held a position at the Dresden Botanic Garden before emigrating in 1848 (the year of uprisings in Germany and across Europe) and eventually obtaining a position at the Botanic Gardens in Melbourne.[7] In January 1870 an article on 'forest culture' appeared, expressing concern over the inevitable conflict between clearing, agriculture and civilisation on the one hand, and the new awareness that 'permanent tree vegetation is of the greatest importance'.[8] The article mentioned most of those elements which were to become staples of the debate, in particular the work of Humboldt, the foremost scientist of the early nineteenth century, who had described the impact of trees on the humidity of the surrounding atmosphere and on moisture in the soil.[9]

In June of that year, tucked away in a report on an amendment to land legislation, Goyder mentioned that in estimating land available for survey he had excluded lands 'requisite as forest or timber reserves, as such must necessarily be secured at White's Forest, Wirrabira [*sic*], ranges on the south-east tiers in the more hilly land, &c., &c'. He advised that as these reserves had been denuded by the occupiers of the open grounds, they should be systematically replanted to keep up a supply – 'as is done in many places'.[10] In practice then, on the basis of Goyder's decision alone, these reserves already existed.

About two months later, in August 1870, Schomburgk presented a paper on the influence of forests on climate at the monthly meeting of the Philosophical Society.[11] Originally from Saxony, Schomburgk was a botanist and an experienced explorer who had travelled in British Guiana with his brother Robert (who had travelled with Humboldt in South America). He had fled Germany after taking place in the failed 1848 uprisings, and had settled on the Gawler River with another brother, Otto. The brothers farmed together and became respected vignerons and winemakers. Schomburgk was appointed director of the Adelaide Botanic Gardens in 1865 and was a vigorous and creative director, one of whose major concerns was the introduction and acclimatisation of economically valuable plants. He also shared with von Mueller, the Victorian Government Botanist at that time, a deep concern for forests and afforestation that went beyond the purely instrumental to invoke the sacred.

Schomburgk began his presentation by recalling humankind's dependence upon forest products. He then outlined the importance of forests to the atmosphere, and the impact of mountain deforestation, illustrating his point with examples of the extent and impact of deforestation from the environmental history of Europe and the Middle East, mostly derived from an examination of classical literature. He concluded this part of the lecture by voicing the hope that:

> times are past forever when the progress of civilization was equal to wasting and desolating the surrounding nature … let us hope that future generations will be wiser than the past ones.[12]

Schomburgk then took a different course. Referring to the example of the Egyptian Delta, he explained how the enormous numbers of trees raised there – Australian species, as it happened, raised from seed provided by the Victorian Government – had dramatically increased the number of rainy days. His argument, from evidence of the impact of deforestation to claims concerning the climate-modifying effects of afforestation, was common at the time. As an argument it had the benefit of an appealing symmetry – if cutting down forests created deserts, then planting them would bring rain, even where trees had not grown before – but it failed to take account of the difference between regional climate and small-scale local effects, a distinction which was not then as obvious as global weather-monitoring and satellite photography has made it now. What Schomburgk's paper demonstrates is that the theory that trees attract or cause rain was not a sort of silvicultural snake oil, spruiked to the ignorant; rather, the theory was part of a developing ecological understanding that was held and promoted by prominent experts.

Within a couple of weeks, George McEwin, another colonist prominent in the general field of plant cultivation and botany and connected with the Agricultural and Horticultural Society, had published his own thoughts. McEwin had come to the colony in 1839 and was an elected life member of the Arboricultural Society of Scotland, an honour conferred for his classification of the plants of Scotland made at the age of 18. He was subsequently awarded a gold medal by the same society for a paper on South Australian forests.[13] An ardent tree theorist, McEwin was deeply concerned by the extent of the destruction of the colony's tree cover and alarmed at what this boded for the future. He was anxious that the already-dry country was about to become even drier. To prevent this, he wanted to see state forest reserves established and arboriculture begun. He also recommended the construction of reservoirs for the collection and storage of rainwater and suggested planting trees in the arid north, to facilitate agriculture there.[14]

The contribution of a Scot was no more a random occurrence than the contribution of the Germans (and Goyder, it should be remembered, had received his later education in Scotland). Influenced by German and French ideas, the Scots were pre-eminent in generating a critique of the environmental impact of settlement in the British Empire, a critique – like the German one – that was based on their own experience of deforestation. In parliament, however, it was the Germans who led the way. In the House of Assembly on 2 September 1870 F.E.H.W. Krichauff requested information on the amount of timber imported by the colony. His question was prompted by advice from von Mueller (a friend) about activities in Victoria, where gathering the same information had led to the conclusion that that colony would be better off raising its own plantations of softwood. A few days later, on 6 September, John Hodgkiss initiated debate in the upper house on the desirability of planting forest trees. The following day, back in the Assembly, Krichauff moved that a select committee be appointed to investigate the whole matter. After being a member of the first parliament in 1857 he had returned in September 1870, evidently determined not to waste time.

Krichauff's motion received general support. The treasurer, John Hart, agreed in principle to a select committee, but thought it would be more expeditious simply to obtain a report from Schomburgk. Krichauff agreed, but asked that Schomburgk be given assistance. The assistance he received was from Goyder. While Schomburgk was mainly concerned with the choice of species, Goyder focused on reserves. He recommended that no reserve be less than four square miles and identified ten sites. These sites constitute most of the forest reserves administered by the South Australian Government today. The first, and now best known of these, was the Wirrabara Forest, in the ranges south of Mount Remarkable, which he had named in the earlier report. Other recommended areas were around Mount Bryan, at several places in the South-East, on the Yorke Peninsula, and at Port Lincoln on the Eyre Peninsula. The margins of travelling stock routes were also included. The area that stands out in this list is Mount Bryan, because now the hills there are bald, but the cover which Goyder was trying to conserve is shown in Frome's watercolours of the area, painted in the early 1840s (see plate 9). The journal kept by one of Frome's companions contains a description of a valley in the area populated by sheoaks, pines and gumtrees.[15] Goyder had evidently been concerned about this area for some time – as deputy surveyor general in 1859 he had recommended the prosecution of timber getters at Mount Bryan (and in the Murray Scrub). The inclusion of this area in his list of reserves must have been a last-minute attempt to protect it.[16] Three years later he described the area as 'grassy undulating country with pines, wattles, bushes in watercourses and a few clumps of scrub; and about three miles [of] stony hills with bushes and a little grass'.[17] He never recommended it as a forest reserve again.

Nothing really came of either Krichauff's or Hodgkiss's motions, and although Krichauff tried to raise the matter of forests again, it was not until 1873, when Krichauff introduced a Bill to encourage the planting of forest trees on private land, that the opportunity was created for Goyder to become actively involved. Krichauff sent a copy of his Bill to the chief secretary, who asked for a report from Goyder before it was tabled. As a result, Goyder was despatched, no doubt on his own initiative, to the South-East and then to the north to obtain information – in fact, to make the observations and decisions necessary to convert his suggestions for the locations of the reserves proposed in 1870 into a complete program to establish state forestry. For advice on the selection of species to plant, he took with him his brother-in-law, Edwin Smith, 'a nurseryman of large experience', as Goyder described him.[18]

The immediate effect of Goyder's participation was to transform vague concerns and proposals into specific programs for action. The report that resulted from the tour of the South-East went well beyond the limited scope of Krichauff's Bill and presented Goyder's ideas about planting forest trees with dramatic directness. It began:

> I am of the opinion that the cultivation of forest trees throughout the Province is urgently required, as, in whatever direction my duty takes me, the rapid decrease in forest trees is brought painfully and prominently before me.[19]

In fact, the forests were all 'rapidly dying', according to Goyder, and the 'indigenous timber' would shortly disappear, 'unless renewed by planting, protection and judicious management'. The report stated that, throughout the province and in Victoria (which he had visited in 1870), large gums in particular were dying off. Remote places where the large timber looked healthy were the exception, and even there 'symptoms of weakness begin to appear'.[20] The situation was dire.

From early in the 1860s the dying forests had also been noted and discussed in Victoria. Goyder dismissed the commonly hypothesised causes – too much or too little water, possums, and maturing of trees – as illogical since it was happening everywhere. His conclusion was that probably 'the soil is becoming exhausted of certain ingredients essential to the growth of *Eucalypti*', and he suggested that an analysis of the soil should be undertaken.

Having delivered his warning, Goyder directed his attention to Krichauff's Bill – the ostensible focus of his report – but only long enough to deem it 'certainly a step in the right direction', and to offer a few practical amendments. To him, the matter was of major importance and concerned public as well as private land. He urged:

> that it will be desirable, if not absolutely necessary for the Government to proclaim and maintain forest reserves on leased and reserved lands of the Crown; and to do this effectually the lands must not only be resumed and protected from the indiscriminate depasturing of sheep and cattle, but inspectors must be appointed and the young stock raised suitable for the locality in which it has to be grown.[21]

The rest of the report contained Smith's advice on what should be done in the South-East. It was recommended that trees appropriate to the surrounding districts be raised in the three nurseries to be established at Mount Gambier, on the Cave Range south of Naracoorte, and near Bordertown. Smith had selected particular species for the eight areas that the pair had identified as suitable for the new reserves. Mainly eucalypts – identified as red and blue gums – had been chosen for the flats and exotic pines for the sandy ridges. Olive and what were described as 'locust' trees were suggested for the government reserves and parklands.

A second report appeared just over a month later, on 28 October 1873. The result of another tour with his brother-in-law, this time to the proposed reserves to the north and south of Adelaide, it consisted of 16 detailed proposals. Goyder had chosen a large number of areas for reserves, which he divided into southern, northern and central districts, comprising in total 777 square kilometres. At the same time he gave the government precise instructions about which sections and clauses of the 1872 *Waste Lands Act* to invoke to enable their establishment and management. One major proposal was the appointment of a conservator of forests, an officer 'thoroughly and practically acquainted with forest culture in all its branches', whose task would be to establish three nurseries, one for each district, so that a beginning could be made 'without delay'.[22] The remaining proposals, apart from recommending that seven nurseries in total be set up, detailed the activities of the new department to be formed under the conservator, who would report directly to the commissioner. As was noted at the time, Goyder's report amounted to 'an elaborate series of suggestions to guide the

Government in establishing … systematic forest culture', a matter which he had approached 'with considerable enthusiasm'.[23]

Krichauff's Bill was passed, but no Regulations allowing for its practical effect were gazetted, and despite the enthusiasm which had greeted Goyder's reports – a public meeting had been held to support the Goyder–Smith proposals and to discuss the details – the government let the whole matter lapse. In March 1874 the editors of the *Register* tried to prompt the government into action with an editorial lamenting that, although the winter planting season was steadily approaching, nothing was being done to effect Goyder's 'grand scheme of State arboriculture'. Meanwhile, Goyder had been doing what he could. His replies to an inquiry from the Colonial Office about colonial timber, which contained a list of questions posed by the Royal Commission on Woods and Forests, the Office of Works, and the Institution of Surveyors, revealed that 'the proposed nursery at Bundaleer Springs has been taken possession of, and a few thousand seeds raised; but more active steps await Legislative action'.[24] The seedlings raised, with some difficulty at first, had consisted of jarrah (*Eucalyptus marginata*, a Western Australian tree), and six kinds of pine, in particular Canary Island pine and what is now known as radiata pine.[25]

One of the questions listed sought evidence on the climatic influence of forests. Goyder replied that the conclusions reached so far – by which he meant those that suggested trees brought rain – were unsatisfactory:

> inasmuch as on the banks of the Murray, and in several places in the interior, and on the sea-coast, where many hundreds of square miles of *Eucalypti* flourish at the expense of almost every other kind of vegetation, the rainfall is known to be less than in treeless districts in the same latitudes.[26]

Goyder was pointing out that forests of mallee thrived in semi-desert, and it remained semi-desert. The trees did not attract rain or produce a humid atmosphere.

In parliament, Krichauff proved relentless in stirring the legislative action necessary for establishing forestry. He had introduced another Bill in July 1874, this time to provide for the appointment of a Forest Board, to which Goyder's proposed conservator would report. The purpose of this was to make use of the expertise that already existed in the colony to sidestep the overworked lands commissioner and actually get something done. Although Krichauff's Bill was received positively, it was nevertheless delayed – the chief secretary wanted a report from Goyder – and the parliament was prorogued before it had completed its passage. The Bill was introduced again in May 1875, and passed on 15 October 1875. The supporters of forestry had finally attained their goal.

The Forest Board

Under the *Forest Board Act*, over 790 square kilometres of forest reserves were created – a little more than Goyder had previously recommended. A five-member board was to manage these reserves, while a conservator would report to the board and be provided with staff to carry out the work. Within a month the members of the board had been appointed. They were Goyder, Schomburgk and McEwin, Finnis, and Colonel William Barber, who had already served on the Railways Commission under Goyder's chairmanship earlier that year and who would later accompany Goyder on tours of the forestry work in progress. The Act determined that the board was to elect one of its members as chairman, making the choice of the creator of the colony's grand and only plan for systematically established and managed state forests inevitable. The first of the board's required monthly meetings took place a week after the appointment of the members, and with Goyder in the chair the board actually met more than once a fortnight until the first annual report – prepared by him – appeared at the end of August 1877.[27] Three nurseries were quickly established: one at Bundaleer Springs on land resumed from the Bundaleer run, another at Wirrabara, and one at Mount Gambier in the South-East. Pines and eucalypts were planted at Bundaleer in 1876, the first planting undertaken by a government in Australia.[28] Within 15 months, 208,144 trees had been raised and others bought. Nearly 100,000 were planted and approximately 30,000 distributed to district councils and others for planting. A large reserve of seedlings remained.[29] By 1881, annual planting was carried out at all three locations, and in the same year a scheme for the free distribution of trees was instituted to replace the provisions of the 1873 *Forest Trees Act*, which had been intended to encourage landowners to plant trees but had not proved effective.[30]

In relation to its own forests on public land, the board made fundamental decisions about how to proceed. Only 'useful timber' would be grown, although 'where ornament and use could be combined, preference was given to such varieties'. Valuable and fast-growing varieties would be preferred.[31] Latitude, altitude, soil and rainfall were given 'due regard'. Goyder's report also documented the board's practical decisions about planting and layout and included the note that, by 'planting different varieties of trees in blocks and each variety together in different localities', it would be possible, 'to mature certain portions of the forest at particular times, and thus be able to prune, thin, cut down, and replant in the most systematic and economical manner'. These plantations were approached with an experimental air, Goyder recording that they intended to 'compare results and to profit by experience' and make their experience available to others.[32]

The useful species planted included those producing timber of various kinds and for different purposes, wattles (to produce bark used in tanning), willow and bamboo, the latter grown to fulfil an unusual purpose. W.J. Curnow, the nurseryman at Bundaleer Springs, had heard that in India large sections of bamboo were used as flowerpots and had hit upon the idea of a composting tube for growing and planting saplings. In 1876, using smaller, narrower bamboo tubes, initially supplied by the Bundaleer homestead, Curnow began experimenting with raising seeds. Eucalypts responded well to this method, and in 1877 he began mass plantings, using a machine which cut the bamboo at the rate of 8000 or more tubes a day. Compared with methods of seed-raising then in use, the system had many practical advantages. Goyder regarded this innovation with delight, praising Curnow and the 'unqualified success' of his new method, and predicted that it was 'likely ere long to become universal'.[33]

The new conservator

With work under way, the board's next major task was to appoint a conservator. For that, they had to look beyond the colony. The man they chose, John Ednie Brown, had been born into the world of forestry, his father having been deputy surveyor of woods and forests in Scotland and an expert on European arbori-culture. Brown had been trained in the management of nurseries and forests by his father and had been employed in Scotland and England. At the beginning of the 1870s he had visited the United States and Canada to study trees and forests, and the reports he had written had won him honours at home.[34] Brown was a generation younger than Goyder – he had been born in Scotland in 1848, the year of Goyder's emigration. When Goyder enthusiastically announced him, and his 'hourly expected' arrival, to the government, he described his recruit as 'the son of Dr. Brown, of Edinburgh, celebrated for his knowledge and work upon forestry'.[35] He was highly recommended, Goyder assured the government members, and 'satisfactory results' could be anticipated.

Brown arrived in Adelaide on 15 September 1878 and set off to examine the forest reserves at Bundaleer and Wirrabara with an alacrity that must have fulfilled Goyder's highest hopes. On 5 October, only three weeks later, he produced a detailed nine-page report. Brown was pleased with the forests and the nurseries he found and pleased with the nurserymen he found in charge of them. He added his voice to the chorus critical of the extravagantly wasteful practices of licensed timber getters and suggested that a real effort be made to control their depredations. Goyder, however, seems to have resigned himself to accepting their activities, at least temporarily, and a letter to the *Register* at the time suggests why. According to the writer, the conditions governing the operations of licensed timber getters determined the price and availability of wood – the basic fuel for cooking and heating – in the city, and the urban poor were those most likely to be severely affected by restrictions.[36] Goyder had previously advised that the existing system must remain intact 'until the timber on the protected portions is ready to be cut, when Regulations will be framed to meet all the circumstances of the case'.[37] In the meantime, only planted trees would be protected.

The remarkable radiata pine

One of the reports Brown had written had been about the trees of California, so he was well qualified to enthuse over the progress of the 8000 *Pinus insignis* seedlings, now a year old, which had been planted at Bundaleer. They were all doing well, he reported; not a single one had died. *Pinus insignis* – now *Pinus radiata* – is a native of California and was commonly known at the time as the Monterey pine. It also came to be known, through the English translation of its then Latin name, as the 'remarkable pine'. As Brown was later to write, these pines were 'a remarkable success, both in the number which … survived the hot weather and in the strong, healthy appearance of the plants'.[38] The species is now predominant in softwood plantations across Australia and there are extensive plantations in the South-East of South Australia.

Although Goyder has been credited with introducing the species, this is one attribution that is clearly incorrect if taken in the most straightforward sense of introducing the tree to the continent. A Monterey pine was reported to have been received by the Sydney Botanic Gardens in 1857, and in the following year the species was included in the list of plants growing at the Melbourne Botanic Gardens. Von Mueller was enthusiastic in his assessment of the pine and its prospects in Australia, and claimed to have distributed it 'extensively throughout Victoria and other some other parts of Australia' from 1859 onwards.[39] He is believed to have introduced it to South Australia, where an avenue was planted in the Botanic Gardens in 1866.[40] It is likely that there were plantings elsewhere, since the species was recommended as suited to the climate and conditions by Heyne in his article published at the beginning of 1870, where he described it as having shown itself, through 'numerous trials … to be thoroughly at home, and capable of withstanding our severest droughts'.[41] (Heyne actually described *P. radiata* and *insignis* as separate but closely related species and recommended them both.) Schomburgk, though, omitted the species from the list of trees recommended in his 1870 report, where he included the Aleppo pine and the maritime pine, which were favoured by all contributors to the debate.

What Goyder did was introduce *Pinus insignis* for official consideration in his first report of 1873. Here it was listed as amongst the species to be planted on the ridges of the Cave Range, south of Naracoorte, and to be cultivated in a nursery there. Goyder's recommendation was on the authority of Edwin Smith, whose preferences had been expressed in the varieties cultivated at Bundaleer by the Crown Lands Department before the formation of the board.

Rain follows the trees

Brown's first annual report was presented in May 1879. It was an extensive document of over 30 pages intended not only to report upon the progress of the 'Forest Board Department', but to provide an expert account of the indigenous South Australian species of trees and the conditions under which they grew; this could act as a reference for all interested parties in the colony and elsewhere. A large part of the report was taken up by his descriptive list. The introductory pages make clear that Brown was familiar with the kinds of trees and the peculiarities of the conditions in his new field of operations. Much of his information had come from the surveyor general:

> whose extensive travels throughout the colony, and known accuracy in forming opinions on the natural resources of the country generally, may be accepted as a guarantee of correctness, so far as possible at present ...[42]

This flattering token of respect was offered at the time when the credibility of Goyder's Line had been crushed by the weight of the harvests from the north. It was probably intended to mollify Goyder, because it immediately preceded a set of statements that, by this time, Brown must certainly have known would provoke him.

Along with the Scottish training and honours that had so impressed Goyder, Brown had brought an equally Scottish commitment to redemption by afforestation. He was another ardent tree theorist like McEwin. Goyder had apparently succeeded in getting him to acknowledge that the climate in the north was both arid and variable, but Brown's faith in the power of trees was boundless. Magisterially addressing the subject of 'the proper proportion under which the area of any country should be occupied by trees', he declared that 'trees regulate a climate ... The want of a due proportion of forest in a country is certain to cause irregularity of climate.'[43] Later in the report, Brown expanded upon his theme for just over two pages under a string of headings that combined micro-climatic effects and macro-climatic claims indiscriminately. The report asserted that trees and forests give shelter, act as fertilisers of the soil, prevent both evaporation and sudden floods, have a tendency to equalise rainfall across the seasons, attract rain clouds, make climate more humid, purify the atmosphere, and 'cause amenity to a country', saving the eye from being wearied by making the land look 'clothed and picturesque'. Brown also credited forests with being able to 'subdue aridity', so that:

> by planting arid tracts of land with properly proportioned belts of timber here and there through them, the result is (1) lower temperature, (2) arrest of hot winds, (3) shelter, (4) more frequent rains, and (5), a more humid climate generally, thus making such tracts of country suitable for agricultural purposes.[44]

Since the idea that trees could introduce major climatic change promoted the possibility of agriculture in the north, Goyder took up the fight again. To ensure that Brown's document would not appear to imply unqualified government approval, Goyder lodged objections, which were attached to the report as an appendix. For Brown, in his mid-twenties, it would have been a frustrating way to begin in a new position, opposed from the outset by the entrenched and powerful senior figure who had recruited him specifically for the professional training or experience he could bring. In his appendix Goyder took issue with Brown over the proportion of forests – 'ligneous growths of all sorts', as Brown defined them – to cleared land, which Brown had calculated from Goyder's own figures.[45] Goyder clarified the figures, explaining that what they concealed was that only a very small area was 'first-class timbered country'. The rest was inferior (for the colonists' purposes), and thousands of square miles of scrublands were excluded from consideration. He also criticised the report for containing inaccuracies.

Where he did agree with Brown was on the many benefits created by trees, including the way in which forests modified their immediate climate. It was only 'from the view that forests tend either to attract or equalise the rainfall' that he dissented.[46] Goyder pointed out again that the large areas of scrublands were in fact low forests and that their presence did not diminish the aridity of the regions where they were found. Interestingly, to update Humboldt's work, Goyder reported the activities of a Dr Ernst Ebermeyer in Bavaria, reviewed, as Goyder explained, by the editor in the *Australische Zeitung* in articles published in early 1875. Ebermeyer had conducted experiments which demonstrated, among other things, that forests had little influence on open plains and did not affect the distribution of the rainfall across the seasons at all. But the argument was not only a theoretical dispute about the relationship between forests and climate; it was also a practical dispute about the reason for planting. Goyder's major goal was to supplement the colony's meagre resources of first-class timber by establishing plantations. He saw Brown as diverting energy and resources from this goal to pursue the futile dream of agriculture in the north.

Brown had also been bluntly critical of the colonists, attacking them in particular for failing to plant trees on private land. Their 'universal apathy' would inevitably lead to a time when 'the State forests will be the only real source from which even a supply of firewood will be procured', a situation that would not be remedied until the population was more settled and less transient. (Brown's prediction was essentially right: forestry was to become and remain largely a state concern.) Brown saw transience and lack of attachment – an attitude in which 'home' was somewhere else – as the root cause of the short-sighted and exploitative attitude of the colonists, a view that was the equivalent of Goyder's opinion of their approach to farming, but Goyder objected to Brown's criticisms as 'uncalled for'. He argued that the existence of the Forest Board and the planting that had been 'going on during the last twenty or thirty years' invalidated Brown's comments. Perhaps he was also worried that Brown would alienate support for the new department, the fruit of a determined struggle by a few, and that the alienation would be expressed when its budget was voted in parliament.

With a planting regime established, the board turned its attention to considering the introduction of a 'system of Natural Regeneration' in the reserves of indigenous forest about which Brown was requested to give a report. In his 1879–80 annual report he concluded that the commercially valuable indigenous species, identified as red, blue, sugar, stringybark and white gums, should be 'reproduced' in the forest reserves, as far as possible. The methods proposed were variations on the cultivation of naturally sprouted saplings. In the same report, under the innocuous heading 'The Progress of Planting Generally in the Colony', Brown returned to his pet theme, dismissing the evidence cited by Goyder, and claiming that:

> In these days of so much statistical record, it is now clearly and undeniably proved, that trees have a very beneficial influence on the climate of a country, and that man can, so to speak, by means of these, through a series of ages, make the climate of a country to suit his own requirements.[47]

Goyder's clear insight into the nature of the inland climate had meant that he appeared to the wheat farmers as overly concerned and pessimistic about the realities of the natural environment. By comparison with Brown, he was cast into the role of an uninspired pragmatist, insensitive to the richness and power of the interactions between trees and the surrounding world, especially since he was not urgently concerned with the need to stop the havoc being wrought by the timber cutters and he was not given to preaching.

Having had his own youthful enthusiasms and certainties disciplined by his experiences in the inland, Goyder may have read Brown's declarations with a combination of practical alarm and knowing empathy. While Goyder had been sure of having discovered a watered inland, Brown, his young colleague, was intent on creating one, declaring that:

> to secure a plentiful and certain rainfall in the colony, and to such an extent that its agricultural interests will be augmented and secured beyond the most ordinary visisitudes [sic] of climatic irregularities, it is absolutely necessary for the planting of trees to be undertaken upon it on a universal scale.[48]

The efforts of the Forest Board, however laudable, were nowhere near on a scale suitable to meeting this goal. Brown argued that plantations needed to be established all over the colony and, therefore, on private land. The farmers had the power in their own hands to ameliorate the 'somewhat dry' conditions of their summer months:

> because it is only by them that the vast bleak plains in the country can now become clothed with trees and so form natural reservoirs for the retention of rain by sheltering and loosening the soil, for the keeping up of a more humid atmosphere.[49]

In early December 1881, when dryness had slowed the speeding plough to a halt, Brown was requested to prepare a report on tree planting in the north. After an inspection, which he admitted was only cursory, of the largely treeless country north of Quorn, he returned to reiterate his already entrenched belief that: 'the dearth of forest land in the district is undoubtedly the chief cause of the uncertain rainfall obtained upon it'.[50] Brown urged that existing trees be conserved, in particular the gums along the creeks in the Flinders Ranges and other trees within and beyond the Wilpena Pound. But he also recommended ploughing down to the subsoil, fallowing the soil throughout summer to destroy weeds, and then sowing seeds, or in special cases planting seedlings, to establish forests. His stated reason for sowing seed rather than planting seedlings was the irregularity of the rainfall. Unlike small plants, seeds could wait unharmed for the unpredictable rain to fall and germinate them, after which, he believed, the rains – inexplicably fickle no longer – would 'nourish the young plants in nature's own time'.[51] Goyder, who only a few years before had warned Carr of the dangers of ploughing up the northern soils, attached a memorandum to Brown's report. His statement of opposition to Brown's beliefs was by now practised, brisk and direct. The memo read:

The large area of the Murray Flats and other large scrubs has but extremely limited rainfall. I do not believe that trees exert any influence on the rainfall, although they keep the air moist within a certain distance of the ground, and the roots open the ground to the admission and conservation of water below the surface, which keep up moisture to vegetation and furnishes water at lower levels in the shape of springs. Timber also prevents radiation or heat, but I do not believe that they [*sic*] attract or increase the rainfall. It appears to me that ploughing, subsoiling, and planting large tracts of country would involve a large expenditure, and that without irrigation or regular watering would prove a costly failure.[52]

Presumably because of the board's earlier decision to allow attendance by newspaper reporters at their meetings, Goyder's reply was published (although it was not appended to the annual report in which Brown's original report was included). It did not win him support. Even the *Advertiser*, the same newspaper which had championed his determined stand over the valuations, now found itself grumbling at his 'excess of assertiveness' over matters about which he was not strictly qualified. When the Forest Board was disbanded for other reasons, not long after the reports appeared, the paper took the opportunity to complain that the government treated Goyder not simply as a surveyor, but 'as if he were a universal genius', and that this had encouraged him to interfere in matters in which he lacked expertise. The paper claimed that the board had 'never appeared to be a happy family', largely because of the clash between the two men, and it was only because Goyder had been allowed to cultivate an inflated notion of his own competence that he dared to consider himself wiser than the conservator, 'in what probably appears to him so simple a matter as forest culture'.[53] But the matter was not 'forest culture' at all. It was climate.

When the three non-official members of the Forest Board resigned in January 1882, the board was disbanded and the powers vested in it were transferred to the commissioner of crown lands.[54] Goyder's clash with Brown was not the reason for the resignations: there had been tension between the board and the commissioner for some time over an area of leased grazing land. But it cannot have helped. Nevertheless, the board had been effective, and in parliament the commissioner commended it for displaying those Goyderian trademarks, 'energy and ability'.[55] Without any ceremony, Goyder's 12-year involvement with South Australian forestry – pioneering both for the colony and the developing nation – was over.

Trees were duly planted north of Quorn, and in 1883 Brown confidently reported that, from the experience they had gained in the previous year, it had been demonstrated 'beyond any doubt that, even with only a fairly ordinary season, we can grow trees successfully in the Far North'.[56] He afterwards fell silent on the subject and by the end of the decade, the only mention of any planting north of Quorn was a plan to establish date palms at Wilpena, Farina, 'Hergott' [Marree] and other places in the 'Far North'.[57]

Regardless of the conflict between them, Goyder had continued to profess faith in Brown's abilities, and when Brown presented the board with the manuscript of *A Practical Treatise on Tree Culture in South Australia* in 1880, the board had taken it up and published it the following year, distributing 2000 free copies and selling another 1000 to cover costs. Brown left South Australia in 1890 to take up a similar position in New South Wales and later became the first conservator of forests in Western Australia.

As something of a footnote to his role in forest conservation, Goyder seems to have prompted the decision to declare Australia's third national park (and the eighth national park in the world). In 1890 he told the Public Service Commission that the government experimental farms had been abandoned and were to be cut up into working-men's blocks, but he suggested that the farm at Belair, which had originally been the governor's country residence, be set aside as a recreational area and for the conservation of flora and fauna. 'It would make a capital picnic ground', he assured the commissioners, evoking in that one phrase the mood of their comfortable world.[58] At the end of the following year, on 19 December 1891, it was proclaimed a national park.

CHAPTER SEVENTEEN

The universal genius

The engineer is in our eyes something more humanizing than the soldier: borne onward by the sublime energy of the thing of his creation; harnessing, so to speak, the very elements to his use, and checking and controlling them as might some magician of a fairy tale, he sweeps from place to place, distributing in his way all the gentler influences of civilization, and knitting more closely together the family of man ...

> – Douglas Jerrold (author of the figurative claim that the Australian earth, when tickled with a hoe, laughed with a harvest).[1]

Performing useful tasks is the delight of everyone's life. Clearly, then, the Lord's kingdom is a kingdom of useful activities.

> – Emanuel Swedenborg, *Heaven and Hell*, n. 219.

The surveyor general's confidential letter book shows that for a period during the mid-1870s he was occasionally called upon to examine and report on patent applications, a small but definite sign that Goyder was regarded as a key source of engineering expertise. In fact, he managed to conduct a parallel career as an engineer throughout his long reign over the Survey Department.

Responsibility for the most prominent of his engineering projects, the drainage works in the South-East, seemed to have left his hands for good when the Public Works Department took over on his departure for England in early 1870.[2] In the years that followed, farming towns had been established on drained land and the effectiveness of drainage had been demonstrated to the doubters, but as time passed, the system was beset by a range of problems, including the technical problems of shrinkage of the soil and loss of gravitational fall, as well as issues of control and responsibility for maintenance. Events there would form a curious mirror image to the pattern of events in the north and would result in Goyder being given responsibility again for this major extracurricular activity.

In 1872, in the face of increasing demand for agricultural land, Goyder, as surveyor general, championed the integrity of pastoral districts in the

South-East and the need for pastoralists there to retain some good land for stock breeding. He warned that it would be a great mistake to grow wheat on any of the lands reclaimed for pastoralism, since the drained soils were unsuitable.[3] By the mid-1870s, Mayurra, one of the two areas in the South-East he had selected to open for agriculture under the Strangways Act, was so badly in need of drainage that the survey and sale of land was suspended.[4] But the pressure on the government for land to grow wheat crops was relentless, and, just as they had agreed to farming north of the Line, the government eventually agreed to sell the swampy inland undrained. A large area was opened up in 1876, and, although not a great deal of land was taken up, by 1878 the purchasers, who had assumed that the government would eventually come to their rescue, began to petition for drains. Once again land was withdrawn from sale. The resident engineer of the South-East, Jonathan Rogers, was given the task of designing a system to drain the inland area. Rogers reiterated earlier warnings that Goyder had given about the unsuitability of the drained land for growing cereals, citing previous crops failures as supporting evidence and explaining that the 'half-decomposed vegetable mould' that was found on the bed of swamps lacked sufficient phosphate for a cereal crop.[5] (The effects of superphosphate were not demonstrated until 1882 and its use did not become general until the 1890s.)

In 1879 and 1880, as the northern areas began to dry, flooding rain fell in the South-East and Adelaide was awash with petitions and demands for further drainage. Playford was unsympathetic and sent the surveyor general to investigate and to discuss the issues with the people. Goyder bluntly informed the South-Easterners that the commissioner did not hold himself responsible for their plight, since the land had only been released because of their repeated demands, and that they 'had only themselves to blame'. But he assured them he would 'represent the case to the Commissioner, who had the interest of the farming population at heart'.[6] He was as good as his word and typically forgiving. The selectors in one area, he explained, had 'realised their mistake in common with others,' and 'in sympathy with them and their losses,' he earnestly submitted to Playford, 'that in all cases of this kind … it will be better to allow selections to be cancelled …'[7]

Goyder also knew that the problem was not just the consequences of the farmers' determination to take up unsuitable land. He later recalled making a journey with Playford which revealed that: 'the thing [had] got into a muddle … the whole country was flooded out', because, as he pointed out, the Public Works Department had failed to ensure that the main drain could carry all the waters the smaller drains could bring to it.[8] A great deal of money had to be spent to correct the problem, and command of the drainage works was returned to the Survey Department. Ultimate responsibility for the drainage remained with Goyder for the rest of his career, although district boards became involved in the last few years.

Because of the spread of settlement, from 1881 the focus of drainage works shifted from draining the coastal flats with cuts across the ridges, to draining the inland swamps by connecting them to watercourses draining to the north-west. In his important 1883 report, 'Water Conservation and Development', Goyder had described the levels of the land and the flow of floodwaters in the South-East in greater detail than in earlier reports.[9] In it he again asserted that growing cereals in the district was a mistake, which had led, for the sake of unprofitable crops, to what was 'almost over-drainage' if the land was to be used as pasture. Worse still, failing cereal farmers who had completed purchase were selling their wasted land to pastoralists and to those who farmed only root crops. 'This is a serious evil,' Goyder pointed out, 'as the soil, if properly dealt with will carry a large population ...'[10] To avoid total monopolisation he advocated leasing the remaining land with limited rights to cultivation of certain portions. He later presented the same views to the Land Laws Commission of 1888.

Transport

Like the South-East, the River Murray was the subject of a voluminous parliamentary literature and a variety of proposals for engineering schemes, and the issues associated with the river had exercised Goyder's mind from his early years in the Survey Department. For the colonists, the Murray was a source of frustration because, as it drained from Lake Alexandrina towards the sea, it split into a number of streams, ending in a shallow, sandy mouth whose main channel was constantly shifting. Although the river could be used for the transport of goods within Australia, it did not connect to the sea trade routes. It posed the sort of problems that naturally attracted Goyder's energies, but it was one area where his ideas would come to nothing.

In 1856 Goyder first proposed the idea of clearing the rocky bar in the Holmes Channel (the Mundoo Channel) to make this the main channel to the sea. He suggested this again in 1870 when a select committee of the House of Assembly was established to investigate the possibility of a port that would enable South Australia to claim the river trade that had been seized by Victoria; and yet again in 1875 to the Railway Commission. There were two options being canvassed at the time: opening the mouth of the river, or bypassing it by constructing a canal from Goolwa to Victor Harbor. In 1870 Goyder assured the parliamentarians that he had frequently returned to the report he had prepared in 1856 and had paid continual attention to the river in the intervening years. He remained convinced that clearing the Holmes Channel would make the sea mouth navigable to vessels drawing 16 feet (nearly five metres) of water.[11] In 1875, as chairman of the Railway Commission, he raised these ideas for consideration again as the commissioners contemplated how a railway might help secure the Murray trade. It was generally agreed that a navigable connection between the Murray and the sea was still required, and a naval navigating lieutenant and first-class surveyor, W.N. Goalen, provided expert advice on the subject, but no resolution was reached. In the following year, Goalen produced a separate report on the subject. He approved of Goyder's proposal, declaring that 'it would have had a good effect if carried out when [it was first] proposed', but since 1856 the channel had filled with sand and was now much less than half the depth it had been then, and the river mouth had changed. Goalen argued that it would not be possible to improve the Murray mouth unless the drift of sand into the Goolwa Channel was halted, and set aside the canal to Victor Harbor as a project 'very much for the future'.[12] The matter was dropped.

Although he was not able to see the mouth of the Murray opened, Goyder, through his position as surveyor general and his early training as a railway engineer, was able to influence the structure of the colony's railways, as well as its roads. Central to his approach to rail transport was his belief, which he came to after a great deal of consideration, that in relation to freight it was better to construct cheap lines of railway than main roads: a rail line was permanent and its maintenance was paid for by its users.[13] The essential task of the railways in South Australia was envisaged from the beginning as the cheap transportation of wheat and wool to the ports.

In 1857, the railway had only extended as far north as Gawler. By 1860 it had reached the copper centre of Kapunda and by the mid-1860s it had become obvious that railway extension would have to match agricultural extension. But, because of the drought, Goyder wanted to see the railways extend even further north, into the remote areas. The interests of pastoralists could also be served: when conditions required, the railway would allow them to bring stock south. As the northern agricultural advance began, a committee was formed to investigate railway extension, with Goyder the first to be called before the committee. Within moments of having been introduced, he was invited to prepare a plan for recommended railway construction in the province. Proposals already existed, but Strangways wanted Goyder's independent view, and asked him to, 'let it be your own plan as Surveyor General'.[14] Goyder was also asked for advice on railway gauges, but he excused himself on the ground that he had 'very little' knowledge, and had not been connected with railway works for the last 20 years. Nevertheless, J.B. Neales persuaded Goyder to agree to give an opinion when he produced his plan – after he had had time to think it over. He did as suggested, but he reminded the committee more than once that on this subject he was only giving his opinion, not tendering expert advice.[15]

Goyder returned with an ambitious plan distinguished by lines reaching into the Far North. One line stretched up the western side of the Flinders Ranges to a point south of St à'Beckets Pond, where it divided into two. One line curved east to Yudnamutana, a place where Goyder had collected many mineral specimens. Using one of his favourite phrases, he complained to the committee that he had not yet had the 'leisure time' to test their value, but he evidently believed that mineral resources worth exploiting existed there: at about the same time he put the idea for a northern railway to the Select Committee on Mineral Regulations.[16] The other line headed north, then traversed west through the pastoral country near the southern edge of Lake Eyre. There were two other lines as well, one stretching north from near Burra to the Mount Craig mining and pastoral area, and another, the Barrier line, curving north-east toward New South Wales.[17] The purpose of this last line was to enable South Australia to dominate trade with the pastoral regions of inland New South Wales – 'to practically annex the whole of the valley of the Darling and the Bogan', as Goyder put it – by providing a reliable alternative to the fluctuating Murray.[18] At the same time, he proposed a line connecting the Murray to Kapunda, and it was on this occasion that put his view that the Murray should be made navigable to a port, and that all main roads connecting to it be macadamised. The plan that the committee arrived at recommended only one line that went into remote country, a line that terminated at St à'Becket's Pond, which they considered close enough to both Yudnamutana and the Lake Eyre South pastoral zone.

But within a few years the introduction of credit selection began to have an impact on ideas about the rail network. Two competing approaches emerged: a plan to connect the wheat lands to ports on Spencer Gulf, and Goyder's idea of lines into the north connected to Adelaide. Popular demand was for short lines feeding to northern ports, but the removal of the line of reliable rainfall as a restriction on agricultural expansion had made the need for lines in the north imperative.

To that point, development had been piecemeal. In 1874, the government led by Arthur Blyth promised public works and established a new commission to consider railways. The seven appointed members of the commission were Lieutenant-Colonel William Barber; Samuel Davenport, the prominent horticulturist and pastoralist; the pastoralist Alfred Hallett; William Boothby, the engineer of the Glenelg railway; Henry Mais, the chief engineer of the

Public Works Department; W.L. Beare, and Goyder. The commission met in Goyder's office for the first time on 1 December 1874 and elected their host as chairman, after which the committee continued to meet there twice weekly. From 13 January 1875 until 19 July, 123 witnesses passed through the office, answering 5740 questions, nearly all of them put by Goyder. The minutes of the commission show Goyder's noted courtesy and sociability operating in conjunction with his intensely focused approach to work. He introduced those called with friendly politeness and an informal touch, explaining to them why they had been asked to appear and what specific questions they could answer – matters that were usually left unaddressed. Witnesses were then issued with a fairly open-ended invitation to give their opinion, which was followed up with a thorough interrogation on the topic. It was an encouraging, almost flattering approach, although Goyder was capable of pursuing a topic relentlessly and sharply. At the end, Boothby and Davenport (one of the leading opponents of the valuations in 1864), passed a motion thanking Goyder for his 'very efficient and courteous' conduct of the inquiry.[19]

The new commission not only endeavoured to prepare a comprehensive plan for a rail system for the colony, with established priorities and a construction schedule, it also formulated basic policies related to engineering, equipment, and operations. As a tool for aiding the investigation and decision-making processes involved in preparing the plan, a density policy was established. In more settled districts it was agreed that it should not be necessary to travel more than 15 miles to reach a railway line.[20] The commissioners then considered possible main trunk lines from Adelaide to Blinman and the Far North, to Kooringa (Burra) and the Barrier Ranges, to the Murray, to the South-East, and to the Victorian border in the direction of Melbourne. Consideration was also given to local traffic lines to carry produce from the various agricultural, pastoral, and mineral districts, to the nearest available port. The needs of the various districts were then considered.

Such a comprehensive, organised and logical approach suggested that the chairman had played a strong directing role; the network of new lines the committee finally proposed also showed the hand of Goyder. As Donald Meinig observed, it is almost possible to infer from the map of their proposals alone that the role of chairman had been filled by Goyder, 'for the whole design carries the imprint of his ideas first outlined in 1866–67'.[21] The 1875 design displayed the same two key features as Goyder's earlier, individual effort: having the capital as the focus of the system and long extensions into the interior to serve pastoralism, mining, and trade from interstate. As usual, for a plan generated by Goyder, it was not only confident and ambitious, but represented a broad view, encompassing a range of interests and a long-term vision. The immediate needs of agriculturalists were only one of several concerns.[22]

Although its specific recommendations were not implemented, the plan and the approach to the structure of a railway network that it reintroduced formed the basis for continuing deliberations upon the subject of railways in the colony for years to come. The line into the north was extended up to Farina and found its use serving pastoralists, and the line to the north-east was extended beyond Manna Hill, the location Goyder had chosen for the government's experimental farm, to the Barrier Range, where silver was discovered in 1884. But the dry years and the depression of the 1880s put an end to calls for further expansion.

Transcontinental

In a more indirect manner, Goyder was also connected with a major, transcontinental communications development: the famous Overland Telegraph line. Communication by telegraph using electricity was a revolutionary innovation, which de-materialised the form of the message and reduced the time taken to transmit it almost to nothing. It had been introduced into South Australia by Charles Todd, who as assistant to the astronomer royal had been offered the post of government astronomer and superintendent of telegraphs in the distant colony. Todd was excited by the new technology and had spoken of connecting England and Australia before leaving for South Australia in mid-1855. Within weeks of his arrival he had connected the port by telegraph to the city.

Telegraphy had been taken up with enthusiasm in Australia, with Sydney, Melbourne and Adelaide linked by 1858. Meanwhile, the international system was approaching Australia through Asia, but it was approaching the north-west coast, where colonisation had not yet taken hold. Stuart had established that an overland route, linking the eastern and southern capitals with the north-west coast, was a real possibility, despite the difficulties, but although proposals for linking Australia with the rest of the world were discussed nothing was done. By the time Goyder was preparing to leave the Northern Territory in September 1869, the line from Europe had reached Java and a commercial proposal for a line connecting Port Darwin to Burketown in Queensland was before the government.

Goyder reported that the line could run to the military reserve at Point Emery, a lookout at Port Darwin, and then travel overland by the surveyed road to Fred's Pass. From there it could go by Mount Charles across plains to the Wickham and Nicholson rivers and the Plains of Promise, and finally to Burketown.[23] But the South Australian Government did not want to concede any advantage to another colony, and the move toward a transcontinental connection began. Goyder had originally planned to return from Port Darwin by an overland route, no doubt so he could search out a connecting route, but he was too exhausted and unwell to undertake such a demanding journey.

After Todd had talked with a representative of the British–Australian Telegraph Company, the South Australian Government offered to build the line at its own expense. The company eventually agreed to this arrangement, with the stipulation that the line was to be open by the first day of 1872. Todd was to construct a line that covered over 3000 kilometres of difficult country, country that no European except Stuart and his companions had crossed, and he had little more than a year to do it. To achieve this, the route was divided into three sections, with the end sections contracted out. There were difficulties with the weather and with the arrangements, and Todd was not able to meet the designated date, but in August 1872 the connection was finally made. Thirty-six thousand poles had been had been set up and nine repeater stations built, all in a period of 23 months. Todd was the hero of the hour.

Goyder's quiet involvement was acknowledged later, at a reunion dinner for officers who had directed and overseen work on the project. Most of the officers responsible for taking the line through the centre of the continent had been selected and trained by Goyder, 'and to that training and the experience gained under him they felt they owed a large portion of their success', one of them explained. He had shown unfailing 'deep interest' and had been, 'a friend to whom they could always apply for advice, and one whose great experience in all kinds of exploration or bush travelling peculiarly qualified him to give it'. Other debts were acknowledged. After a lengthy celebration of the efforts of Stuart, John Chambers (one of the pastoralist brothers who had commissioned Stuart's work south of Lake Eyre) lightened the proceedings by observing that it was a 'noticeable fact that this great work had been performed by three little men': Stuart, Goyder and Todd. Goyder was acknowledged again and designated 'their great little man', to general laughter.[24] Many years later, in a speech at Goyder's retirement, Todd himself stated that the successful completion of the overland line 'was largely due to Mr. Goyder', a remark that drew a chorus of 'Hear, hear' from those assembled, many of whom would have been involved.[25]

With the telegraph line in place, Todd, who had taken responsibility for the collection of meteorological data from Kingston as part of his position, was now able to use the relay stations along the line to collect rainfall and other observations. The presence of the telegraph also generated the notion of a railway to run beside it. In 1887 Goyder was called to an inquiry on the transcontinental railway. He stated that the line would be of use in shifting stock and would promote settlement in the interior – to the extent that such settlement was possible. He reiterated his belief that eventually through minerals and tropical products the Northern Territory would be a source of revenue to the government. But while the commission had stated that the main lines of railways should be constructed by the state, Goyder evidently considered a north–south transcontinental line as a service to a limited group, since, although he favoured the construction of such a line, he strongly opposed the use of public money to build it.[26] Eventually the Overland Telegraph led to the construction of the famous Ghan line to Alice Springs, providing the line to the south of Lake Eyre that Goyder had long recommended. (This line was ripped up in the 1970s, and the first transcontinental railway to come into existence runs east–west, not north–south. Adelaide and Darwin were later connected using a line further to the west of the original Ghan line.) The line Goyder had initiated to the Barrier Ranges was eventually extended over the border to the booming mining town of Broken Hill.

Mining and geology

Goyder had further exploited his background in engineering by combining it with his interest in geology to involve himself in mining. Perhaps inspired by Selwyn, he had taken the position of chief inspector of mines while still deputy surveyor general, and retained it when he was appointed surveyor general. His position had resulted in his visiting Wallaroo in 1861 and the North Flinders Ranges the following year. Simultaneously holding the positions of inspector of mines, valuator of runs, and surveyor general, he consolidated a presence in all three areas of major economic importance to the colony: copper, wool and wheat.

Goyder's interest in geology and his encounter with Selwyn also seem to have affected the way in which geological knowledge developed in the colony. Although there had been geological reports prepared in 1846 by the deputy surveyor general, Thomas Burr, and in 1856 by Benjamin Babbage, who had been sent to tour the central highlands in search of gold, Selwyn's report of 1859 was the first geological report of any significance. Selwyn had been invited to conduct his inspection in search of coal, water and gold as a short-term response to a suggestion made in parliament, that a geological surveyor should be permanently appointed to search for artesian water. The general response had been that this was the business of a surveyor general and that experience in the colony was the most important factor. The 'Surveyor-General [Freeling] could see quite as far below the surface as the Geological Surveyor', according to John Barrow, and it was this belief which won out.[27] The subsequent journey Goyder took with Selwyn did more than increase Goyder's knowledge and skills: by making him aware of the limits of his geological insight, it inspired him to see surveyors as collectors of geological information. Once he was in charge, surveyors became the chief source of geological information until the appointment of a geologist in 1882 – the same general period when Goyder was detaching himself from forestry and water conservation, as new administrations were established in these areas.

The search to find gold had not ended with Selwyn, and five years later another attempt to find minerals and gold was undertaken on behalf of the government by Edward Hargraves, the pursuer and promoter of goldfields. Hargraves was to follow a route determined by Goyder, and inevitably he was directed to investigate the central ranges, but Goyder's instructions also sent him out across the north-east plains toward the Barrier Ranges and the as-yet-undiscovered Broken Hill mining field. However, on the basis of a fossil he was shown, Hargraves used his discretion and omitted this section of the inspection.[28] Apart from these tours and the knowledge built up by Julian Tenison-Woods and geologists Ralph Tate, professor of natural sciences at Adelaide University from 1875, and G.F.H. Ulrich, who had worked under Selwyn and lectured at Melbourne University after the Victorian Geological Survey was disbanded in 1868, the chief source of geological information came from the work of the government surveyors, who collected specimens and made observations in the field.[29] The first geological map appeared not long after the first geologist, H.Y.L. Brown, had been appointed.

Goyder's formal involvement with mining finally ended in 1889, after a commission investigating mining followed his suggestion and recommended the establishment of a separate department. The Mines and Goldfields Department was created under the inspector of mines, who remained within the Crown Lands portfolio.[30] The new Mines Department was finally formed in 1893, with Brown as its head, under the lands commissioner as minister of mines. The role of inspector of mines was perhaps the least memorable of all the roles that Goyder took up as part of his overall career. Nevertheless, geology, minerals and mining consistently interested him and it is not inappropriate that it is in that context that he makes his only appearance in police inspector Alexander Tolmer's colourful reminiscences, testing a sample of black sand with a blowpipe and producing 'several globules which he pronounced to be tin'.[31]

Quediable

Mitford had hissed spitefully that 'Goyder was the most talented energetic little man that ever was or ever will be – everything he did was right and everything he didn't was the same', but the government's universal genius was not always competently in charge and at ease.[32] Tucked away in the parliamentary papers is the transcription of an unfortunate comic encounter with the government counsel, Josiah Symon, before the arbitrators appointed to settle a dispute between contractors and the government over the construction of piers at Rivoli and Kingston bays. Goyder's performance was perhaps influenced by the fact that he had been called as a witness *against* the government. In any case, invited to examine scale drawings of piers and provided with a small-scale ruler with which to calculate actual sizes, Goyder used the wrong side of the scale. 'I must have turned the scale around while using it', he offered when his mistake was pointed out. 'My eyes are not as good as they used to be.' In his summing-up, Symon dismissed all the witnesses against the government as useless in their different ways, and mocked Goyder for having 'protested too much' at being one of them – as well as for his unfortunate performance with the ruler. Evidently in full rhetorical flight and enjoying himself, Symon professed himself sorry to see Goyder 'in such company', adding: 'the question forces itself irresistibly upon one, *Que diable allait-il faire dans cette galère?*'[33]

Part four:
ENDURING
MARKS

CHAPTER EIGHTEEN

A house in the hills

Goyder's upbringing, ambition and abilities had combined to generate in him a passionate dedication to giving physical form to a new society, a society that would be, as far as he could ensure it, prosperous, fair, and sustainable. His profound self-confidence and willingness to take on responsibility meant that he never doubted the value of his own contribution, and he rarely stepped away from any role, even when it had been thrust on him unwillingly. Life, he believed, was about duty. Over the decades, government after government benefited from this happy coincidence of personal qualities in their leading public servant. At the same time, they exploited his willingness to overwork himself – just as Light and Frome had been overworked. Inevitably, illness began to affect and then to dominate his life.

Goyder had never really recovered his health after the rigours and stress of the Northern Territory expedition and the shock of the events that followed. The distress in his personal life had been followed by the reversal of his initial success in having agriculture confined to the reliable rainfall zone. That reversal had its own impact on his health. In August 1875 he requested leave because the pressure of trying to keep up with the demand for land had brought on severe headaches and 'giddy turns' (dizziness).[1] In March 1876 he tried again to obtain leave, this time on the advice, once again, of Dr Allan Campbell, who insisted that he needed a change from his official duties. Campbell, one of whose brothers-in-law was the Supreme Court judge and deputy governor Samuel Way, seems to have been an ideal friend and physician for Goyder: a similarly devoted, spiritually inspired, hyperactive, and unrelentingly thorough counterpart in the field of public health and also in societies for promoting the arts, literature and science. Campbell had grown up in a village near Glasgow and had studied architecture before fragile health prompted him to take up medicine instead. In South Australia he put both skills to work as one of the founders of the Children's Hospital (now the Women's and Children's Hospital) in North Adelaide – his name still appears in large letters on the front of the old building – and was prominently involved with other institutions to benefit sick children. As Goyder was 'Little Energy', Campbell was 'the children's friend'.[2] Campbell was a Bible Christian, not a Swedenborgian, but like many members of the New Church he was also a supporter of homoeopathy and this fact, combined with the presence of homoeopathic medicines in the story about Jones of the Survey Department and the inland sea, suggest that Goyder probably was too. In his youth, Campbell recalled, he had given himself: 'to God and to the service of humanity', and took it as a fixed principle 'to enter every open door of usefulness'.[3] While Goyder had chosen the public service because of the unique scope the position of surveyor general had then offered, Campbell had maintained a professional practice and became a member of the member of the Legislative Council. He was an active supporter of federation.

For his prescribed holiday, Goyder had wanted to visit New Caledonia with William Morgan, then the chief secretary, but Campbell was concerned about 'brain symptoms' and, in the tropical heat, a recurrence of the problems that had debilitated Goyder in the Northern Territory and for some time after.[4] He advised that Tasmania or New Zealand would be more suitable and recommended that Goyder be granted three months leave, on the understanding that more time might be required. Goyder proposed to start for Tasmania at the end of the month (typically, going via the South-East to confer about the survey there), so that he would be back by the time parliament reassembled. In response, the Cabinet offered to pay all his expenses if he would visit New Zealand and provide them with a confidential report on the state of the lands, public works and the workforce, the effect on the soil of permanent settlement, and anything else he thought relevant or worthwhile.

Since Goyder seems to have later taken another holiday in New Zealand, it must be supposed that he enjoyed himself. Certainly, when the government made their proposal, he declared himself 'most happy' to comply, no doubt in part because he continued to struggle with financial problems.[5] Only three years before, in 1873, he had warned that he intended to resign at the end of the year because his salary was 'wholly inadequate to meet the requirements of my family and position'.[6] Despite 'utmost economy' he had been forced to spend three hundred pounds in excess of what he had earned, and was heading towards insolvency.[7] The *Register* had reported that he complained of an income that was: 'totally inadequate and utterly disproportionate to the salaries paid to the heads of departments in other colonies'.[8] In doing so, he had opened up the problem of civil service salaries in general. A new *Civil Service Act* was passed the following year.

Goyder had ended up with a salary of £950 (£1000 had been agreed on) and, after the Act, new Regulations quickly followed. Some of these restated old Regulations; others introduced entirely new principles. The short daily working hours (10.00 am to 4.00 pm weekdays and 10.00 am to 12.00 pm on Saturdays) were left intact, with mandatory unpaid overtime as required and no mention of a lunch break (which the *Register* thought 'rather hard'[9]). The Regulations endorsed the established principle that an officer should hold no competing posts or receive any reward from outside the service for official duties. Measures were introduced to ensure a minimum standard of competence for candidates for cadetship.

In 1878 Goyder resigned again, this time because of a financial disaster he felt he could not hope the government would address – or so he claimed. The company to which he had been making life insurance payments for 20 years had failed. At 52, and with damaged health, he could no longer afford the premiums required to establish an equivalent policy and he believed he had only 10 years work left in him. If he remained with the service until his retirement, he would be left with a retiring allowance 'that would barely afford the simplest necessaries of life to my family however favourably the same may be invested'. He had decided instead 'to endeavour to form a private practice before the increase of age and its attendant infirmities place such a step beyond my power'. What made his situation really pressing was that he was not the only one affected by illness: 'several of my family are invalids and can only hope to obtain suitable and remunerative employment where more than ordinary consideration is extended them'. He intended to solve this problem by employing them himself. He conceded that his prospects of success were uncertain, but any chance of success seemed preferable to the certain penury of leaving things as they were. He asked for six months leave of absence before his resignation became effective to enable him, while still enjoying a secure income, to set the business up and make connections. The need for this was inescapable:

> as I have not only separated myself from many likely to prove of service to me, but by my official acts have frequently made enemies instead of friends. I have also failed to invest in any way – not alone from want of means but because the adoption of such a procedure, tho' leading to large profits – might have swayed my judgement in the performance of my duties which were doubtless simplified by the absence of such influences.[10]

Goyder wrote this confidential letter on 8 March 1878, but it was not sent. Eleven days later he wrote another version, only slightly modified but more urgent in tone, and accompanied by an assurance that he would continue to monitor departmental affairs and report on official correspondence. The government responded with a salary increase of £300 – more than was earned by most of his staff in a year – taking his basic salary to £1250. Together with forage allowances for two horses and a house allowance of £100, which he was already receiving, this came to a total of £1454, making him the most highly paid public servant in South Australia.

Warrakilla

With this increase Goyder immediately set about working out a scheme that would enable him to profit from his knowledge and secure his family's future, without generating a conflict of interest. In 1879 he bought the Wheatsheaf Hotel, a hotel originally licensed in 1842 and with land attached. The old inn, which settled into the shoulder of a slope on the road through the stringybark-covered hills between Mylor and Echunga, overlooked flats leading down to the River Onkaparinga. Gold had been found there in 1852 and over the next decades it continued to be found, but despite the initial excitement, the Echunga field had proved very modest.

Goyder employed one of the leading architects, David Garlick, to design additions and alterations to the building.[11] These seem to have consisted of adding large rooms for entertaining at either end of the old inn, including a ballroom with chandeliers. The series of drawings Goyder had made in the back of his field notebook while in the Northern Territory suggest that he had been considering a house with this plan for a decade. At that time the house on Robe Terrace had only just been enlarged, so it is unlikely that he was rede-signing it from scratch again and, overall, the plan in the notebook seems to match his later house: the dining room and the room behind it even have the same dimensions as those in the sketch plan. Since he never wasted time doing things twice, it would be typical of Goyder to employ the plans he had made a decade before. According to a local historian, the name he gave the house, 'Warrakilla', was a term acquired during his stay in the Northern Territory. The spelling first given was 'warekilla', and the meaning was defined as 'place of the changing winds'.[12]

The inn had been surrounded by an acre of garden and a 40-year-old orchard of cherry trees. Edwin Smith had already expanded into the hills, leasing Leawood Gardens, at Devils Elbow on Mount Barker Road, near Glen Osmond, sometime around 1874. Smith took advantage of the cooler climate there to grow European and other trees in the open. Leawood Gardens was taken over by Edwin's second son, Harry Arthur, who expanded the gardens and extended the fruit plantings. His cherries and strawberries were known for their excellence and were popular with tourists to the area and, like his father, Harry grew award-winning roses.[13] Goyder believed in the value of the hills country for producing not only fruit but also fodder, and on the flat section of his property (between 50 or 60 acres, or 20–25 hectares) he set about putting into practice the ideas he had espoused in his major report on water, developing an elaborate irrigation and drainage system to make intensive farming possible. The system was planned to support 17 acres of apple, pear and cherry trees and to produce fodder for a large number of cattle; the cattle would be milked to make cheese, as well as fattened for market.

Goyder had the land around the house landscaped to form a semi-circular terrace planted with a couch grass lawn (which stayed green almost all year), bordered by hedges of hawthorn and laurel. Other areas of the garden were planted with decorative American catalpas and alternate plantings of plane trees and oak trees, with pine trees nearer the road. Below this extended the cultivated flat. Pigs roamed about, feeding themselves on peas and acorns. Basket willow, which could be profitably harvested, grew on the river banks.[14] The garden, and probably the estate as a whole, was perceived at the time as 'experimental'.[15] The author of a series of articles on the stately houses and gentlemen's country seats being established in the hills captured Goyder's purpose with clarity and tact:

> To the professional man, who looks forward to retiring from business and spending the latter portion of his life in liberty and comparative ease of mind, an estate in the hills, gradually improved with an ultimate view to profit, is one of the best possible forms of investment. Such an estate is Warrakilla, the residence of G.W. Goyder ...[16]

Goyder's will makes clear that the estate was also intended to provide a residence and income for his wife and unmarried daughters after his death.

But after beginning the transformation of the old inn and its grounds, Goyder found himself unable to work because of a severe attack of 'lumbago' (pain in the muscles of the lower back). The trip to England and America to buy boring and pumping equipment in 1882 was actually, like the trip to New Zealand, another journey which was meant to double as a recuperative break. The *Register* commented that it was a pity that the surveyor general was unable to have a period of absolute rest, and that it would be better for his leave to be too prolonged rather than too short, but contented itself with the conclusion that it seemed:

too bad to map out an extensive field of enquiry for the most hard working of the Government officials when he is leaving on a holiday; but Mr. Goyder has such a superabundance of energy that what would be hard labour to many will probably be a pleasant recreation to him.[17]

A note in Ellen's autograph book suggests that he did indeed need a real break, and that he succeeded in having one. Under the poem 'Mizpah', after recording the dates of his journey to England, Ellen had written: '19th May 1883 – New Zealand – returned 24th July 1883'. The departure date is only nine days after the date of the major report 'Water Conservation and Development', in which Goyder had delivered a picture of his vision for the colony.

Apart from encouraging a belief that nothing was too much him, his reputation for 'superabundant' energy and invincible vigour has led to some mistaken claims after his death. Goyder has repeatedly been described as insisting on driving a buggy from the hills to Victoria Square and back again every working day after he moved to Warrakilla – a distance of nearly 65 kilometres.[18] In reality, he had driven (or been driven) just over seven kilometres of winding and undulating road to the Aldgate railway station.[19] The station had opened in 1883, and Goyder seems not to have taken up permanent residence until then. A letter from Mrs P.E. Horn – although undated, it appears to have been written in the 1890s – suggests that the Goyders continued to maintain a city base. The Horns, who were friends of the Goyders, had been enjoying themselves so thoroughly at Warrakilla that Mrs Horn enthused that it was no wonder 'when you get up here you are *such* an age before returning to Medindie again'.[20] A letter from Ellen gives a glimpse of life at Warrakilla, with members of the family, guests, and Goyder himself, coming and going. An invitation to dine at Government House could not be accepted because he was out of town.[21] (A property in Medindie which continues to be associated with Goyder is the mansion, The Myrtles, on Hawkers Road, which has been described as having been built by Goyder. In fact, the connection is distant and slight. According to a heritage survey undertaken for the Walkerville Council in 1987–88, the house was built on a lucerne paddock he had owned and sold to William Wadham, the land agent, in 1877. After changing hands again, this property was bought by the wealthy merchant H.C.E. Muecke in 1880, by which time a house called The Myrtles had been built.[22] However, in 1880 Goyder did own a residence on Hawkers Road, with a value assessed by the council as £100.)

Although Goyder had departed for the hills, Edwin Smith remained at Clifton Nursery, and worked there until his retirement in 1898. Two of his children had also moved to the hills, Harry taking over Leawood Gardens, and his eldest daughter, Ellen Hannah, who was married to Frederick Wood, running the Woodlands nursery at Kyneton, south-east of Angaston.[23] After Edwin's death in 1907, Clifton Nursery was taken over by another son, but it eventually changed hands, and was subdivided for housing after the First World War.

Maturity

The change from being entirely dependent on his salary and insurance to establishing a hill resort and productive farm had paralleled changes in Goyder's professional life. As the northward movement of the wheat farmers finally came to an end and the heavy demands on the Survey Department ceased, Goyder's own administrative dominion began to diminish as he relinquished responsibility for departments he had created to administer forests and the search for artesian water. Prompted by his failing health, he seems to have traversed the period from 1878 to 1883 well aware that he was moving into the final phase of his career and life and determined to see both his projects and his children provided for or standing on their own feet, independent of his presence. The report on water conservation, a plan for the future of the whole colony and the fruit of decades of observation and reflection, was written at Warrakilla in the year that he began to live there. That his life was moving on was pressed home by the death in 1878 of his father David; in 1884 Ellen's 94-year-old mother, Susannah, died at Warrakilla and, two years later, Goyder's mother Sarah died in London, at a similarly advanced age.

At this time, too, the life of the family matured into the phase of marriages and grandchildren, although Goyder and Ellen still had young children of their own. The year the old inn was purchased, 20-year-old Isabella had married George Frederick Hallett, the son of Alfred Hallett, who had been a member of the Railways Commission chaired by Goyder. Alfred Hallett had come to the colony in 1838 to manage a copper-mining venture, but eventually became a partner with his older brother, John, in pastoral enterprises. The brothers had been among the leading pastoralists in the colony until the drought of the 1860s. A few years later 'Gerty' married, and both daughters soon had children. George and David married at the end of the 1880s, at the ages of 33 and 27 respectively. Norman married in 1891 and Alexander in 1894, and the young twins later. Of the 'several' members of his family described by Goyder as invalids, only Florence, the eldest daughter, can be identified: Goyder wrote a letter in 1874 requesting leave so that he could travel to Melbourne by steamer to attend a consultation about her continued treatment for 'disease of the eye'.[24] Florence remained single, as did Mary Ellen.

In 1885 legislation enabling the surveying and sale of working-men's blocks was passed, and in the same year the town of Mylor was surveyed, on the way from the house to Aldgate. Early the following year the first working-men's blocks were released in the area. Mylor has been described as 'Goyder's little village', and Warrakilla does seem to have functioned much like a manor house: 'many local people were employed to help in the stables, kitchens, and as waitresses and maids.'[25] Local histories record that Goyder lived 'in grand style entertaining celebrities from all over the world' – presumably on a semi-official basis – and, along with local people working in the house, members of the family may also have helped.[26] In 1882, in a competition for the most artistic table laid for a stand-up buffet held by the Horticultural Society at the Adelaide Town Hall, one of the Misses Goyder had scored a 'runaway victory for after-noon and evening tables', with nearly twice as many points as her nearest rival.[27]

While Warrakilla was certainly one of the grand houses in the hills, it is noticeable that it is only local histories that extol the exceptional splendour of life there. The ballroom with the chandeliers was 'the talk of the area', and certainly must have provided a contrast to the slab huts occupied by the surrounding community of 'blockers', many of whom needed to seek employ-ment in towns or in the city – or at places like Warrakilla – to make ends meet. However, the local reputation of Warrakilla should not obscure the fact that Goyder was dependent on his salary, and although the house was a grand one, neither it nor its residents were in the same league as the solidly wealthy fami-lies and their dwellings. Goyder's obituaries only recalled that he liked to gather with his old cronies for late-night rubbers of whist.[28]

In 1887 Goyder's health broke down again, and, once again, he resigned. Since the situation could not be resolved with money, Goyder's indispensability worked against him to some extent. In November he was granted eight months leave, but he was not given official notice of this until after a major problem with pastoral leases was resolved. He began his leave early in 1888 and returned to work on 1 September.[29] He later described the threat to his physical well-being as having been 'incipient paralysis'.[30] If this was correct, he was probably lucky to have his physical capacities intact.

Whatever medical condition 'incipient paralysis' indicated, it can neverthe-less be said that Goyder was lucky still to be alive. By his own admission, he had only just escaped from the Northern Territory with his life, and he had nearly died years before, when illness had forced him off his horse while travelling ahead of his party in the salt lakes country.[31] He had been in serious danger from lack of water on another occasion south of Lake Eyre, as well as in the country behind Fowlers Bay, and anecdotes published after his death show that he had come close to death twice in the South-East, in both cases because

Goyder and John Harvey, Ellen's first child, at Warrakilla. Since John Harvey is still in short pants, the photograph was probably taken in the early 1880s.

he could not tolerate being unable to complete the task he had set himself.[32] On one occasion thirst nearly claimed his life after he left the track in the Ninety-Mile Desert and run out of water, but had been determined to carry on. Fortunately, the man accompanying him had found water and saved them both. On another occasion, he almost drowned when, in the absence of a boat, he paddled a packing case into the middle of one of the lakes in the South-East to take soundings. The crate had capsized, but, once again, Goyder had been rescued by the man accompanying him, who was a good swimmer and able to drag him ashore. There is an element of recklessness in both tales that does not cohere with the thoughtful planning, attention to detail, and concern for the wellbeing of men and animals that characterised his early explorations. These incidents and those recorded in Milne's 1863 diary, in which Goyder ran several miles to retrieve a tobacco pouch, forged a cross-country 'shortcut' with an axe, and, in his determination to get back to his gravely ill wife, took off on an exhausted horse only to continue on foot in the darkness when the animal could no longer go on, present an aspect of his conduct that was not preserved in any official source.

The incidents in the Ninety-Mile Desert and with the packing crate in the lake were recounted in the popular paper *Quiz and The Lantern*, where Goyder was described – after his death – not as dedicated, capable and energetic, or courteous and kind, but as a man of intense determination, intrepid, impulsive and abrupt.[33] The evidence suggests that he was all these things and that there were two sides to his character: the perceptive observer, thorough administrator, prescient planner, and master of detail who was genuinely considerate of others and devoted to the public good, and the hot-tempered, impulsive, ambitious, determined and passionate individual. Usually both sides of his personality worked together, one setting the course, the other providing the drive, but occasionally they disengaged. Since the two incidents in the South-East seem to have been the recollection of staff in the department at the end of his career, it is likely that they took place in the later 1870s or the 1880s, and, if they were accurately reported, Goyder evidently had not become less driven or more cautious with age. In his early years, exploring and surveying in remote country, his survival had been largely due his intelligence and iron constitution. In his later years, he seems to have owed a great deal more to good fortune.

For all that, it would be wrong to imagine Goyder as so self-confident that he did not know or would not concede his own limits, or so courageous and intrepid that he never acknowledged fear. In 1866 he commented to a committee that crossing Lake Alexandrina to Milang by steamer in rough weather was 'no joke', and soon after, having been asked about MacDonnell Bay as a harbour, he replied: 'Well, I was horribly frightened there, for I thought that we were going to the bottom. I know I thought it was a frightful place.' Immediately after he showed that he was not tempted to give advice beyond what he considered the limits of his expertise. Asked if he thought the harbour could be improved, he responded, 'I could not form the slightest opinion about that'.[34]

Colonial gentry

The possession of a house in the hills was typical of members of the 'old gentry' of South Australia, not all of whom would have been considered gentry in England, many being of more humble origin than the Goyders and Smiths. Along with the house, Goyder exhibited other characteristic features of this group. He was a member of the Adelaide Club, the family lived in Medindie, and their eldest son (at least) was a pupil at St Peter's College, 'predominantly the school of the landed and professional classes'.[35] But, compared with others in this category, Goyder lacked the substantial wealth that distinguished most. Nevertheless, in the early 1890s, the Goyders enjoyed the distinction of being included in *Burke's Colonial Gentry*, where the family was said to be an offshoot of the family Gwydir (meaning, presumably, the Wynns of Gwydyr). Arms with a Welsh motto ('*heb dduw, heb ddim*' – literally, 'without God, without anything') were provided. The shield, showing three rampant lions on a background half-silver and half-black (*per pale argent and sable three lions rampant countercharged*), had been conferred on the baronet Sir Henry Etherington in 1775, suggesting that the connection was through Goyder's mother's family, although the crest of a lion holding a sword was not that of Sir Henry.[36] In any case, the family's sense of its Welsh background was evidently maintained with some pride. George Arthur Goyder earned an entry in the early twentieth-century *Cyclopaedia of South Australia*, where he was described as residing at a house named 'Gwydyr' in Gilberton. He had earned this entry as an analyst and assayer in the mining industry and his innovative work has been described more recently in a history of colonial technology.[37] His publications indicate that he shared his father's interest in water, and a family interest in botany, especially trees.

CHAPTER NINETEEN

Steward of all the Crown Lands

The name of a Surveiour is a French name and is as moche to saye in
Englysche as an Overseer.

– Book of Husbandry and Book of Surveying, 1523

A journalist visiting the expanding northern areas in the spring of 1874 saw the
camps of the surveyors dividing the new hundreds into sections as charmingly
picturesque Bohemian oases, with their 'clean white bell-shaped tents' dotted
about, and their kitchen 'caboose', reminiscent of a Gipsy wagon, nearby. The
writer had no difficulty in identifying the resident 'Bohemians' as 'smart young
fellows from Adelaide, who, although once "doing the block" [promenading] as
dapper members of swell society', were now 'roughing it', to the evident benefit
of their health and wellbeing, as they went about their 'pleasant and useful
work'.[1]

What these once-dapper young bohemians were really doing was imposing
a deliberately created design on the land, an 'intricate pattern of roads; fences,
paddocks, towns and, eventually, farm boundaries and administrative areas,
which formed the framework for all subsequent geographical activity'.[2] Goyder
had overseen their work from a straight-backed chair behind a large desk in his
office overlooking Victoria Square, at the centre of the colony's neatly planned
capital.[3] Like the founders, he was giving conscious geographical expression
to the ideals of the new society. Light had given basic shape to the colony by
locating and designing its capital; Goyder was responsible for much of the
framework that followed. Thus the sense of clarity and order that had inspired
the layout of the capital was extended into the country beyond (or imposed on
it), but not as an unvarying pattern. Goyder's survey was an evolving work that
responded to economic, social and political changes, as well as to technological
developments in farming and to the different types of land that were colonised.[4]

Hundreds and townships
Of the three major administrative units of the sectional survey, the most
important were the intermediate unit, the hundred, and the sections into
which the hundreds were divided. The largest unit, the county, was relatively

unimportant. Up until 1860, the size and form of the section had remained fairly rigid, while the hundred had been flexible and the subject of experimentation. During that period, the function of the hundred had been to relate section holders to pastoral commonage, and there had been a preference for using the distinguishable features of the landscape to define hundreds. As commonage declined, the need to have easily recognised boundaries was less pressing.

The sketch of a 10-by-5-mile block in the recommendations Goyder made for pre-survey examination in 1855 suggests that he was already thinking in terms of the hundred as the basic pre-survey planning unit at that time.[5] In 1860, new legislation established the hundred as a tool with the potential to be used to control the pace and direction of the survey and settlement, and when Goyder became surveyor general he set out to clarify the situation, as well as to improve the quality and accuracy of the surveys.[6] From that point on, the hundred became firmer, while the size of the section varied, increasing in response to technological developments in farming, to different soils, and to an underlying concern to see that farming families could create a decent livelihood for themselves. Under Goyder, sections increased in size as the survey moved out over the rough country on the eastern side of the Mount Lofty Ranges and onto the sandy plains of the mallee country, until the 640-acre, or square-mile section became common. In 1865, Cabinet agreed to a proposal that each hundred should contain one township.[7] The hundred was then used to address important social needs, through the system of roads and the location of urban centres with educational, social and religious facilities.[8]

The agricultural land rush of the 1870s, which saw 92 hundreds proclaimed between 1874 and 1879, inevitably affected procedures. In 1875, in response to pressure from the squatters, the House of Assembly had moved that before the declaration of any new hundreds it was advisable to obtain a report on the suitability of the land for agriculture.[9] (It was these reports that Goyder had used to warn against unreliable rainfall.) The effect of this reporting requirement (which also enriched the content of the surveyors' diagram books) was to increase the usefulness of the hundred as a unit for planning and making landscape. At the same time, the hundreds became increasingly similar in design, as the surveyors hurried to churn them out at the rate required. The model the new hundreds increasingly resembled eventually made its official appearance in the first edition of the *Handbook for Government Surveyors* in 1880.[10] Half of all hundreds in South Australia were proclaimed and surveyed during Goyder's administration.[11]

The requirement that each hundred contain one township meant that government-designed townships tended to be scattered across these landscapes at approximately equal intervals, although not all townships marked out by the Survey Department achieved actual existence. Before 1875 the only government-surveyed townships had been in pastoral areas, where speculators were unlikely to undertake such development. (The site of Willochra, beside the Mud Hut, was among the first batch of these.) Goyder had been doubtful about the value of these pastoral townships because he believed inns and public houses would inevitably be established there, luring the shepherds and station workers with 'ardent spirits, rum and beer'.[12]

Like the hundreds, the townships were laid out according to a model, in this case the one Goyder had first made public when he had included it in his advice to Finniss on the survey of the Northern Territory, given in 1864. A grid straddling a river with a central square and a neatly rectangular margin of parkland, the model modestly emulated the capital and resulted in little adulatory echoes of Adelaide – sometimes stretched around topographical features, sometimes straight from the plan – being dotted across the country.

Charles Reade, the first South Australian town planner, criticised the way in which the standard design had been applied, often with insufficient attention to drainage, grade of roads and local features. He also criticised the ring of parklands as disconnecting the centre of the township from the suburban development beyond.[13] But Goyder had himself grown up in the industrialising north, where booming populations of working people crowded into dank little dwellings huddled around narrow streets and lanes. By the mid-1850s, the worst areas of even the colonial capitals had become 'gothically squalid in their wretched stench and putrefaction'.[14] Adopting a model based on Light's plan meant that, in whatever way these townships were to develop (if they did), their design would at least ensure that the inhabitants of their original centres would not be denied access to space, air, light and unbuilt land.

Straightened out

As aerial photographs show, the most characteristic and enduring imprint of the South Australian survey is the straight line, an outcome of what Williams has kindly described as Goyder's 'quest for order and regularity'.[15] In his report on the triangulation of the north, Goyder described how the unclaimed country was divided into runs by extending straight lines from one trigonometric point to another, so that 'quadrilateral or trilateral figures may be produced at pleasure, according to the area required'. A vision of the surveyor as William Blake's *Ancient of Days*, reaching down with a pair of dividers, generating tidy figures 'at pleasure' is hard to escape, and his regret that the entire landscape could not be dealt with in this way is almost palpable in his comment that:

But for the claims that existed prior to my arrival, this system might have been adopted throughout, and with advantage; as it is, it has been carried as far as the triangulation permits.[16]

Enthusiasm for straight lines was not just an expression of map-dominated geographical tidy-mindedness. Frome had instructed that natural features were to guide the layout and boundaries of sections, but where no such guides were present, the sections should be square. Since the landscape of much of South Australia tends to flatness and is noticeably free of rivers and lakes, there was plenty of opportunity for the straight line and the right angle to flourish, and they had begun to do so in Light's time. In the absence of topographical features, working to the grid of latitude and longitude enabled the country to be divided relatively cheaply and efficiently, especially on the mallee plains where movement was difficult.

One area where surveyors have demonstrated a frequently troubling fondness for the straight line since the days of the Roman Empire is in the layout of roads, and the surveyors of South Australia were no different, although the temptation to produce impassable roads because of a fondness for the straight line was a hazard that Frome had warned against. District councils had also made their objections known. In his satirical history of the colony, Mitford took the Survey Department to task, no doubt drawing on his experiences as a teamster. With their 'chains, and measures, and theodolites, and staves and little flags, and sundry other gimcracks', he wrote, these 'wise men':

> went forth into the wilderness rejoicing greatly; and they measured the four quarters of the earth, and all the surface thereof; but in straight lines only did they exercise their skill, so that the earth became like unto a chess-board, and the Governor saw that it was good.
>
> And it came to pass that as they measured the lands only in straight lines, turning neither to the right nor to the left, lest their calculations should become disordered and themselves be lost; so did they mark out beautiful roads thirty cubits in width that went *nowhere*; others ran over perpendicular mountains, or through impassable swamps – many went into deep water-courses, or hideous caverns – and some did terminate in the depth of the ocean, or hung suspended on lofty gum trees; and the people were exceeding glad, but the working bullocks lifted up their voices and wept.[17]

Mitford's criticism prompted others to complain, and Goyder responded by warning that any surveyor discovered to be the author of such a road would be dismissed.[18]

Method

Along with developing the sectional survey of the agricultural country, it was he, Goyder claimed, and not Freeling, who had been the initiating force behind the drive to triangulate the pastoral country. In 1857, maps of the pastoral country and the runs had been transferred from the office of the previous commissioner, Bonney, to the Survey Department. These were the plans compiled from information given by pastoral lessees which had been at the centre of the Moonta dispute. Most of the diagrams of the leases had been made only with the help of a prismatic compass, with the distances estimated on horseback. Very rarely were any marks made on the ground to establish boundaries in a definitive way, and to complicate matters further the lessees had the habit of making their own arrangements for transferring portions of their runs, without informing the Land Office. Goyder claimed that after he'd got to know the country he realised that the runs were so far out of position that if the plans were followed, many settlers would lose the head stations where they lived and from where things were managed – as well as the most valuable portion of their runs. He had urged Freeling to make a general survey of the entire province, 'getting the starting points, and laying the runs down actually on the plans for the information of the Government, and then dealing with each case separately'; Regulations were framed enabling this to be done. The Regulations included a clause which made the surveyor general the final referee in a boundary dispute, but, as Goyder put it, Freeling 'still refused to incur the responsibility'– implying that he had refused it at least once before – 'and a special Act had to be passed to allow licensed surveyors to deal with these disputes'.[19]

It was as a result of this that he had been sent exploring and, eventually, triangulating in the northern country, and it was in response to the demands of this work that he had developed the method which transformed the preliminary work necessary for triangulation and the inevitable search for water into a system of crisscrossing of the country that added informally located detail to create a map. A simple plan which accompanied Goyder's report on the northern triangulations shows the area investigated during the expedition, but the hand-drawn map reveals the detail with which the central area, just below Lake Eyre South, had been mapped. It was because he was so pleased with this map that he had recommended the use of this method for the Northern Territory in 1864 and later in 1880 in the *Handbook for Government Surveyors*. Under the heading 'Rough Triangulation (Preliminary)', he stated that this technique had enabled large areas of the Lake Eyre country to be mapped, 'with a considerable degree of accuracy in an incredibly short time'.[20] A sample of a journal entry was provided to illustrate the systematic method of making and recording observations while moving between points. It reads like an entry from one of the valuations books.

By 1890, near the end of his career, South Australia had taken much of the shape, in terms of land use, that it would have throughout the twentieth century. Pastoralism had continued to expand into the arid areas, and the only land still to be converted to agriculture was in the Murray Mallee and on the Eyre Peninsula. All but the desert region, the area around Lake Eyre North and the country along the inland border with New South Wales – over half the area of the province – had been connected by triangulation. That task had become more and more difficult the further the surveyors distanced themselves from settlement. Outside the settled districts, Goyder explained:

> all provisions must be carried with the party, and frequently the want of water produces the most terrible sufferings. This was especially the case before camels were used in connection with the conveyance of survey parties. In consideration of the privations which they have to undergo we pay the members of such parties a somewhat higher remuneration than the ordinary staff we receive, and on their return to town they receive also some consideration in the shape of extended leave.[21]

The trigonometric survey was suspended during 1890–91, but resumed in the following year. In typical late-imperial style, Goyder described the staff and equipment accompanying the trigonometrical surveyor and his assistant as: 'a party of eight white men, an Afghan camel-driver, and two black boys, with equipment of camels, camp equipage, rations, instruments, &c'.[22]

By 1890 the defects introduced into the sectional survey through the use of contract surveyors had been remedied, and as settlement extended and transport became easier and cheaper, the initially high cost of the survey decreased rapidly. Costs had also been reduced by gradual changes in the system of survey: although transport distances had increased as the sectional survey moved out, the increase in the size of sections as the South Australian environment was better understood meant that less work was entailed in marking out larger blocks. Another gradual change reported by Goyder was that 'roads were defined with greater care'.[23]

Although Goyder had not begun his career as a surveyor, he ended it as a prominent and authoritative figure in the profession in Australia. In England, surveyors had been slow to form an enduring professional association – 1868 is a comparatively late date – and they were even slower in South Australia. An association was not established in Adelaide until 1882, at which time – during an urban land boom – it was felt that some form of official accreditation was needed to separate the capable from the unscrupulous. Goyder was the founding patron of the resulting body, the Institute of Surveyors, and the founding president was Charles Todd. Four years later, a system of licensing by examination was introduced through legislation, with examinations to be conducted by a board of five examiners, including of course, the surveyor general. Three members were to be nominated by the newly formed institute, and in the years that followed, Charles Todd was always amongst them.[24] The board held its first meeting in Goyder's office in March 1887 and inevitably Goyder was elected chairman.[25]

When he finally retired, after 41 years in the Survey Department, 33 of them as surveyor general, Goyder left an organisation recognised for its efficiency. The quality of the Survey Department's cartography had received international acknowledgment much earlier, when a map submitted by South Australia was awarded a gold medal at the Paris Exhibition in 1877. After his death, William Strawbridge, his successor, commented that he had 'introduced and continued a system of survey and recording which was unequalled in any other part of Australia'.[26]

Surveyor and steward

For all his spleen, Mitford had posed a pertinent question when he had wondered what to make of the position of surveyor general as it had been filled by Goyder. The contempt he expressed was founded on the presumption that surveyors were men with theodolites, and therefore Goyder's major excursions beyond those limits were nothing more than the expression of his personal sense of omnipotence. But the extraordinarily expanded role that Goyder and successive governments had together created can be viewed in an entirely different way – as an unexpected reversion to the roots of his role as surveyor. In a history of surveying commissioned by the Chartered Institute of Surveyors to celebrate the centenary of their professional association in 1968, the historian F.M.L. Thompson argued that, although the history of surveying is usually presented as an unbroken chain of development from the ancient world to modern times, in reality the skills accumulated and developed in various cultures of the ancient world were lost with the collapse of Roman civilisation.[27] According to Thompson, the profession of surveying in the Western world has its origins in the Dark Ages, and the real ancestor of the modern surveyor is the medieval *seneschal*, the estate manager or steward. These manorial stewards were usually little involved with boundary definition and map-making, since boundaries tended to be stable during this period; they were more likely to be occupied ensuring the maintenance of good agricultural and husbandry practices and in overseeing the exercise of customary rights of usage and access.

The modern surveyor began to emerge when social and economic changes turned land into property, transferable by sale. These transactions created a need for a clear definition of boundaries and hence for cadastral surveys and plans. The development of instruments for measuring space and time and determining location, which were necessary for navigation on the open ocean in pursuit of the trade that was developing at the end of the Middle Ages, meant that tools were available for the creation of charts and the development of the skills of mapping. The combination of needs and means produced the evolving phenomenon of the surveyor.

What reintroduced the stewardly function to the role in the colonies – especially in the smaller colony of South Australia, where the effect of Wakefield's theories was amplified by an expectation that the surveyor general might function as an expert in all scientific and engineering matters – was the need to consider and protect the public interest in a new society that had first to establish itself physically on the ground. In 1825 instructions had been issued in England which defined the public purposes for which land was to be set aside in New South Wales. Reservations were to be made for:

> Public roads and internal communications ... the scites [*sic*] of towns and Villages ... for the erection of Churches, schoolhouses, Parsonage houses and burying grounds ... [and of] vacant grounds, either for the further extension of Towns and Villages, or for the purposes of health and recreation ... [and] in the neighbourhood of navigable streams or the sea Coast, which it may be convenient at some future time to appropriate as Quays and Landing Places ...'

In general, land was to be reserved for 'every object of public convenience, health or gratification ... before the waste lands of the Colony are finally appropriated to the use of private persons ...'[28]

First in the line of responsibility for converting these instructions into reality were the surveyors in the field. One historian has commented that they were assumed to be able to 'convincingly integrate the skills of the field scientist, spatial planner and Crown real-estate agent', and that 'any consideration of the public interest required a keenly developed sense of imagination, objectivity and public spirit' – it was 'all a question of striking a balance, of shaping the landscape appropriately'.[29]

In the all-encompassing way in which he had filled and enriched the role of surveyor general, Goyder had undertaken all of the modern technical work of the surveyor while unintentionally returning to the origins of the profession in the work of the medieval *seneschal*. At the beginning of the 1890s, a commission of inquiry that had been working its way slowly through the public service finally reached the Survey Department. After investigating the extensive range of activities undertaken there, the inquiry reported that, in addition to its fundamental role, the department had responsibilities for working-men's blocks and special provisions to encourage fruit growing on them, pastoral lands and their irrigation, scrub (mallee) lands, drainage and irrigation in the South-East, Aboriginal reserves, gold, university and education lands, the agricultural college, land boards, roads, forests reserves and travelling stock – and this was after Goyder had shed forestry, water conservation and mining. This was a greater range of concerns than occupied the survey departments in other colonies, the commissioners noted, before arriving at the succinct conclusion that: 'the Surveyor-General really has the stewardship of the whole of the Crown Lands'.[30] The unique stewardly function of his role as surveyor general had finally been acknowledged and expressed.

Guiding lights

In advising governments and managing the vast estate of South Australia's crown lands, Goyder had been guided by a framework of values, attitudes and beliefs that remained stable throughout his career. These principles were articulated, along with aspects and details of his vision for the colony and its development, in his many reports and in the evidence he gave to numerous commissions and committees. Two of the reports are actually histories, intended for the guidance of the public, including intending immigrants. The first, *The Past and Present Land Systems of South Australia*, for which he is the presumed author, was published by the Surveyor-General's Office in 1881. The Public Service Commission later directed him to replace this with a new pamphlet explaining the nature and conditions of land tenure in the colony, despite his pointing out that, since all the good agricultural land was taken up, a pamphlet intended to assist and encourage immigrants was no longer required. The *Report on Disposal of Crown Lands in South Australia* of 1890 which resulted was welcomed as timely because the parliament was currently concerned with land legislation. Although the report was judged to be 'carefully compiled', it was also found to be dauntingly lengthy, 'very elaborate', and virtually impenetrable without the aid of an index and summaries.[31] Although Goyder did not see the document as necessary, he nevertheless brought the full weight of his detailed knowledge to bear on the writing of what was, in effect, a history of land legislation in the colony. His account of the repeal of the Line and the events that followed is distinguished by a distinct change of tone, from the bureaucratic to the colloquial and narrative, and by a pervasive air of regret.

The goal of all Goyder's work was settlement, and settlement that was as dense as possible. In 1888, before the Land Laws Commission, he spoke of aiming 'to get the greatest number of people on the land that the country will carry if it is to be developed'.[32] In 1870 he had described with awe how in Victoria, after only 35 years, 'that wonderful colony has attained a population of three-quarters of a million'.[33] But, sold or leased, land was the government's greatest source of income, and politicians tended to view it purely in that light. Goyder was unwavering in his belief that establishing settlement, not extracting income, was the goal. In 1865 he had told the committee investigating land sales that he questioned 'whether Government should care much about the highest price being realized, so long as they get *bona fide* farmers to settle on the land'.[34] In 1888 he reiterated this message to the Lands Laws Commission: the government's aim should not be 'to secure the highest price for the land, but to put the most suitable persons as permanent settlers upon it', and for that purpose he favoured the creation of a land board.[35]

Goyder expected that the pattern of settlement would be structured in such a way as to promote fairness, the public good and an enriched quality of life for those settling. When arguing against free selection in 1870, he had been concerned not only with its likely impact on pastoral lessees, but also with the neglect of the public interest and the welfare of the community that he was certain would result from the dispersal of settlement. He pointed out that there had been claims that allowing settlements to become dispersed in the other colonies had promoted bushranging and lawlessness. Settlement that was not thoughtfully and logically contained could also be expected to damage individual lives, the community, and even the land itself. Distance effectively denied education to the children of remote settlers, he argued, and the result was, 'hardy men and women, brought up to a practical and laborious life ... but it is the development of muscle, without a proportionate expanse of mind'. The combination, he warned, was a 'growing evil'. Children denied the 'inestimable advantage which education confers become less useful members of society, and enjoy a less rational life than those more favourably situated'.[36]

His evidence to the 1888 commission reveals a constant concern for the situation of the rural poor, for the battlers and strugglers, and the hard lives of their families. He criticised the credit land leases because: 'a man is required to do certain things and he has not the means. A bad season or two throws him into arrears, and the penalty of forfeiture hangs over his head ... The conditions are objectionable.'[37] He did not support making rabbit-proof fencing compulsory, although he believed it would be effective and the best thing for the state 'perhaps', because the cost would break the small leaseholders, and when asked if lessees should be given the power to recover the cost of rabbit-proof fencing from their neighbours, he replied that it was a question he could scarcely answer: 'it might ruin a hard-working but poor man to compel this'.[38]

Goyder's demand of settlement itself was that it should be permanent. Permanent settlement required genuine settlers, people who intended to establish a way of life for themselves and their children, and who had the skills and resources to carry out successfully the task they had set themselves. In 1870 he had warned that credit arrangements that were too liberal worked against the goal of expanding permanent settlement and encouraging fresh immigration by rewarding abuse of the land with the opportunity to move on.[39] In 1890 the same concerns underpinned his explanation that the pastoral country in the north-east and west was:

> not a poor man's country. Large areas are required to carry anything like a paying quantity of stock, and any attempt to subdivide these lands below a proper area will result in disappointment to the State and ruin to the individual who holds them, or if not ruin, at least a precarious existence, coupled with the maximum amount of anxiety and toil.[40]

The fundamental requirement for permanency then was appropriateness. The way in which land was divided and disposed of had to be appropriate, just as the people and their goals, resources and capacities had to be appropriate. And, of course, the country and climate had to be appropriate. But 'appropriate' land use meant more than merely distinguishing between cattle or sheep country and wheat lands. Goyder recommended that South Australians make an active investigation of agricultural activities in similarly arid climates around the world, and later in his career suggested not only kangaroos, wallabies and rabbits, but even ostriches as a better alternative to sheep and cattle in certain pastoral country.[41] The kangaroos, wallabies and rabbits were presumably intended for meat, and rabbit fur was used to make the felt from which hats were made, but ostrich feathers were in great demand for hats and decorative plumes, and fortunes had been made in South Africa farming ostriches.

Permanent settlement required that the land be treated with care. In 1879 a committee on credit land laws had as their 'special object' to determine 'whether we wish to have a permanent settlement on the land', and asked Goyder's opinion on how things could be improved. They were told:

I have already stated in evidence that I believe that as soon as we have alienated all the land within a certain distance of market which is fit for agriculture – as soon as it will be impossible for farmers to break up new ground – their attention will be drawn to a better system of agriculture, which will make settlement permanent. I am quite certain of that. Farmers must be taught that they must put back on the soil what they take off, and then they will have good crops, as in former days.[42]

One of the complaints Goyder had about free selection was that, ultimately, the lack of education resulting from dispersal of settlement was felt by the land itself: 'it is the want of this element which conduces to bad farming and the utter exhaustion of the soil'.[43] Not surprisingly, he also regarded the opportunity to own land as an inducement to improve and care for it, and evidently admired those who farmed their small allotments intensely.[44] Goyder's own background would have disposed him to a generally 'enlightened', scientific approach to agriculture, but his induction into the horticultural world of the Smiths must have added content and force to this general outlook, transforming it from a general disposition into a detailed, concrete understanding.

That Goyder had been at one with the ideals and methods of South Australian settlement from the beginning is indicated by his early unquestioning acceptance of the practice of putting all land up for auction at the minimum price of one pound per acre. In 1865, when confronted by the daring suggestion from Lavington Glyde, who had sent him out to value the runs the previous year, that land could be roughly valued and put up at between one and three pounds, Goyder answered: 'I should like to think over that question before replying; it is one I have not thought of'. He did think about it, and, at a second appearance before the same commission a week later, agreed that it might be a good idea to increase the minimum price of rural town allotments and suburban blocks – to deter the squatters from buying them all up – but did not agree that it would be worthwhile with other rural land.[45]

Although at first he had seen agricultural land as freehold and pastoral land as leasehold, by 1879 he was one of the early voices supporting perpetual leases for agricultural land.[46] He even considered that under certain circumstances it might be acceptable to sell pastoral land, since in some cases a leaseholder:

gets fond of a locality, and makes everything he does very substantial [and] I am conscious he would do what he undertook with more heartiness with the knowledge that all would ultimately become his own property, and the home of his children. That is one motive. Another is, that in this country we have incurred a very heavy burthen by borrowing money, and this is likely to be a very serious tax on the farming community as well as on the community generally, and if there was a right to purchase the results would help to remove this incubus.[47]

Goyder's acknowledgement of the fondness leaseholders could develop for the areas they had settled in stands out for its injection of warmth into a usually dry realm of discussion. He also had no hesitation in describing the attachment of the Aboriginal people to their land, not as some sort of primitive response or 'tribal instinct', as his contemporaries were prone to do, but as love.[48] His willingness to recognise and respect these feelings in others indicates that affection and respect for the land was something he had found strong in himself.

CHAPTER TWENTY

Final years

Throughout the 1880s, following the trend introduced in the *Scrub Lands Act* of 1867 and the Strangways Act of 1868, the beliefs and assumptions about land and rural settlement on which the colony had been founded continued to be superseded. Inspired by the settlement of the Victorian Wimmera, legislation was passed in 1884 which enabled farming land to be held on lease. Within a year, 3000 farmers had transferred from credit selection to leasehold.[1] Two further changes, which Goyder supported, met with resistance. One was the notion of perpetual leases, which were thought to be dangerously alien to English culture. The other was the idea of local land boards, which had been suggested by the farmers. The government investigated this possibility, but nothing was done on this occasion.[2]

By 1888, after six years with insufficient rain in the north for agricultural purposes, it was inescapably obvious that the failure of farming there was due to more than a poor choice of sites or unfavourable financial arrangements. This led to the formation of the Land Laws Commission under chairman Clement Giles. (Later, another prominent commission member was F.E.H.W. Krichauff). The commission, which met in the first half of the year and heard from nearly 290 witnesses, most of them farmers, represented the serious attempt of a community to come to terms with the realities of attempting to farm the north. Rainfall was the only aspect of the natural environment that the commission was directed to investigate.

When his time came to be questioned, Goyder – predictably – explained the area's unreliable rainfall to Giles and assured him that he had issued warnings repeatedly and that he had never considered the northern country fit for agriculture, although he still adhered to the belief that mixed agriculture and pastoralism could be successful. He also suggested that it might not be too late to implement his proposals of 1882, providing for resettlement south of the Line, where even on inferior country the farmers would at least be sure of crops.

The region Goyder suggested was the South-East, where he had set aside a large area for 'miscellaneous and special purposes'.[3]

The commission was a public vindication of Goyder's position on the Line. The very first finding was that:

> a large number of farmers have for years been struggling to make a living by growing wheat where the rainfall is insufficient, and too uncertain to allow of this industry being carried on at a profit; whilst on a considerable area in the south-east the same want of success has resulted from other causes.

The second finding recorded the vast losses involved, highlighting that many people had been reduced to destitution, while the third noted that a large number of selectors north of Goyder's Line had converted their selections to leases and taken up grazing. A further finding added dolefully that the leases were too small, because in much of this country the capacity to carry sheep was also severely limited by the rainfall. There were seven findings in all, to which the commission added 13 recommendations, all aimed at putting an end to the hopeless situation and enabling the failed farmers to remain in the colony.[4] These included the introduction of local land boards and perpetual leases.

Pastoralism

The year 1888 was also the year in which many pastoral leases, first taken out in the late 1850s and early 1860s and subsequently extended, were due to expire. (These were the leases that Goyder had been required to deal with before he was granted leave to deal with his threatened paralysis.) For pastoralists on leases beyond the agricultural frontier, the 1870s had been good years, with the wool clips resulting from good seasons and improved practices attracting better prices than anticipated. Since many of the original leaseholders had retired at the end of the drought, their leases had been scooped up and consolidated into very large runs.

Although urban politicians and farmers had accepted that agriculture could not be taken further, there was a continuing concern for the future of the second generation on the land. The result had been a call for these large, consolidated runs to be subdivided into smaller runs and auctioned at the expiration of the leases in 1888, and an Act, which included the creation of a pastoral board, was passed in 1884 to achieve this. Goyder was opposed to the principle of auctioning leases on their expiry, on the grounds that the original discoverers should be allowed to benefit from their discovery, although this hardly described the situation where runs had changed hands.[5] The pastoralists were too disunited to mount any effective opposition, and the situation in which they found themselves deteriorated even further when an Act was passed in 1887 which interpreted the value of improvements on a run as the cost of those improvements – as Goyder put it simply: 'They said the value of a well means the cost of the well'.[6] This put the pastoralists in a difficult situation, because in reaching their agreements with the banks, they had chosen to interpret the value of a well as the impact the water would have on stocking rates and the value of the run as a whole. The banks became reluctant to extend credit. When the runs were finally auctioned in 1888, prices were higher than had been anticipated. Despite that, most of the original lessees managed to hang on, but both Goyder and the Pastoral Board were concerned about the survival of the squatters, even though the legislation of 1884 that had directed the dividing-up of the runs had also stretched the period of tenure to 35 years.

A major problem that had come to bedevil the pastoralists during this period became a preoccupation in Goyder's last years. Rabbits and wild dogs – dingoes, feral domestic dogs and crosses of the two – were ravaging the pastoral industry. Rabbits in plague numbers had stripped the country of vegetation and the dogs attacked stock. The method Goyder favoured for controlling rabbits, which were also a problem for farmers, was to enclose areas with wire-netting and to require landowners to kill all vermin on their properties.[7] He had promoted this idea from the beginning and was obviously proud of having done so, because in 1891 he embellished one of his remarks to the Pastoral Lands Commission with the information that:

> it is now fifteen years ago since I recommended the adoption of wire netting, when the inroad of rabbits first took place from Victoria to South Australia; and Dr. W.J. Brown was the first to introduce wire netting into the colony, and now its use has become general.[8]

As Goyder foresaw, since the government was now the proprietor of the leases abandoned in 1888, the rabbits were the government's problem. He suggested that the land be let at a low rate, with a vermin destruction provision in the lease, and that the ranges, where he expected most of the rabbits would remain, be fenced off. His hope (inadvertently anticipating the introduction of myxomatosis and calicivirus) was that, when their numbers were large enough, an epidemic would break out. He also invested considerable amounts of time and energy investigating means of destroying the invaders, with his office publishing a pamphlet with advice on the use of bisulphide of carbon and another dire method, phosphorous poisoning.[9] In the last year of his career, as the first witness before a commission investigating vermin-proof fencing, Goyder testified that it was useless to send out government parties to destroy rabbits and dogs, reiterating his belief that the only way to get rid of them was to fence blocks in the occupied areas and close off the unoccupied land.[10]

While rabbits remained a problem to farmers, by 1893 they posed much less of a problem to the pastoralists. Their numbers had been reduced, not by epidemic but by drought. The Pastoral Lands Commission set up that year focused more attention on dogs. Along with others, Goyder assured the commissioners that the only way of getting rid of dogs – he believed the wild domestic dogs to be the most troublesome – was to fence the outside country.[11] One of the commission's recommendations was the immediate erection of a wire-netting fence from Morgan, where the River Murray bends almost 90 degrees to run south, to Winnininnie, north-east of Teetulpa.

During the 1890s other dog-proof fences were erected in different parts of South Australia to prevent dogs from attacking stock. These were eventually replaced by a fence which runs from the New South Wales border east of Lake Frome, through the salt lakes between the northern Flinders Ranges and Lake Eyre South, to the north of Fowlers Bay. Over 8000 kilometres long, it is the world's longest unbroken barrier. The Dog Fence divides the pastoral country, with sheep being grazed within the protection it offers and cattle grazed beyond. Goyder's connection with the creation of this 'line' extends no further than the support he gave to fencing the country to control rabbits and keep out dogs, but he may well have helped to determine its form. Always looking for a solution to a government problem, in 1893 he told the Vermin-Proof Fencing Commission that he had devised a way to reduce the cost of fencing from 70 pounds to 40 pounds per mile. The bottom of the netting was to be fastened into a ploughed furrow in the ground, and the top selvedge fixed to the two strands of barbed wire stretched above, advice which appears to have affected the form the fence took.

After helping to 'open' the country in his early years, Goyder had gone on to do everything he could to ensure that pastoralism was maintained in the inland. It is now recognised that the effect on the rich but fragile ecology of the arid zone from grazing animals and introduced feral species has been the loss of over half the mammal species. The question that inevitably arises is how Goyder would have viewed his efforts in the light of this knowledge, but it is the sort of question which cannot be answered simply. Nothing in Goyder's writing reveals that he had any interest in indigenous mammals, but given that there was no concern about them at the political level, and that he was never asked to give his opinion, it cannot be assumed that his silence represents complete indifference. With human settlement as his goal, he obviously took for granted the replacement of larger mammals with stock, but personally he may not have been indifferent to the impact of introduced feral animals. At the very end of his career he included the conservation of fauna as a reason for establishing a national park at Belair, and since his foresight was based partly on a sensitivity to interconnection and interdependence there can be no doubt that he would have appreciated the insights of the science of ecology, including the larger implications of the loss of biodiversity, had they been available to him.

The South-East

In the South-East, the attempt to drain the inland by directing water into existing drains had led to renewed flooding on the coast.[12] In 1890 Goyder produced another comprehensive plan for draining the region, one that, unexpectedly, did not incorporate all his previous proposals.[13] Instead, it was 'an uneasy compromise between opening-up the north-flowing natural water courses and cross-cutting the ranges'.[14] If put into effect, it would have involved 530 kilometres of major drains, affected half of the area of the South-East, and cost nearly a million pounds, but the economy was in depression and the government was not interested. In 1892 a royal commission was established to deal with the issue of drainage in the South-East and it produced a 170,000-word report, to almost no effect. Goyder recommended to the commission that responsibility for drainage be given to the Public Works Department, provided it appointed people competent to do the work.[15] This eventually came to pass, but only with his retirement. The engineer who took over came up with his own plan for draining the South-East at only a fraction of the cost of Goyder's plan, but it was no better accepted.

Another phase of drain-cutting occurred in the first quarter of the twentieth century, focusing – as Goyder had suggested back in 1863 – on main drains that passed at right angles through the ridges to the sea, although the effects were not fully experienced until networks of local drains were added in the middle of the century. In the late 1960s, conservationists fought to have the last original undrained swamp, the Bool Lagoon, separated from the drainage system and preserved. Like the Coorong, it is listed under the Ramsar convention as a wetland of international importance. It is a breeding habitat for waterbirds and a refuge during drought. Curiously, the landscape that now surrounds it is not widely recognised as a drained landscape now vastly different from its original form. In the promotional brochures found at the tourist information centre at Millicent – described as 'the city of drains' in a much earlier public relations exercise – drainage is simply not a topic that is mentioned. Goyder's early visions of a grand navigable canal, with branches reaching out into the surrounding country, and the South-East as a sort of pastoral Venice with steam cars puffing along a network of embankments were never realised and, on the whole, the drains on the South-East are unprepossessing, neither picturesque nor dramatic. The smaller ones are little more than ditches, entirely lacking in marketable romance, especially at a time when it is wilderness, not artifice, that is the increasingly scarce commodity. A traveller from outside the region can pass through the South-East without being aware of the nature and origins of the country they see around them, although this could change now that some restoration of the wetlands has begun. Goyder's early plans would have created a man-made landscape with a distinctive, and perhaps engaging, character, but the water would have been drained with ruthless effectiveness, and the wetlands would have vanished altogether.

The Murray Mouth

In 1890 the Commission on River Murray Waters was formed as a result of concerns about the riparian rights of the three colonies involved: New South Wales, Victoria and South Australia. One issue troubling South Australians was that the massive quantities of water planned to be pumped from the river for irrigation by New South Wales and Victoria would jeopardise the river's navigability. Goyder was a member of the commission as well as a witness before it. Addressing his fellow members in an early hearing, he directed their attention to the river mouth, quoting the reports he had made in 1863 and 1856, and encouraging the commissioners to inspect the ridge he had examined in 1856, which they did. He was no longer interested in utilising it as the foundation for a road, but, having taken soundings along its entire length and located a bar, he wanted to use it in another scheme to solve the problem of the Murray mouth. He told the other commissioners:

> I think that it is now desirable for the Commission to ascertain whether this bar cannot be utilised in the way of enabling an embankment to be thrown up from a point on the main land at Pelican Point right on to Hindmarsh Island, in such a way as to close all the shallow channels between the island and the Coorong. Then there might be constructed a lock at Goolwa, which would retain a certain quantity of water and be a means of communication between the sea and the river.[16]

The object of the lock was to secure free navigation of the river and access to Goolwa for fishing on the Coorong, while the embankment was intended to raise the level of the water behind Hindmarsh Island and exclude the salt water. Goyder saw this as a necessary response to the reduction in flow resulting from the expansion of irrigation in the other colonies, expansion that was far from complete. Settlers below the bar would still have access to fresh water, he claimed, and, 'the blacks would not be interfered with in their fishing operations, for the different classes of fish would be obtainable as at present'.[17] (This was an important consideration in the context of a plan he had in mind to declare the Coorong and surrounding lands an Aboriginal reserve.)

Goyder was not alone in suggesting locks and embankments, although proposals for weirs to be placed across the channels usually located them further down and in fewer places. Later, a plan prepared by the government engineer, Moncreiff, was tabled before the parliament, but despite all the proposals, the issue lapsed. It was not raised again until the early 1920s, when soldier settlers were placed on drained land in the area. By the 1930s, the parliament was constantly presented with questions on salinity, water levels and barrages, and by then, as in 1870, there were large numbers of unemployed men, providing an incentive for the government to undertake a major scheme of public works. By 1940 there were six weirs and locks on the river below Blanchetown. A series of barrages was eventually constructed that followed the line Goyder had investigated in 1856, the Aboriginal route across the bar. These barrages serve exactly the purposes he had identified in 1890: they separated the fresh and salt water, preventing the intrusion of seawater during periods of low river flow, and stabilised the lakes. (They also support pipes carrying Murray water to Adelaide.) The effect of these works was an overall rise in the water level of about one metre, which enabled gravitational irrigation of the flats of the lower reaches and permitted safer navigation. But it also drowned thousands of river red gums and flooded large low-lying areas above the locks, increasing the levels of salinity.[18]

Water

Goyder continued to think about water, as the cluster of testimonies and reports from the beginning of the 1890s shows. In his *Report on Disposal of Crown Lands* of 1890, he continued to put forward his plans for making use of the floodwaters of the inland, explaining that pastoral country could be improved by diverting water from watercourses out across the plains. Often while travelling on the arid plains to the north and north-west of the Flinders Ranges, Goyder had come across a line of trees marking a watercourse, which: 'proved to be bank high with flood waters from the ranges … whilst very slight works placed at the foot of the ranges, sufficiently high up to direct these waters to the crowns of the spurs or low watersheds, would divert and distribute the waters over the plains, filling waterholes … resulting in rich and succulent vegetation wherever the water reached'.[19]

He continued, too, to advocate irrigation for agriculture on the Madrasi model, telling the Barossa Water Commission that any scheme he recommended:

> would be of a simple kind, which would be carried out by the occupants of the land. I altogether object to the Government undertaking large works for the purpose of irrigation, and the scheme I propose would be to show that the waters at present running to waste could be conserved by all the occupants of the land one after the other and utilised for irrigation purposes.[20]

His belief that the government ought not to involve itself in large schemes encouraged him to view favourably the establishment of large-scale irrigation systems on a commercial basis. Goyder was also keen to see the low-lying land to the east of the great bend in the Murray and the flats south of Morgan embanked, drained and irrigated. In the *Report on Disposal of Crown Lands*, he outlined a scheme for doing this, pointing out that he had suggested such a scheme as far back as 1857.[21]

Goyder's role in the development of water resources in the colony was acknowledged in his lifetime in Edwin Hodder's *History of South Australia*, published in 1893. After stressing the critical importance of water to the colony, Hodder presented Goyder as the first person to be entrusted with any active work in connection with water conservation (that is, collection), in the context of his role as one 'to whom the colony is indebted for the prosecution of many important works'.[22]

Original custodians

Even though Goyder had come to acknowledge that the country had already been occupied by people whose lives, like his own, were governed by laws which regulated relations between people, and between people and country, it is not likely that he ever grasped that these people were the stewards of this country. In his later years, in his own stewardly role, he seems to have made a serious attempt to provide for a viable way of life for the people dispossessed by the settlement that was his primary concern, by allocating good land for their use. At the same time, like many others, he believed that Aboriginal children, for what was considered their own good, could be relocated without any apparent concern for their family and community relationships.

While he was still deputy, Goyder had been one of the officials involved in the selection of the site of the Point McLeay Mission Station – now known as Raukkan. In early 1859, George Taplin, a young missionary appointed by the Aborigines' Friends Association, had made a journey of reconnaissance to the country around the lower Murray and the lakes, where the people now known as the Ngarrindjeri (or Narrinyeri) were suffering from disease, social breakdown and the impact of alcohol. Taplin intended to establish a mission. At Wellington, he was advised on the choice of an appropriate site by the local inspector of native police, by one of the leading squatters of the district, and by Goyder. The advice was basically good. Although there was criticism that the site was exposed, located as it was on the south-eastern shore of Lake Alexandrina, it was ideally placed in relation to fishing and other resources, and at the time was relatively isolated. Moreover, it was in the heartland of the Ngarrindjeri and was already an important meeting place. It was to become a secure refuge.

In the 1870s Taplin outlined the impact on the Ngarrindjeri of the destruction of wildlife and useful vegetation in the area of the lower Murray. By the 1890s, the degradation of the environment of the lower Murray, Lake Alexandrina, Lake Albert and the Coorong was so well advanced that it was not only the Ngarrindjeri who were affected. The numbers of Murray Cod and Mulloway were depleted (the fishermen blamed pelicans and cormorants), the salinity levels in the lakes and the river were changing, and, to benefit the diners of Adelaide, virtual war was being waged against ducks and other waterbirds, using large-calibre guns, sometimes mounted on punts.[23]

There was also a threat to the reserved land itself. In the second half of the 1880s the strong movement to make small blocks of land available to working men for growing their own food and supplementing their income resulted in legislation that left all Aboriginal reserved land potentially at risk of acquisition. Point McLeay, along with two other major reserves, Poonindie, on the Eyre Peninsula, and Point Pearce on the Yorke Peninsula, were specifically excluded by the *Crown Land Act* of 1888, but another reserve – the Needles, on the Coorong – remained unprotected. The reserves around Adelaide were available under the scheme, and the Working Men's Blocks League, which funded an inspector to locate suitable land, had applied to divide land near Lakes Alexandrina and Albert into small blocks.[24]

When the Public Service Commission reviewed the Aboriginal reserves in 1890, Point McLeay was found to be in a very different situation from the other mission stations. Most of its 1740 hectares were remote from the mission or rented – in fact, a substantial proportion of all Aboriginal reserves constituted small blocks that were rented for farming – and the approximately 320 hectares that remained had to support 136 people. Attempts to irrigate and work the land had failed, and the people there complained of their enforced idleness. The result was an expanding population of people dependent upon the station. When interviewed by the commissioners, Goyder stressed to them that the land near Lakes Albert and Alexandrina was 'peculiarly adapted for the purposes of aboriginal reserves, the lakes being full of fish and wild-fowl, and the land providing suitable camping places ... [the] Commissioner of Crown Lands has therefore given instructions that the lands in question shall be kept for the use of the aborigines'.[25] In fact, these lands were so rich that, prior to European settlement, they had been the most densely populated area on the continent. Goyder also recommended that 'every piece of land abutting on the waters [of the Coorong] should be reserved for the use of the natives', and that non-Aboriginal people should not be permitted to engage in commercial fishing or shooting there.[26] A Ngarrindjeri witness made the same point to the commission, explaining that it would be: 'a very great help if the Coorong were kept as a sort of reserve to supply food and employment for the blacks'. That witness (identified in the minutes only as 'another native') also recommended that no one should be permitted to engage in commercial fishing.[27] Goyder advised that, if the reserved land near Lakes Alexandrina and Albert was expanded to include the Coorong and surrounding land, the Point Pearce Mission on the Yorke Peninsula might be amalgamated with Point McLeay. One of the commissioners responded by drawing his attention to the fact that: 'our evidence shows that the tribal instincts and habits of the aborigines are strongly against their removal from any part of the country to which they are accustomed'. 'Yes,' Goyder agreed with alarming ingenuousness: 'only the children can be moved about in that way'.[28]

The new history of land settlement he had been directed to write in 1890 gave Goyder another opportunity to put forward his ideas and he did so with lively enthusiasm. The most important Aboriginal reserves in the colony were Point McLeay and the Needles, he asserted, and these reserves:

> should be made the resort of all the aboriginals whose love of the country round their birth-place does not form an actual bar. These lands possess all that can be required by aboriginal life. There is timber, grass, herbs, shelter, game waterfowl, fish, and water; in fact a black turned out with a blanket and tomahawk and a few lines and hooks can get a living at once, and no further portion of the lands abutting on these lakes, or on the Coorong should be alienated, but be held inviolate so long as they are needed, and there are aboriginals to resort to them; nor should the timber be destroyed, as the grubs for bait are obtained from them. There are salt and fresh water fish, from cod to coongulty, crayfish, cockles, mussels, wallaby, kangaroo, and smaller marsupials: turkeys, geese, ducks, & c., too numerous to mention, as well as swans and pelicans. I cannot place too great emphasis on the value

of these reserves, and trust that they will be properly cared for, and as the landholders who reside thereon show them the greatest kindness and attention, as well as find work for them when they wish to be employed, careful oversight on the part of the trustees appointed or acting in their behalf being all that is required to ensure their comfort and happiness, as far as such can be attained.[29]

It was typical of his foresight and attention to detail that he would point out the necessity to maintain trees to ensure bait for fishing.

About 18 months later, in November 1891, he used a report on land transactions to campaign again. Reserves in the area had been slightly increased as leases in the vicinity expired, he explained, and:

a proposal has been made, I am informed, not only to increase the area set apart for the use of the blacks, but also to make the institution self-sustaining, part by the establishment of fisheries, various trades and also rabbit-preserving and skin-curing.

It was the best site available, he reiterated, and if the reserve was extended into the desert, it would be possible to:

utilise these lands for the preservation of kangaroo, wallaby, and rabbits, and also ostriches, thereby extending the usefulness of the natives and increasing their domain to an extent calculated to give comfortable homes and living to the present natives of the district as well as to the young born in other parts of South Australia, where it is impossible they can be so well looked after as on the lands adjoining the lakes. Were this suggestion put in force the other reserves might be disposed of, leaving the Point Mcleay Mission the only aboriginal reserve in South Australia, and which, if extended as opportunity offers, will prove ample for all purposes connected with South Australian aborigines.[30]

The enthusiasm of his tone suggests that he believed he had found a real solution, one that would secure a viable future for these people, based on a continued use of traditional skills coupled with the development of a local industry that would engage with the settler economy. Point McLeay remained South Australia's oldest Aboriginal reserve until, in 1974, it was returned to the Ngarrindjeri people, who restored its former name.

On the other side of the Gulf there was another important early reserve, the Poonindie Native Institution, which had been established near Port Lincoln on the Eyre Peninsula, initially on land purchased in 1851 by an Anglican archdeacon. Its function was to take children of the dispossessed and devastated Kaurna people of the Adelaide region and to raise them in isolation, with the purpose of retraining them. Goyder was never involved with Poonindie, but it is possible that his influence helped ensure its survival for the time that it did. A highly successful agricultural operation developed at Poonindie and the reserved crown land was regarded by the settlers with barely concealed envy. In 1865 Goyder told the inquiry into crown land sales that there was a considerable area suitable for agriculture in the Port Lincoln district, 'but the best portions are included in the aboriginal reserves; by far the best land there is included in the land set apart for the Pooindee [sic] Institute'. Immediately after, apparently realising the implications of what he had said, he tried to correct this emphasis by stressing that the Poonindie block was 'not available at all, being set aside for another purpose'.[31] And so it remained until 1894, the year of his formal retirement (although in fact he effectively retired at the end of 1893). In 1894, the Poonindie Institute was brought to an end, and the farming land divided up for sale to the settlers who had been demanding it. It seems unlikely that this was merely a coincidence.

CHAPTER TWENTY-ONE

A gentleman of the Civil Service

Failure on the part of private attempts … [is] sympathised with by most, if not all; but failure by a Government officer is a very different affair. Doubtless public criticism does good; but many able, painstaking men are sensitive, and retire from a position which renders them liable to censure so publicly administered, and those who can bear unmerited blame are few.

– George Goyder, 1883[1]

When Goyder had become surveyor general in 1861, activities for which he was responsible – principally the division and sale of land for agriculture and the administration of pastoral leases – generated nearly half of the total government revenue from all sources, while only costing the government just over three per cent of that total income. It was therefore critically important to the government that the crown lands were administered efficiently and effectively, meaning that whoever occupied the position of surveyor general was undoubtedly the most important public servant in the colony. Given Goyder's drive and ability and the constant flux of governments forming and then collapsing, it seems virtually inevitable that an influential 'king' of the Land Department should appear. But it would be wrong to suppose that Goyder had set out to become that king. With his characteristic foresight, Goyder would not have expected this state of affairs to go on forever, and even when he applied to become surveyor general, he had probably been anticipating a reduction in the importance of the role, not just because of Freeling's departure and a more complete transfer of authority to the elected commissioner, but because of the decline in activity that would inevitably result once all the suitable agricultural land had been surveyed and sold and the northern pastoralists were securely settled on their leases. Certainly, this scenario was what he hoped for, and by 1870, despite increases in population and in total government revenue, income from crown lands and the costs of the department were beginning to decrease. However, the force of the mania for land that erupted in the early 1870s was something Goyder could not have foreseen and it resulted in a dramatic reversal of the decline in the Survey Department. New teams of surveyors and their assistants had to be recruited and despatched ahead of the agricultural frontier as, for the rest of the decade, the frenzy of the farming expansion drove the survey across vast areas. When demand for land dried up during the 1880s, the survey parties were largely withdrawn and, by the time of the Public Service Commission, revenue from lands had crashed, with the cost of running the department diminishing accordingly. Nevertheless, because of all the other activities for which he was now responsible, Goyder was still surrounded by 'a perfect swarm of clerks', as he put it, who were summoned by electric bells to attend his desk, and such were the complexities of his still-extensive realm that it was necessary for him to guide the commissioners through its workings.[2] No other department had to deal with 'such a variety of intricate matters of detail', they noted. There were, 'no less than twenty-eight Acts of Parliament affecting the covenants of current agreements between the Crown and its lessees, and … absolutely no less than sixty-seven forms of lease now actually current'.[3] Since Goyder must inevitably have advised on their content, he can be seen as stretched on a rack of his own creation, but he was not necessarily in agony. One of his contemporaries testified that he was regarded as one of the few men who had 'the ability of the American lawyer' required to deal with the land laws and Regulations.[4] Overall, the commission judged that the department functioned 'with efficiency and economy'.[5]

Goyder had been concerned about record-keeping systems and information from the very beginning of his career. In his application for the surveyor generalship he had offered to ensure that the commissioner was kept informed, and he was true to his promise, passing on a 'welter of reports and correspondence'.[6] Interestingly, no annual report from Goyder as surveyor general was ever presented to parliament, although even Freeling had managed to make a rudimentary gesture in that direction in both 1859 and 1860, and Strawbridge promptly presented one after Goyder's departure. Given that he had already attempted to ensure that the commissioner was fully informed, Goyder probably saw an annual report as an abhorrent duplication of effort. Even when writing the numerous reports he did produce, Goyder claimed (early in his career, at least) that he avoided wasteful duplication by abandoning the usual practice of drafting a rough copy and then following it with a 'fair' one. Instead he made halves of the usual sheets of foolscap paper (longer than the now-standard A4), and wrote on those, only replacing half-sheets that needed altering.[7] Nevertheless, his reports were so well informed, succinct and authoritative that they were usually printed without alteration as parliamentary papers and often accepted as government policy in the same way.[8]

When the colony was planned in the early 1830s, little consideration had been given to establishing a civil service – an institution which, in England at that time, was based on patronage and commonly acted as a refuge for gentlemen incapable of making a living in the professions. Of necessity, a government administration developed and, as soon as partial self-government was achieved in 1851, the colonists set about addressing it, passing the first legislation in Australia to attempt to regularise government administration: the 1852 *South Australian Classification Act*.[9]

In the mid-1850s a major reform of the civil service in England ensured that entry into the service and promotion within it would be based on ability and dedication, not class or patronage, and Goyder's rapid rise from chief clerk to surveyor general in less than 10 years demonstrated that the South Australians were in complete agreement with the reforms in the civil service in England. His success, clearly attributed to zeal (that great Victorian virtue) and ability, was not just a matter of personal characteristics, but said something about South Australian society, something of which its members could be proud. Goyder's rise to the top showed that the South Australian public service was not only 'open to talent' – as the cry went – but open to a degree commensurate with the talent displayed.

The sense of a moral purpose that informed the times and the English reforms to the civil service gave birth to the notion of public service as service rendered to a greater good, a noble vocation above mere personal profit that demanded not only ability but integrity. In the face of managerialism and changes in society, that ethos has faded, but Goyder exemplified this attitude, and it is impossible to understand him without reference to it.

The 'Big Boss'

Lionel Gee, the surveyor whose recollections of the South-East in the 1870s were published decades later in the *Register*, described 'the Big Boss' as 'a gentleman of cyclonic energy', a leader who was able to draw the best out of his men, and 'a wonderful organiser. Instructions from him left nothing to be surmised, clear, concise, complete.' Despite Goyder becoming the highest paid public servant in the colony, Gee, writing in 1928, also saw him as underpaid, commenting that 'his recompense was not within coo-ee of what it would be now for the same amount of work'. But Gee also described him as 'officially a martinet'.[10] Since Goyder was strongly driven by notions of duty and had a famously determined approach to discipline, it is a claim that deserves consideration.

One person with whom Goyder can directly be compared in this context is B.T. Finniss. In his role as leader of the failed attempt to establish a settlement in the Northern Territory, Finniss had also been described as a 'martinet'. Goyder's success in undertaking the same project was attributed by many at the time to his being a trusted leader of his men, and their diaries and writings bear this out. In fact, his firmness, coupled with a trusted fairness, was seen by the men as protecting them and their interests. (Finniss, it should be pointed out, had not been able to choose his own officers, and it was no doubt in reaction to his experience that Goyder insisted on being able to choose his.)

Another famous 'martinet' with whom Goyder might be compared was the head of the Department of Mines in Victoria, Brough Smyth. Like Goyder, Smyth was acknowledged to be unremittingly energetic and zealous, but, unlike Goyder, subordinates accused him of tyrannical and overbearing conduct, which included abusive outbursts that reduced men to tears. Clerks worked under the threat of fines for spelling errors, or dismissal if they used erasers. The situation became so bad that an inquiry was held, resulting in the claims of the abused being upheld. By contrast, an anecdote published shortly after Goyder's death illustrates only the claim that he was a 'great stickler for etiquette'. Goyder had summoned one of the clerks; the man, who had been working in shirt sleeves because of the heat, had hurried straight to his office and had asked if Goyder wanted him. Goyder reputedly put on his glasses and studied the man stonily before replying, 'Yes sir. I want you to return to your room and put your coat on, sir.' The cool approach was obviously a favourite tactic. Another anecdote described an incident in the Northern Territory involving a party of men on a boat in the river. Frightened of getting into the water because of crocodiles, they had began to quarrel about who would unload the provisions. Goyder, who had been watching from the shore, 'waded out to the boat, with the water up to his arm pits, and quietly enquired, "Well, what are you waiting for?"'[11] This is undoubtedly controlling behaviour, which asserts Goyder's superior position through a display of superior courage, and it would have left the men feeling uncomfortable and embarrassed, but it is of an entirely different order from threatening a subordinate with unemployment for the use of an eraser, not least because Goyder had used his own body to make his point. It is also obvious that both stories were remembered because they contain an element of humour and express an attitude of amused affection and respect. In her book on the Northern Territory survey, Margaret Goyder Kerr described her grandfather as a strict disciplinarian, but immediately and unselfconsciously placed that strictness in the context of competence, adding that 'if you had survived the rigours of a couple of trips with Goyder, you had a passport to almost any field job in the colony'.[12] Goyder might have been firm, but he was not an irrational tyrant. While Gee's use of the term 'martinet' undoubtedly tells us something about Goyder, it also reveals Gee's own limited experience of real 'martinets'.

Although he was known for his quick temper it seems never to have caused any serious problems. When asked on one occasion if delays in the drainage project had been caused by friction between officers of his department, Goyder (tellingly, perhaps) interpreted the question as referring to friction between himself and officers of his department. 'I never have any friction with any of my officers', he answered. 'If I am wrong I give way at once; if right, I put some other person to carry out the work.'[13] The evidence supports his claim about giving way when wrong. Apart from his evident willingness to admit mistakes, one of the field notebooks from the expedition to Lake Eyre South recounts that late one night, after a 'dull and threatening' evening, he found the camp in disarray, with guns and rations lying on the ground. Goyder got all the men up, but two of them remonstrated with him. One of them, Tom, who objected to being woken up, was threatened with dismissal. The other was the cook. Despite his anger, Goyder listened to the cook's complaints, decided that he 'really had too much to do when all the men are at camp', and selected an assistant for him.[14] This was the same decisive responsiveness that Goyder demonstrated later in the Northern Territory, where the men were clearly not afraid to bring problems directly to him.

However, there was one case where there was a serious misunderstanding between Goyder and his staff, and that was over the bonuses specific to the Northern Territory work. This emerged once the party had returned to Adelaide, and left the men, who had not understood that the extra money they had earned constituted their bonus, feeling cheated. However, this was very much the exception. Goyder was normally seen as concerned about employment conditions and loyal to his staff. In a further example of his care for those under him, in early February 1870 he had warned all officers, from his deputy down, of the likelihood of their being retrenched at the end of the month because the government was short of funds. The usual practice in situations like this was to leave employees in ignorance of their impending dismissal and then give them only 12 to 24 hours notice.[15] And in 1874, while conditions in the service were under review, Goyder strongly objected both in person and in writing to having officers in his department, who were evidently needed, kept on the provisional and temporary list, where they could face dismissal without compensation, for as long as 10 years.

Apart from the tendency of successive governments to keep substantial numbers of staff on the 'pro. and tem.' lists rather than employing them under proper conditions, employees of the department faced an additional problem: those taken on as juniors during the great expansion of the 1870s ended up languishing in the promotional doldrums because the department had not continued to expand at the same rate, with many remaining on their junior salaries as they grew to adulthood and established families. In 1889 Goyder strongly recommended to the government that their salaries be increased. He warned that they risked losing some of the best staff, and threw the whole weight of his presence behind his appeal:

> from me on behalf of my officers, who are fully and most completely deserving serious consideration, and in the interest of the service which they have entered, and in which I hope they will be enabled and find it profitable to remain, by due and fair consideration being given to their services and merit.[16]

In his *Report on Disposal of Crown Lands* in South Australia, Goyder saw fit to include the information that the field surveyors and chainmen:

> are all highly efficient, but paid at very low rates, a defect that I have striven to remedy over and over again, but, I regret to say, not successfully.[17]

Other factors, such as the large-scale surrenders of land and the increased public demand for information, created enormous amounts of work for the office staff, despite the decreased land sales. So much overtime was being worked that Goyder judged it a threat to the wellbeing of those concerned. He praised the dedication, competence and efficiency of his draughtsmen and the office staff, and stressed the need for more staff and proper remuneration.[18] When the surveyors and draughtsmen made their own suggestions to the Public Service Commission in 1890, they referred to Goyder's attempts to improve their situation.

Earlier in his career, Goyder's own working experiences had ultimately affected salaries and benefits across the South Australian civil service. His situation and his complaints had led to remuneration being received for travelling expenses, and to a complete review of service salaries and conditions in 1874. In 1881 the government had dispensed with the retiring benefit to civil servants, at the same time extending the hours of work and cutting pay. Goyder led the heads of department in opposing loss of the benefit.[19] Eventually this resulted, in 1885, in the formation of the Civil Service Association, with membership open to all civil servants – the first public service union in Australia.[20]

Goyder's long tenure was characterised by generally peaceful and harmonious relations with his staff and the records show that most staff remained with the department for many years. Despite his strictness and his intense focus on his work, there was 'nothing unkind about him', Strawbridge explained. 'He was genial in disposition, and won the confidence of his subordinates'.[21] Gee, despite also considering him a martinet, agreed that he had 'a kindly helpful nature that strengthened the steps of many a young man in sickness or trouble and in his journey through life'.[22] The confidential letter book reveals that Gee had every reason to praise Goyder's fatherly concern. As chief clerk, Gee had suffered a fall from grace, evidently through drink, and would have been discharged. Goyder wrote a note to him in confidence offering to recommend that the commissioner overlook whatever had happened if Gee would remain a strict teetotaller, and take 'the usual pledge to that effect'. Goyder urged: 'for your own + yr family's sake – I do hope you will be man enough – with God's help – to keep to this determination'.[23] Six months later he penned another private note, telling Gee he had arranged for him to take up his old surveyor's position, '+ I think it may lead to your complete restoration to health'. Goyder considered that a quiet life surveying in the country would benefit Gee and had already discussed these arrangements with Gee's wife.[24]

Relatives and friends

Despite his obvious fair-mindedness, the suggestion that he favoured members of his family for positions in the civil service dogged Goyder. In 1872 Captain Samuel Sweet, a keen photographer and a pilot on the Roper River, was instructed to provide the government with a plan of the river. He complained that the result had been rejected because of the report of 'some relative of the Surveyor-General', meaning Goyder's nephew George MacLachlan.[25] MacLachlan died at Port Darwin, but Goyder's nephew Edwin Mitchell Smith and brother-in-law Arthur Henry Smith remained with the department. (A.H. Smith had joined in 1859 when Goyder was only deputy.) His eldest son, George Arthur, had become an assayer and analyst of minerals with the department and, later, when it was formed, with the Department of Mines. His younger son Frank joined as a surveyor, and Goyder's son-in-law, George Frederick Hallett, was employed as inspector of scrublands. An embittered clerk, who had been retrenched, voiced his complaints to the Public Service Commission.[26] When Frank Goyder was examined, he explained tersely that he had been with the department for eight years, was still on the 'pro. and tem.' list, was paid only £120 a year, did not know if he would receive a recommended increase, and did not think that any men in charge of survey parties received less than he did. The only allowance he received beyond his ration allowance was 'a free copy of the *Government Gazette*'.[27] He eventually left both the department and the colony, finally settling in Western Australia. There is no suggestion of favouritism in the promotion of nephew Edwin to an important position: Edwin Mitchell Smith was praised elaborately for his competence and long hours by the Public Service Commission, and had his salary increased by them.

Along with relatives of the surveyor general, the Survey Department was also home to other members of the New Church. In the early 1860s, even its minister, E.G. Day, was gathered into the departmental fold in the role of storeman, and during the frenzy of activity at the time of the Moonta claim, he was perceived to guard the surveyor general's door, much like a private secretary.[28] But as Goyder's relationship to Chauncy shows, the connection was in place before Goyder's arrival, and the small membership of the New Church included a number of men in engineering and related technical professions. Strawbridge had been with the department for nearly as long as Goyder had been its head, and had been promoted from chief draughtsman to deputy surveyor general in 1886.

In 1893 Goyder was asked about the employment of relatives in general and he replied that, although he generally preferred not to do this, he had no objection to relatives being employed in the same area because 'traits of character and ability may run in families'. As he explained: 'in a small community like this, where connections frequently are formed by marriage, they are not always to be avoided'. He was not merely justifying his own situation: the civil service lists, published annually in the parliamentary papers, reveal the same family names as borne by cabinet ministers and parliamentarians, business leaders and squatters, indicating that Goyder also employed the offspring or relatives of his political masters. In fact, the civil service list was a list of men with better than average education: members of the middle class. In a small colony, this was a small group, and the same names occur again and again.

Commissioners

The character of Goyder's relations with his superiors was dominated by the character of the political process at the time. As deputy surveyor general, Goyder served under six commissioners (including Bonney). During the 33 years of his service as surveyor general, there were 42 appointments to the position of commissioner of crown lands and immigration. Some men were appointed more than once, and some served successive terms, but there were still 39 actual changes of office – an average of more than one for each year of Goyder's tenure – involving 22 different men. Only Strangways in the early 1860s, and Playford in the 1870s and 1880s, had significant careers as lands commissioners and each was also able to influence this area as premier. Goyder's relationship with Strangways is hard to read: some of Strangways's comments suggest a considerable respect for Goyder, and even a pride in being a confidante, although some of his remarks must have been annoying to their object. It was under Strangways that Goyder was seen, and remembered, to have been 'king'. 'Who is the person who practically rules in the Land Office?', Boucaut had wondered (on Mitford's behalf) at the inquiry into the Moonta. 'The Commissioner of Crown Lands', Goyder had answered dutifully, but Boucaut had continued with his questioning, elaborating the insinuation for comic effect with a flurry of similar questions.[29]

The situation with Playford was very different. Two sequences of correspondence show Playford supporting Goyder strongly against charges of partiality in relation to pastoral land. In one case Goyder regarded the threats and allegations as 'too contemptible to notice' and 'activated by base motives'. In the other, he was surprised and explicitly stated that his vindication was up to the 'Honourable Commissioner'. In both cases Playford informed those involved that he would not consider their requests unless the threats or allegations were withdrawn, and withdrawn unconditionally, commenting, in the second case: 'any person who is at all familiar with that officer [Goyder] must know that he is quite incapable of conduct such as that attributed to him'.[30]

But all was not well between them, and the so-called 'Playford resumptions' may have been the cause. Thomas Playford, who led his own government for the first time in 1887, and again from 1890 to 1892, was a leading figure in the colony's politics in the late nineteenth century and had held the office of lands commissioner from February 1876 to June 1881, interrupted only by the short occupancy of John Carr, and again for some months in 1885. Playford was an orchardist – his desire to become a lawyer had been frustrated by his father – who had entered parliament in 1868. As lands commissioner he was a reformer, and in a collection of brief biographies of 'representative men' of the colony published in 1883, he was proclaimed the best crown lands commissioner to that point.[31] He was also exceptionally tall, usually dressed in ill-fitting and ancient garments, and with comically large feet. (A consultation between the lanky commissioner and the economically packaged surveyor general must have amused more than one onlooker.)

In 1888, before the Land Laws Commission, Playford testified to the accuracy of the assessment of the country Goyder had expressed in his valuations and in the Line. He also recalled that the members of the House of Assembly had been generally of the opinion that the Line was mythical, 'that cultivation had sensibly increased the rainfall, and the dry country no longer existed as a fact'.[32] He did not acknowledge, as both Goyder and Strawbridge had testified, that government members had gone ahead even though they had guessed that the northern farmers might be doomed, and in an extraordinary exercise in mean-spiritedness even attempted to shift some of the blame to Goyder. In direct contradiction to Goyder's evidence, 'Honest Tom', as he was known, claimed that Goyder had been: 'so greatly impressed with the fact that some of the hundreds beyond his rainfall line grew such good crops that he advised that, as long as the people were willing to take up the land and would live upon it, it should be thrown open to them'.[33]

> 'You don't then remember his advising that the land should not be cut up for farming?', the chairman queried. 'No,' Playford repeated, 'not whilst I was in office.'[34]

And it was certainly true that Goyder's plea to delay further expansion had been written during the brief Crown Lands Office interregnum of Carr.

The resumptions had aroused a great deal of anger and criticism of Playford, and since they had been introduced on Goyder's advice – although that advice consisted of nothing more than doing the arithmetic required to facilitate government policy and acting as required on behalf of that policy – Playford appears to have taken the opportunity to use Goyder as a tool for denying his own role in the events, while at the same time exacting some revenge. Goyder was clearly uncomfortable and made several attempts (before this commission and later elsewhere) to exonerate Playford, on the ground that ploughing up the pastoral country was the will of the people. He also had to defend his own position from the perception that when he identified the 'need' for massive resumptions it reflected a change in his beliefs about the climate.

Finale

In 1889 Goyder's contribution was formally acknowledged when he was made a Companion of St Michael and St George, joining his friend Charles Todd, who had received this honour for the Overland Telegraph in 1872. In 1893, however, Todd received a knighthood, which invites the question of why Goyder was never similarly honoured. One possible answer is that, since his marriage to Ellen was virtually incestuous according to English law, such an honour could not be considered.[35] Another is that he had made enemies at the highest political level.

In his last years Goyder was beset by ill health, with Campbell in 1892 writing to him warning that his symptoms of severe nervous prostration indicated that he needed a prolonged and complete rest. Later, he was confined to bed for a period by influenza and acute bronchitis.[36] He made his last trip – in a buggy with a driver – early the following year, returning to the South-East, although his health was a cause for concern throughout the journey. The end of his career came not long after this, but ill health would be only the official reason for his retirement. The real cause would be that a new and explosive element – a politician with a personal vendetta against him – had been added to the standard mix of squatters, leases, and controversy that had been a part of Goyder's career from the beginning.

The politician was Charles Cameron Kingston, the son of George Strickland Kingston, a man who had made his fortune through the Burra mines. Born into the small and interconnected group of the wealthy and influential who dominated South Australian society at the time, C.C. Kingston was admitted to the Bar in 1873 and eventually made Queen's Counsel. He gained his experience in government as a protégé of Playford. Despite the circumstances into which he had been born, Kingston was true to his father's political attitude and came to power with the support of the Labor Party as a champion of urban working people at a time of economic depression in all the colonies. The United Labor Party won its first three seats in elections for the South Australian Legislative Council in 1891.

Kingston became premier in June 1893, holding the portfolio of attorney-general, and led what was regarded as a brilliant government of unprecedented talent. Under his leadership a new continuity was introduced into South Australian politics, with his government remaining in power for a record six years. At the outset Kingston announced his desire to reduce the number of public servants and introduce a sliding scale of reductions in salaries which would cut the earnings of those on £600 or more by 10 per cent.

Kingston possessed a personality viewed by even friends and admirers as flawed, although it was agreed that he had many worthwhile and likeable attributes. He was certainly a person whose impact would not be forgotten in a hurry, with a journalist on the *Register* for many years remembering Kingston, after his death in 1908, as:

> a veritable gladiator … He gloried in strife and revelled in his triumphs. His vehemence and vituperative attacks, his desperate tactics, his unrelenting attitude, made him many enemies … But the fiercer the opposition the greater his determination to pursue the battle.[37]

The same writer was also impressed by Kingston's many gifts, his nationalist spirit, and his vigorous commitment to reform. (In South Australia Kingston extended the franchise to women and was a leading figure in federation. He became a federal politician.) Alfred Deakin even found him quiet, gentle and considerate, in private. Nevertheless, it was Goyder's misfortune, when he was ageing and in poor health, to be seen to have crossed him.

At the centre of the dispute was a group of pastoral leases near Lake

Alexandrina which, through a mistake in the Land Office, contained a clause specifying that they could not be resumed except where required by 'public utility'. Playford's last government, in which Kingston had been chief secretary, had wanted to resume these lands and lease them again as perpetual pastoral leases, and had sought advice from Josiah Symon, the Queen's Counsel who acted as government counsel, over whether they had power to do this. Symon had been articled with Kingston under Samuel Way, and like him would become a federalist and a member of the Australian Parliament.

Symon at first advised that they could go ahead, but then changed his mind. The government proceeded anyway, and the three squatters involved – Bowman, Harvey and Macfarlane – went to the Supreme Court. Disapproving of what he now perceived as the unfairness of the government approach, Symon handed in his brief as government counsel and took up the squatters' case. At the time legislation being presented by the government included amendments to the *Crown Land Act*, designed to remove the leaseholders, and when this was queried in parliament, Kingston accused the member who had asked the question of having been instructed by Symon, now perceived as an enemy to be overcome at all costs. Kingston spoke so scathingly of Symon that a war of insult and accusation between them broke out in the daily press.

Kingston was further enraged and frustrated by the way in which the Leader of the Opposition, Richard Baker (the son of the pastoralist John Baker, who had many years earlier opposed the valuations in the parliament), was delaying legislation. In the Legislative Council, Baker advised the chief secretary that the government should consult the surveyor general on the matter of government powers in relation to exchanging lands, and ask for a report from him clarifying which interpretation was correct. On the same day, 28 November, Goyder was called in to see Kingston in Cabinet, and then again later in his office, in relation to the Bowman and Harvey cases.[38]

In his bullying way, Kingston accused Goyder of acting on behalf of the lessees. Certainly Goyder would have known the squatters, and Bowman in particular may have been a good acquaintance or friend – he had been attempting to farm ostriches. Goyder denied partiality and left Kingston's office 'very much upset by his manner'. In his own office he went over the paperwork with P.P. Gillen, the commissioner. Meanwhile, in the Legislative Council, his old friend Dr Allan Campbell was moving a motion that would defeat clauses the government had prepared aimed at the disputed leases. Afterwards, Goyder prepared a memorandum about the Bowman and Harvey leases, and had it taken to Kingston's office by his nephew and chief clerk, Edwin Mitchell Smith, because he still felt 'too much upset' by his previous encounter.

The next day, Saturday, Goyder saw Kingston himself. The 'Hon. Attorney General', he later noted, appeared to think that he had not been loyal to the government. In fact, Kingston believed Goyder had given orders to Smith to supply information to Symon's clerk, apparently by allowing him to see a letter regarding a Land Board meeting. Goyder agreed that the letter had been seen, but it had been supplied because access to such a document would normally have been granted to a member of the public (an explanation that is entirely consistent with his early determination to make as much information as possible available to the public). It had not been 'leaked' with the intention of undermining the government's case. The argument was wasted on Kingston.

On Monday 4 December, the last day of the case, the Cabinet summoned Goyder again and interrogated him over details of the leases. The claim of partiality was extended to include the accusation of not supplying information to the crown solicitor. In court, the government not only lost the case, but was seen to be so determinedly unjust that costs were awarded against it. Even the pro-Kingston *Advertiser* opposed the manifest injustice of the government's approach. Goyder was called before Cabinet again. This time it was Playford who accused him of favouring the leaseholders. Goyder was deeply offended and called on their long history together as proof of his loyalty. No one could have been more loyal than he had been, he remonstrated. But Playford gave no sign of being moved and Kingston took up the attack again, referring to papers seen by Symon's people. At this Goyder 'took exception'. The papers had been seen in court, he pointed out, and he had notified the crown solicitor that application had been made for them. 'Yes,' Kingston agreed nastily, 'on the day before the trial'. 'On the day the subpoena reached my office', Goyder retorted.

Goyder had stood up to vociferous opposition, public denigration, and attacks on his integrity over the years – he was one of the few who, in his own words, could bear 'unmerited blame'. But the attacks by Kingston, and now Playford as well, struck at the heart of his sense of himself as a public servant, at the meaning of his entire career. It was not the substance of the attacks that was ultimately troubling, but that they were being made at all. Motivated by an ethic of duty and service, Goyder's understanding of his own value resided in his being a faithful and loyal servant of a succession of governments and ministers. But to Kingston he was clearly just an overpaid old bureaucrat, a friend of squatters rather than urban working-class people, and, unforgivably, someone Kingston saw as having aided an enemy. Playford, whom Goyder had served for many years, had abandoned him as well. The fact that he had not made any special disclosures to the pastoralists, or withheld anything from the crown solicitor made no difference: innocence was no defence from the fury of Kingston's malevolent attack. No doubt personally wounded and deeply alarmed as well, Goyder went back to his office, called for a new diary and recorded a brief account of these meetings.[39] The daily entries that followed were perfunctory – 'usual office work' – except for one recording the referral of a decision about another pastoral lease to the commissioner as a matter of policy. Before long, the entries ceased, but not because he had gained enough confidence or composure to consider a record unnecessary.

On 13 December Goyder wrote and delivered what would be his final resignation. It was made on the grounds of ill health and accompanied by a medical certificate from Campbell. He proposed to be absent from the end of the month (and year), and asked for six months leave, making the resignation formally effective from the end of June the following year. Goyder pointed out that during his long career he had only taken a total of 22 months leave over three occasions. The message was plain: they were morally obliged to give him the leave – in effect half a year's salary – in addition to his retiring allowance. He also revealed that the situation had not changed much since his last resignation:

> I venture very respectfully to express the hope that the Govt. may favourably consider my case as, though the salary received by me during the later part of my service has been good, my expenses have been large and the settlement of my affairs will absorb the greater portion of the retiring allowance coming to me – so that I shall have – before long – to enter upon some new business, and to form new connexions which my long and arduous career in the service of the Govt. has in a great measure separated me from – [and] during which – the strict performance of duty has not enabled me to make many friends.

The end of the letter was pure Goyder:

> I need only add that any experience gained by me during my long tenure of office will be heartily at the disposal of the Govt. should it at any time be required.[40]

The same day, the lands commissioner announced the news in the House of Assembly, where the members had been unaware of Kingston's attacks. Gillen made the usual remarks about Goyder being a 'faithful servant' and a 'true friend' to ministers he had served under, and regretted the ill health that had provoked his retirement. The first response was from Jenkin Coles, a former lands commissioner who had known Goyder for over a quarter of a century and who considered himself a personal friend. No officer 'was more absolutely honest, and earnest in the performance of his duties, and no one in the service was more loyal to his Minister and to the Government in power', he told the members. Goyder's loyalty 'went sometimes to the extent of self-sacrifice'. Coles was followed by Sir John Downer, who had been replaced in his second spell as premier by Kingston. Downer focused on the difficulties of Goyder's position. He considered that there had been 'no man in a public position in South Australia for so many years who has been in the fire so much'. Goyder had held 'immense responsibility and [displayed] immense energy' and had performed his duties 'with the most remarkable ability'. J.H. Howe, another former lands commissioner who had known Goyder as deputy and had chased him across the north trying to deliver the dispatch containing news of Freeling's resignation, testified to Goyder's enormous capacity for work and leadership:

> He spared neither himself nor those under him, but at the same time he was a good general, and all his officers and men acknowledged that they had an honest and just man over them, and they served him as faithfully as he served South Australia.

But one member, Henry Grainger, announced that he was not surprised at Goyder's resignation and made it clear he did not believe that ill health was the real reason. Grainger was basically a supporter of the Kingston Government, but his distinct political style was to criticise any government vigorously. He announced that he had noticed for 'some considerable time' that there had been 'a sort of set' growing against the surveyor general, 'as it always did grow against an officer who was independent'. Grainger considered that Goyder 'had always been on the right side, notwithstanding that he had the weight of certain parties against him', and in the course of his career he had been 'blackguarded and accused of all sorts of crimes', ultimately to the discomfiture of his accusers. If he had been a 'timeserver', Grainger told the members, Goyder would have been 'a very popular and wealthy man': as it was, he had strong enemies and strong friends. (The claim that he had loyal friends, as well as enemies, is supported by the observation, made in an obituary, that the 'personal magnetism of the man drew around him many true friends'.[41]) He admitted having advised Goyder to retire some years ago, when, in private practice as a surveyor (and perhaps a land agent) his knowledge and reputation would have enabled him to completely dominate land affairs in the colony; he characterised Goyder's enemies as mean-minded people who 'disliked to see men getting large salaries or with independent minds', people who could be found in every small community, 'very pious respectable men, who went to church twice on Sunday – people who never had a good word to say for such a man as Mr. Goyder ... there were a lot of narrow-minded men who would rejoice at the fact that Mr. Goyder had been worried, when he was ill, into resigning'. Grainger repeated that he was sure that if Goyder had not been worried, he would not have resigned.[42]

'Honest Tom' replied by invoking the old dictum about being saved from one's friends. Grainger's depiction of Goyder as retiring because of worry was false, Playford assured the House. Not content with outright lies, Playford cynically and disparagingly manipulated Goyder's reputation for courage. The surveyor general had never cared a 'snap of his fingers for what was said about him, and as to anything being said in Parliament about him it was years ago', Playford declared. (He was perhaps referring to attacks which had included the charge that Goyder secretly owned a run in the South-East against which Coles had defended him in 1887.[43]) Of late, 'everything as far as Mr. Goyder is concerned is as quiet as possible'. Did they think that the man who had fought the battle against both the squatters and the selectors would retire, 'having lost courage', when nothing had been said about him? The real cause, the House was assured, was not that Goyder had been driven out in any way: it was his health. 'He had to put up with you, and that was enough to kill any man', Grainger interjected. Playford settled for declaring that nothing done by either the ministry or the parliament had caused Goyder's retirement, and then joined the hymn of praise, although in his own less-than-candid way, proclaiming that he had 'always spoken in the House in the highest terms' of Goyder, a claim which could be true whatever he'd said elsewhere.[44]

The following day both major dailies announced the retirement as their second leading article. Grainger's charge and Playford's response were completely ignored, perhaps out of consideration for Goyder's pride and dignity, as well as his health. The *Advertiser* conferred general praise according to the already well-developed themes. The *Register* adopted a more considered and honest approach. After regretting his illness, lauding his 'inexhaustible energy and amazing versatility' and listing his various achievements, the writer picked up the theme of his 'least enviable' position. 'He is in the place,' the article explained, 'where motives are peculiarly liable to misconstruction and decisions to suspicion. Mr. Goyder could not reasonably expect to escape hostility and vilification, and in actual experience he has had his full share of both.' Worst of all, the paper considered, he had been maligned in parliament by legislators who had failed to recognise that defending a public servant obliged to carry out difficult and unpopular tasks was 'one of their most sacred duties'.[45] The *Quiz and The Lantern*, which specialised more in snippets of news and reports of goings-on about town, took the side of Goyder's critics, commenting that he was an admirable officer, in many respects, 'although he made mistakes, some of them costly ones'.[46] The reader was presumed to know what these were. Perhaps reflecting on the same unnamed errors (which probably included the extensive resumptions of the late 1870s), the *Observer* remembered Goyder as 'absolutely loyal to his Minister even to the extent at times of enduring blame in silence at the hands of the public in cases where the blame was not perhaps personally deserved'.[47]

Farewell

The other heads of department organised a formal presentation. Although some had wanted to raise money for an expensive gift, it was decided to give him an illuminated copy of the address to be delivered on his retirement. A younger civil servant – 'Disgusted' – wrote to the *Public Service Review* to complain that:

> our fathers have told us what he was in the old times, a brave, energetic, useful servant of the Crown, and at his retirement he deserves something better than the illuminated address, that goes to every good young man who has taught in Sunday-school ...[48]

Goyder was eventually presented with a copy of the address given by his friend Sir Charles Todd. Inscribed in large letters, the address was long enough to take the form of a volume bound in red morocco and gilded. The ornamental work around the text showed South Australian flowers, and there were representations of places significant in his career: Port Darwin, a surveyor's camp, Mount Lofty, Mount Gambier, Mount Remarkable, and Termination Hill. The family crest was displayed, and the address was signed by 35 prominent members of the service, representing all departments.

The presentation took place in Todd's office at 5.00 pm on 31 January 1894. There was a large crowd, although the many apologies received included those from the acting lands commissioner, the lands commissioner (who was ill and sent a telegram) and the treasurer, Playford. Todd read his address and two others spoke briefly. Even in this situation, which should have been a celebratory occasion, there was an undercurrent of unease and awkwardness. In his reply, Goyder commented that, unfortunately, 'his very disposition and temperament had frequently separated him from those with whom he should have liked to be on more genial terms', echoing the feelings in his letter of resignation.[49] It was an honest and curiously vulnerable admission. In a separate gathering some time later, Goyder's staff presented him with another address and a silver tea and coffee service.[50]

The day before the presentation the *Register* had led with an article,

'Worrying the Civil Service', in which it repeated the claims that Grainger had made when Goyder's retirement was announced and which had been ignored at the time. It was understood, the paper stated, that Goyder had been 'harried into resigning before he intended to do so', and that it was not the minister (Gillen) who was to blame. The article then dwelt in some detail on the fates of other senior public servants, seen as the victims of the Kingston Government's enthusiasm for cutting public service costs. The chairman of the railways commission, it was reported, had opened his newspaper one morning to learn that he was thinking of leaving the service and that the government heartily recommended him to any other colony wishing to benefit from his abilities.

The subject arose again when Goyder's resignation took formal effect at the end of June 1894. The *Public Service Review* published a long 'Personal paragraph' recognising the final departure of someone who was still 'in our minds and hearts'. Goyder had evidently made enemies in the public service as well, because the paragraph concluded with the observation that, for him, leaving the service meant escaping from 'envy and uncharity, which, like the toad, are ugly and venomous'. (Of course, the sources of these venomous feelings were not named, but it can probably be assumed that his long-term control of the drainage of the South-East, forced by the absence of competent staff in the Public Works Department, is unlikely to have won him friends.) In an apparently oblique reference to the line of reliable rainfall, the writer described him as possessing a 'singular judgement', adding that few would now not admit that he was 'about right after all'.[51] Some months after his retirement, Goyder was presented with a purse of a thousand guineas (over a thousand pounds) on behalf of the very section of South Australian society, the most prominent and well-established members of which Goyder had once done so much to enrage: the pastoralists. The money was presented in gratitude for his efforts in opening up the country.[52] It was a gesture that expressed far more accurately the nature of his relationship with the pastoral industry, which he had worked long and hard to assist and develop, than the battles of the mid-1860s.

After Goyder's official departure, William Strawbridge was appointed surveyor general, with Edwin Mitchell Smith as his deputy. Smith would succeed Strawbridge in 1911.

Retirement

Goyder continued to be seen around the office 'not infrequently' after his retirement because he had taken up private survey work.[53] The *Public Service Review* reported the general view that it was a pity he had to 'commence life again' at his age.[54] But Goyder also took the opportunity of retirement to fill, if only

for a year, a position he might be expected to have held at some time, that of
President of the South Australian Branch of the Royal Geographical Society
of Australasia. He had been a member of the provisional council to establish
the society, which began in 1885, with early meetings held in his office. After
holding the presidency for 1894–95, he was made a life member of the society.[55]

At Warrakilla, life seems to have continued on, perhaps with Goyder now
having more time for direct involvement in the orchards and the running of the
estate. The file in the Mortlock Library contains only one complete personal
letter from Goyder, written on his seventy-second birthday, to thank one of his
grand-daughters for her 'kind letter'. It also describes his birthday:

> a graced day. Nearly all the members of the family wrote to me – + we had a
> quiet jollification at home. Jack proposed my health – + the family responded
> with musical honors – whilst I sat beaming all over.

Four months later, on 30 October 1898, Goyder's old friend Allan Campbell
died of a series of heart attacks. Goyder himself had been confined to the house
by illness for a short time previously and had been growing weaker, despite
the visits of a doctor. In the days immediately after Campbell's death, Goyder,
too, had a heart attack, which was followed by a stroke. He died in the evening
of 2 November 1898, at Warrakilla. Strawbridge believed his death had been
hastened by the death of his friend.

On 4 November, the *Advertiser* devoted its leading article to a reflection on
his career. 'Fortunate are the Governments who attract to their service men of
the high calibre of the late Mr. Goyder', it began, and went on to declare that
it was not 'too much to say that the civil service had gained reflected glory by
the presence in it of men like the late surveyor-general – able, conscientious,
and devoted to their duty'.[56] When he had addressed the celebratory dinner
for officers of the Overland Telegraph in 1872, Goyder promoted duty as the
talisman that led to success in every venture and which was even 'the link
which connected the two extremes of life [birth and death]', effectively defining
duty as life itself.[57] The implicitly ethical and spiritual aspect of duty informing
Goyder's approach to life was prominent in the *Advertiser*'s assessment of his
attitude: 'Public servants of that type add to their ability a qualification the
value of which can hardly be overstated', the paper proclaimed:

> They give practical expression to what is really, whether they are conscious
> of it or not, an exalted ideal of duty. It is this which, above all things, lends
> dignity to labor. Quaint George Herbert has said: –

'A servant with this clause
Makes drudgery divine;   ·
Who sweeps a room as for Thy laws
Makes that and the action fine.'[58]

'Quaint' George Herbert was a seventeenth-century Anglican clergyman whose poetry was later set to music and absorbed into the Anglican hymnody. Given Goyder's lifelong spiritual seriousness, it is probable that the newspaper accurately articulated Goyder's conception of what he was doing, although, as the writer noted, his task was not the modest one of sweeping a room.

Goyder was buried in Stirling cemetery (which he had helped to establish) on a tranquil Saturday afternoon. Parties came up from Adelaide for the funeral, but the number of people attending was affected by a public service picnic being held on the same day. Playford was among the ex-commissioners of lands who attended, but Kingston didn't trouble himself, or bother to send an apology.

The estate that Goyder left was valued in his will at just under £4000. His having 'died a poor man' was seen as 'his best epitaph in these days of hypocrisy and chicanery'.[59] His death had a marked effect on Ellen, who survived him by less than seven months. In the following year she was admitted to Parkside, the large public asylum in Adelaide, where she died on 24 May. Her death was attributed to hemiplegia (paralysis of one half of the body, perhaps the result of a stroke) and diabetes. A story in the family reports that she had scolded Goyder for being out in the orchard in the middle of the day without a hat, and had gone back to the house to bring him one. She had returned to find him collapsed.[60] The trauma of this, echoing as it did finding her sister years before, ultimately overwhelmed her. The property was divided, but Warrakilla remained in the Goyder family until 1911.

There is a photograph of Goyder taken in what must have been his last years – perhaps on his last 'graced' birthday. He is lounging on a verandah seat, very much the country gentleman, wearing a wide-brimmed hat. There is what appears to be a fountain pen in his breast pocket, next to what might be a pipe. Despite this slightly business-like touch, the portrait is also one of a white-bearded grandfather in embroidered slippers. He sits still, relaxed but poised, fixing the camera lens with a level gaze – the same acute, steady gaze with which he looked upon the world.

CHAPTER TWENTY-TWO

Remembering

The prediction made at Goyder's death, that his name was 'never to be forgotten' because of his 'famous line', has so far largely held true.[1] The Line has perpetuated Goyder's name, while the 'other reasons' – compelling reasons – why it might have been remembered have slipped into obscurity.

For a biographer, Goyder as a person was elusive, and without diaries, reminiscences, or personal letters I had no choice but to pursue him through the public record, looking for clues in the reports and field notebooks, which were his chief written expression, and in the pages of the major newspapers. Like the parliamentarian who, as a young trooper, had chased him across the outback in an attempt to deliver mail, I found him a hard man to overtake, and when I did catch up with him physically, at his grave in the Stirling cemetery, it was to find him as I would have in life – surrounded by other people, in this case the members of his large family buried with him. His friends and his colleagues of the Survey Department were also there – on the stone obelisk they had erected over his grave. My experience of that encounter was an awareness of being an interloper, forever peering over shoulders at the edge of a crowd, not quite able to see and just missing what had been said. The situation did not begin to improve until persistence unearthed his role in the Moonta dispute, and serendipity led me to the story of his involvement in a shipwreck. The minutes of the inquiry into the dispute over the Moonta claim provided a glimpse of Goyder at work in the field and in town and clarified the circumstances surrounding his succession of Freeling, while, in his account of the wreck of the *Queen of the Thames*, I was able to hear his personal voice for the first time. The angry description of his experiences confirmed Lionel Gee's recollection of a practised and competent storyteller, but here, to express his outrage, he also used the sarcasm that occasionally appears in his other writings. The story revealed a person of courage and integrity, although someone not afraid to disclose moments of anxiety and vulnerability. Above all, the incident shows a person who constantly considered the wellbeing of others. It was a mixture that I had encountered enough in my research to recognise as distinctly Goyder. But the pleasure of finally hearing his voice was muted by a grim realisation of the harshness of the events he had to endure during the middle period of his adult life – the death of his wife and how he received this tragic news at a time when he was exhausted and ill, the shipwreck and the legal problems which followed, and the death of his niece on the journey back to Australia. Perhaps the most startling document, however, was the diary in which Goyder recorded, tersely, the encounters with Kingston that so cruelly ended his career, although no patience or good fortune had been required to uncover that. It was in the archive, listed as Goyder's office diary, apparently untouched, except by archivists, since the day Goyder had left it – or, perhaps, had deliberately sent it to be filed.

In South Australia...

As I investigated the whole range of areas and concerns to which he had devoted his immense energy and considerable intelligence, I was not surprised that some semblance of a personal life for Goyder would have to be constructed from the little pieces of evidence I accumulated along the way. But, as I began to recognise the unique role he had played – a remarkable fit between individual and circumstance – I was certainly surprised to realise how much that he had done had been forgotten and how little acknowledged he had been. Had it not been for the scholars already mentioned – Michael Williams, J.M. Powell, D.W. Meinig and K.R. Bowes – it would have been impossible to form a sense of the scope and depth of Goyder's contribution to the way in which South Australia has taken form, especially given that, outside academe, others, at least in the past, have found it possible to write about the history of forest and water management in the state or exploration of the salt lakes region without mentioning him at all. And yet it is not just the landscapes of the Line – the ruined houses or a sudden transition from farming to pastoral country – that give silent testimony to his presence. The looming pine plantations and drained flatlands of the South-East owe their presence largely to his foundational effort. To picnic or camp in the Wirrabara forest is to enjoy a natural environment that, without his dedication, might well have been destroyed. To drive through many rural areas or to cross the country by train is very likely to travel on a route proposed or approved by Goyder. The landscapes of the older wheat lands and their towns are the product of his management and supervision. The vast pastoral areas of the inland also owe much to his exploration and subsequent decision-making: as the pastoralists themselves recognised, without his promotion and support of the industry through the creation of watered stock routes and the introduction of artesian bores, it is doubtful whether the industry would have survived. Even to return from the Flinders Ranges with an album of stunning photographs is to follow in his path. But, apart from the country around the Line, Goyder's presence is not felt in these environments, and his absence is especially striking in comparison with that of the ubiquitous Colonel Light, the acknowledged author of an urban plan. Considering them side by side caused me to reflect not only on those we choose to remember and why, but how we identify human agency. In an earlier part of *The Making of the South Australian Landscape* Michael Williams noted that in the early seventeenth century William Penn had written that, 'the country life is to be preferred for there we see the works of God, but in the cities little else but the works of man'. Penn should have known better, Williams commented:

> for all around him pioneers were hacking out a new landscape from the woodland. Similarly, Cowper, an English poet, well over a hundred years later wrote, 'God made the country and man made the town', which again was strange as the open fields of Olney, Buckinghamshire, where he lived and wrote those lines had been enclosed a few years before, and a new landscape created.[2]

Cowper and Penn were probably content to see God's handiwork in trees, grasses and unobscured skies, but now it is wilderness that is celebrated for its spiritual value. At the same time it has become apparent that many of the landscapes perceived as untouched by European observers had actually been cared for and managed. Nevertheless, for the average city dweller at least, the tendency persists to perceive both these and settled rural landscapes as just having been there since time immemorial. Apart from highly structured landscapes, such as irrigated terraces, we do not generally acknowledge our impact on the natural environment in the same way that we claim our other creations – our cities, our technology, our thought systems and the artefacts of our cultures. Even when we do think about it, it will not be to identify 'landscape authors'. The rarity of 'practical geographers' with the reach and impact of Goyder – individuals whose work cannot be recognised or understood without this acknowledgement – goes some way to explaining this, but only some way. Perhaps, as we are increasingly compelled to confront the devastating impact that humankind has had on the natural world – and we are forced to search for ways of ameliorating this impact – we will cease to see landscapes without histories or authors (collective or individual), and people like Goyder will have a more significant place in our historical narrative.

To describe Goyder as a landscape author is not to applaud or approve his every decision. Quite apart from the vast and often complete destruction of existing landscapes and ecologies that was, and is, the basic reality of European settlement in Australia, I was often aware, while reading the original documents, that I was witnessing the genesis of what are very real problems now. Pastoralism in the inland has resulted in damage to many mound springs and contributed to the pressure on vanishing indigenous animals. The drained soils of the South-East have proved problematic. Even on a much smaller scale, this was true. Goyder was keen to see a small steamship converted to a dredge to clear snags from the river and keep the waterway open for trade, but this has had to be reversed in places because it is now known to be a form of habitat destruction that affects the breeding of native fish. Nevertheless, it must be understood that Goyder's chief concern was the establishment of stable, permanent settlement of communities that were at least respectful enough of their environment to be perpetuated indefinitely, and he did this on the basis of the best knowledge and understanding available to him, much of it, of course, the fruit of his own observation and reflection.

As surveyor general Goyder was also instrumental in converting the traditional lands of the Indigenous people into colonial real estate, and the landscapes he helped to shape were superimposed on those already given meaning through an elaborate belief system conferred by the original owners. Goyder, however, was unusual in that, following his experiences in the Northern Territory, he did not conceal from himself or others the implications of what was being done in the name of the Crown, identifying the surveyed land according to the peoples to whom it belonged. Rather than resorting to retaliatory violence after an incident where a member of his party was speared by Aboriginal people, he recognised that the underlying reality was that the original inhabitants were being dispossessed of their land, in this way choosing to acknowledge Aboriginal ownership and law. How he reconciled his role – if he ever did – with his stated belief that the Indigenous people and their possessions should be held sacred is not known; if he succeeded, it would probably have been on the basis of a belief in a Divine injunction to develop the resources of the earth. To a large extent, he was as clear-eyed in his perception of the people of the continent as he was in his perception of the land and the climate. His insight and perspicacity were based on an uncommon ability to put aside the expectations of his culture and simply to see, and respect, what was there.

Other reasons have been put forward to explain why Goyder has not been remembered in a way that matches the extent of his contribution. Michael Williams argued that Goyder was sparsely commemorated on maps (apart

from his line) and far less well remembered than 'the Stuarts, the Sturts, the
Burkes and Wills, the Wentworths and the Leichhardts', not only because his
explorations lacked memorable drama, but because he was the 'model of a faceless public servant' and, consequently, little was known about him.[3] But Goyder
was not faceless in his day. Perhaps merely being a public servant was enough to
trigger collective amnesia. In a society in which representative government was
a relatively recent achievement and universal franchise was still to be achieved,
it was the legislature, not the executive branch of government that was the focus
of interest. Being a public servant certainly set him at a great distance from the
heroes of Australian myth, who have been characterised as:

> outlaws. They are Ned Kelly, or they are non-functional or dysfunctional
> white males who can't bind with anyone – like the Man from Snowy River.
> They are people who do not respect public authority …[4]

The practice of commemoration through the naming of topographical
features had been actively pursued in the nineteenth century, although preservation of indigenous names was officially encouraged. Frome had directed
that a particular effort was to be made to determine, as accurately as possible,
the existing names of 'remarkable places, streams and districts', which he confidently expected 'will generally be adopted in preference to any modern appella-
tions'.[5] Goyder's report of 1857, in which he highlighted the names of mountains
he had been able to elicit from the local Indigenous people and explained that
he had named an area after its residents, suggests that he adopted this approach
with some enthusiasm. In 1874 he instructed the party heading north to explore
Lake Eyre to consult the people who lived there, whom he advised would be
found, 'extremely useful in giving local information and names of places, which
in all cases should ever be availed of when obtainable'.[6] But even if the names
of all the significant topographical features had been gleaned from the locals,
there was still plenty of opportunity for naming as new towns, streets, and
elements of the survey and political structure were created. The names of politicians, bureaucrats and surveyors and their families dotted the newly formed
landscape.

Goyder's own first name and the names of his family, especially Frances and
Ellen, appear here and there – in the country east of Lake Eyre, and in Smith
Street, the main street of Darwin, and nearby Frances Bay, and in the streets of
Moonta and elsewhere. In October 1872 parliament debated the resolution that
it was inappropriate to assign hundreds and townships 'the names and surnames
of ladies and gentleman', especially while 'euphonious native names' abounded.[7]
The motion was defeated, but the point was not entirely lost. A few months

before the debate had reached parliament, the *Register*'s satirical columnist (the editor J.H. Clark, posing as 'Geoffry Crabthorn' and his correspondents) had stirred up the issue of naming by posing the riddles: 'Who are James, George, and Laura?' (of Jamestown, Georgetown, and the township of Laura) and: 'Who are, or were, Isabella, Mary, Ellie, Ellen, Gertrude, Florence, George (No. 2), David, and Alexander?' The latter question was purely rhetorical, as the reference to George No. 2 indicates. Since the names of the Goyder children graced the main streets of Port Pirie, Clark suggested that if their bearers could see 'that picturesque spot' they might consider the compliment dubious. After attacking the 'ludicrous system of naming places after "nobodies"', he found an entirely novel use for the Line, suggesting that its name be changed to 'Snobs' Line', and that it become in future:

> the boundary between the 'natural' and 'unnatural' names, the 'Snobs' keeping that portion of the country south of 'Snob's Line', (the map of which is already disfigured by many of their names), whilst in the country north of that line the aboriginal names be retained intact. Having taken everything else from our sable brethren, at least do not let us 'filch from them their good names'.[8]

The name 'Goyder' itself was commemorated in the survey structure: it was attached to a hundred above the top of Gulf St Vincent at a time when every leading figure in South Australian public life seems to have acquired a commemorating hundred, and later to a parliamentary division. More recently, the regional council of the area around Burra has been named 'Goyder'.

Goyder's Lagoon was the first feature in the natural world to receive his name, and this has been followed by a waterhole and 'lake' in the region, and even a bird – the Eyrean Grasswren, although named after Eyre (through the region, presumably), has the scientific name *Amytornis goyderi*. Goyder's name can now be found elsewhere as well. The channel between Lake Eyre North and Lake Eyre South is the Goyder Channel, and South Australia's highest peak, Mount Woodroffe in the Musgrave Ranges, bears his middle name. In the Northern Territory, the Goyder River flows toward the Arafura Sea (very appropriately, it joins the Glyde), and a Mount Goyder sits to the north of the Arnhem highway, west of Kakadu National Park. Bores and waterholes bearing the name are scattered about.

Changes in our understanding of the physical world have brought Goyder's work into a new and clearer focus and there is evidence of the continuing relevance of his contribution to South Australia. Goyder recognised that water was South Australia's most precious, but most precarious, resource. Water was central to almost all of his activities and was the subject of a master plan presented to government. He never lost sight of its importance – not surprisingly, for a person who had more than once come close to dying of thirst. Goyder was out searching the South African landscape for drinking water even while passengers of the *Queen of the Thames* were being ferried ashore. When a tropical downpour disrupted the unveiling of a commemorative plaque in Darwin, Margaret Goyder Kerr reassured those involved that her grandfather would not have been displeased: 'If there was anything G.W. Goyder liked, it was fresh water!', she later explained.[9] In 2010 an independent institute supported by all three South Australian universities, the CSIRO and the South Australian Government was founded to address the critical water problems faced by that state, and appropriately named the Goyder Institute for Water Research.

... and Australia

Goyder did more than help to shape and establish the state of South Australia. For many decades Darwin was a remote tropical township, and his position as its founder was of marginal historical interest, but Darwin is now a modern city close to Asia. Goyder must therefore be viewed as a figure of early significance in two of the seven major regional entities that make up the Commonwealth of Australia. (The Australian Capital Territory is a separate administrative entity, of course, but regionally, it is part of New South Wales.) And Goyder did not just select, design and survey Darwin; he was responsible for establishing non-Indigenous settlement in the Northern Territory.

However, his relevance and significance to the nation as a whole derives from his early, unrecognised encounters with climate and, most crucially, his extraordinary early understanding that the climate of the inland of the continent was one in which extreme and unpredictable variations in rainfall were the norm, and that this was the major determinant of the ecology in those regions. Realising that the presence or absence of particular species of plants identified areas where the rainfall was unreliable enabled him to take the first major step towards adapting European settlement to this new climatic reality. This aspect of Goyder's contribution to the national story has not been celebrated, not because it has been forgotten, but because it was not recognised in the first place. The story of the settler exploration of the Australian continent as it was told in the nineteenth century did not include a chapter on the extreme variability of the climate, because that climate was not understood. Goyder reported vast areas of fresh water and lush vegetation in the north of the salt lakes, but the country itself seemed to prove him wrong, and by the time his claim was vindicated, the narrative, from which he was excluded, was already determined.

Throughout the twentieth century Goyder continued to be presented as the misguided victim of floodwater and mirage – if he was presented at all. His explorations in 1859–60, when he saw a drought break and the salt lake country flood again, confirmed his belief that flooding did occur in the inland, awakening him to the realisation that the phenomenon was critical to life there. However, these journeys were never perceived as 'exploration' in the recognised sense, simply because he was not the first European to set foot in the region, and his stated business was surveying, not exploring. So there was no story.

But people continued to live with the weather. The drought that took hold during the last years of Goyder's life, impressing on his contemporaries the validity of his line, exerted a lasting effect on another South Australian, directing him, too, towards an interest in weather and climate. Sir Hubert Wilkins, internationally famous during his lifetime as a war correspondent and photographer, a polar explorer, a naturalist, geographer and climatologist, as well as an aviator, was born in 1888 and grew up near Mount Bryan, just outside the Line. 'Our living was precarious because we never knew what weather nature held in store for us', he recalled. In good years they were prosperous and happy, but the horror of watching, as a young child, tens of thousands of animals die during the drought, and the realisation that his family's life was lived at the mercy of a variable climate initiated a lifelong interest in weather forecasting.[10]

Not many people had such a clear and straightforward experience or response to the weather as Wilkins. In the year following the 1901 drought a royal commission attempted to determine whether it was worth building a railway to the Pinnaroo area. The surveyor general, William Strawbridge, and his deputy were queried about the variant mappings of the Line. Given that Goyder had not examined the country before 1882, Strawbridge ventured to suggest that the modification which excluded Pinnaroo was 'just a curved line' and an 'accident' – prompting one of the commissioners to respond with alarm that it seemed as if the rainfall line had been drawn 'at a venture' – more or less at random – in the first place. Strawbridge had to hastily assure the commissioners that this was not so, and it was then that he offered the observation that Goyder had monitored the presence of bluebush and saltbush, pointing out that no saltbush was found at Pinnaroo, which was therefore properly placed south of the Line.[11] Despite Strawbridge's assurances that the original line was the one used in the Survey Office, the commissioners were unimpressed by his inability to interpret what were evidently deliberate changes. Edwin Mitchell Smith, who knew nothing of them either, had already been despatched to gather information.

Although the commissioners were clearly determined to discover Goyder's opinion of the Pinnaroo country and whether or not he had considered it to fall within his line, the discussion that took place displayed how completely his real point had been lost. Rainfall continued to be spoken about, not as unreliable, unpredictable or variable, but as 'fair' or 'poor' or 'deficient', and the focus of attention was on annual averages alone – surprising, given that Strawbridge and Smith had worked with Goyder for years and were connected to him by religion and family respectively. The deliberations of the commission only served to demonstrate how entirely at the mercy of their inadequate conceptions and vocabulary its members were.

The commission eventually decided that prospects for the area were good (although some commissioners dissented) and the railway went ahead. The land surrounding it was soon taken up, and wheat farming began in the Murray Mallee area, as well as in the south of the Eyre Peninsula. In both places it spread rapidly, and in the Murray Mallee it had soon advanced well beyond the Line. As Michael Williams has put it, over-confidence engendered by new technology and farming methods eventually led yet again to a 'recklessness that was shared by settlers and government officials alike'.[12] When the next drought arrived, in 1914, it replaced the great drought of 1865 as the most severe since settlement. No area in the newly settled mallee yielded more than a half a bushel an acre (less than 35 kilograms per hectare), and in many places nothing was produced at all.[13] There were rains and bountiful harvests again around 1920, but drought had visited once more by the end of the decade. Returned soldiers from the First World War who had been settled by the government on small blocks in the drier country fared disastrously.

In 1908, ten years after Goyder's death, Dorothea McKellar published the poem that was to be memorised, at least in part, by generations of school children. 'My Country', with its passionate honouring of 'droughts and flooding rains' (among other things), would become the poetic equivalent of a national anthem, and that particular phrase was to enter the Australian idiom. But a deeper, more thoroughly integrated recognition of the reality and impact of a highly variable climate was still a long way from finding a place in the consciousness of Australians, as the work of another poet of the period demonstrates. Six years after 'My Country' appeared, the great folk poet A.B. 'Banjo' Paterson wrote 'Song of the Wheat'. The poem is a paean, not only to the grain itself – grain that would be shipped to grateful 'teeming millions' across the world – but to the spread of wheat farming in Australia. Paterson, who was seen to express the experience of Australia's rural base, could be expected to know something of the trial-and-error experiences of farmers moving into

marginal country, not only in New South Wales, but also in Victoria and especially in South Australia, the leader in wheat production. (In New South Wales
in 1900 a royal commission had investigated the problems of settlement in the
drought-prone west and, in a strategy akin to Goyder's Line, had defined the
area as the Western Division, pastoral country with special leasehold provisions.) But in his determination to celebrate a successful industry, Paterson,
apparently unaware of what he was doing, turned history, and environmental
reality, on its head. The problem is not that the poem celebrates the triumph of
wheat, but that it presents wheat as victoriously succeeding sheep, not on good
agricultural country, which in the course of settlement was first taken up by
graziers, but on 'grim grey plains', where even the pastoralists had struggled
hopelessly in 'a ceaseless fight with drought'. According to Paterson, after this
land had been cleared of trees ('Yarran and Myall and Box and Pine', which
indicate semi-arid, drought-prone country), work began on the 'surface-mine
of the grain'. Remarkably, the wheat proves itself:

> Better than cattle and better than sheep
>   In the fight with the drought and heat.
> For a streak of stubbornness wide and deep
>   Lies hid in a grain of Wheat.
> When the stock is swept by the hand of fate,
>   Deep down in his bed of clay
> The brave brown Wheat will lie and wait
>   For the resurrection day:
> Lie hid while the whole world thinks him dead;
>   But the spring rain, soft and sweet,
> Will over the steaming paddocks spread
>   The first green flush of Wheat.[14]

Like the forester J.E. Brown, whose response to the climate of the inland was to
replace tree seedlings with seeds, Paterson proclaimed drought while denying
its realities: having extolled the wonderful, drought-resisting hardiness of the
wheat grain, he brings on the spring rain anyway. The poem was published in
1914, the year of another massive drought, but nobody responded by declaring
it nonsense. In the 1920s an illuminated manuscript of the poem in book form
appeared to elevate it to something akin to Holy Scripture.

A promising moment of insight came in 1918, when the geographer T.
Griffith Taylor published *The Australian Environment (especially as controlled
by rainfall)*. The account that Taylor gave of Goyder's Line demonstrated that
he had approached the heart of Goyder's achievement – and he introduced

appropriate technical language to describe it. Taylor saw the Line as practically agreeing with the southern boundary of the 'salt-bush, mulga and dwarf mallee country', and understood that, although 'an immense amount' of wheat could be grown beyond it in good years, 'the safe farming country lies south of this ecological isopleth'. But, taking advantage of what was then the most recent information, Taylor went on to write that, once the isohyets of average annual rainfall were charted, the Line was found to agree closely with the 12-inch isohyet (except along the coast north of Wallaroo). Having apparently equated the Line with that isohyet, he encouraged 'this logical method of estimating the possibilities of new country' – referring to the 12-inch isohyet, presumably – as one that ought to be carried out in arid regions across the country.[15] Across much of Australia, aridity and variability change together, increasing aridity being matched by increasing variability, a reality that may have led Taylor to equate the 12-inch isohyet with the limit of the safe country. If Taylor believed this, he did not make that understanding explicit. Instead, he seems to have helped to establish a trend towards understanding the Line in terms of average rainfall, which, paradoxically, obscured the very point Goyder had struggled to make clear.

But it is hardly appropriate to criticise Griffith Taylor. In the same year that Taylor's work was produced, Edwin James Brady published *Australia Unlimited*, a solid volume, full of illustrations, which sold well. As part of his promotion of the country's unlimited potential, Brady argued that local artists and writers had misrepresented the place, embracing 'accidental happenings' such as dry seasons and bushfires, in the absence of such dramatic material as wars. Brady argued that the explorers, unaware that Australia's water was underground, had created a 'Desert Myth'. The 'Dead Heart' was actually a fertile 'Red Heart', which would one day be full of people.[16] For Brady, the Australian climate was a misleading illusion, there to be ignored. The view that climate should not set limits on the future of settlement or the size of the population was expressed by others as well as Brady and eventually this approach predominated. It was this attitude that led to the banning in Western Australia, in 1921, of what has been described as the first competent textbook on the climate and weather of Australia, which Griffith Taylor had produced in association with two others.[17] Taylor, like Goyder, believed that the land should be settled in a rational way, guided by scientific understanding, rather than by often destructive 'testing'. In the debates that followed other expressions of his views, Taylor was reviled in the press for his pessimism.

In South Australia in 1938 the situation as a result of the clearing and ploughing up of the north had become so dire (as Goyder had known that it would) that a Soil Conservation Committee was formed to investigate the extent of soil erosion. The committee found that much of the moderate soil drift, and all of the severe drift, occurred beyond the Line. The following year another committee reported on marginal lands, including country around the Murray and a large area in the west beyond the Eyre Peninsula, which were inside the Line. This committee found that there was still a need to transfer many farmers south to safe agricultural country so that the size of properties and the number of livestock carried could be increased and the extent of wheat cultivation in the north decreased.[18] Perversely, it was at this time that the geographer John Andrews asserted that the Line was nothing but the border of a particular drought, damning it as a 'vanished frontier' that had become a stultifying 'fetish', frustrating economic development in the state.[19]

What was lost when a line that had been conceived as a line of reliable rainfall became known only as Goyder's Line is demonstrated by the use of the phrase 'imaginary line', as if there were no better way to describe it. In the 1870s this was a term of dismissal, although by the time of Goyder's death, it had been rehabilitated, making it a 'most useful imaginary line'.[20] The term even made its way into cyberspace: for a time during 2005 the *Wikipedia* entry began by boldly announcing that 'Goyder's Line is an imaginary line ...' A tourist handout distributed at Orroroo at that time also made use of the term, and in April 2006 a reporter on the ABC's rural affairs program *Landline* described a sheep and grain property as being on 'tough farming country ... just short of Goyder's Line, the imaginary boundary across South Australia which, for 140 years, has separated farming land from pastoral land'.

Since the 'line of reliable rainfall' had been stillborn, the word that came to carry Goyder's message was 'safe': the country south of the line was 'safe' agricultural country, safe from drought (if you thought about it). That this was not enough to preserve the Line's full meaning is demonstrated by Meinig's unsuccessful attempt in the late 1950s to determine from the existing studies what the Line expressed. It was not until the second half of the twentieth century, as understanding of the climate and ecology developed and the analysis of meteorological data became more sophisticated, that the real meaning of the Line began to be restored. The historical geographer J.M. Powell appears to have reintroduced Goyder's terminology, always referring to the Line as defining reliable rainfall, although he continued to draw on the then-standard historical account, which ascribed its origins strictly to the 1865 drought. Inevitably, misunderstandings have continued to abound. The 2009 edition of the *Macquarie Dictionary* offers a definition in which Goyder's position as surveyor general is assumed to have played a mysterious role. The Line, it explains, is 'a surveying line which separates areas of good rainfall from areas in which rainfall diminishes rapidly thereby becoming unsuitable for farming, especially wheat farming'. The image of Goyder peering through a theodolite to determine where the rainfall ends begs a cartoon from *Pasquin*.

Ruins

Early in my research I had been very excited at finding a photograph in a book that was virtually the image of the isolated ruin I had first seen, the ruin that ultimately led me to the story of the Line, and to Goyder.[21] The only difference between it and the image in my memory was that the photograph was a close-up with a vertical orientation to show the bare sandy soil and the pebbles, whereas the house lodged in my memory sat in the horizontal expanse of the landscape. The caption bluntly characterised what was shown as the result of two hundred years of environmental degradation. Sometime later, I came across a wonderful postcard of an abandoned house, a classic South Australian stone house, faced at the corners and around the frames with brick. It stands in a field of golden stubble, under a blue sky, across which high-altitude winds have drawn a single gauzy line of cloud. Just behind, bare hills form the horizon. A moment's reflection reveals that the house has not been abandoned for a great period of time. The roof, although rusty, is still complete, and two tanks, no longer connected but roughly intact, stand behind it. Although the windows are empty, the wooden frame remains in the doorway and not a stone is missing from the walls.

Of all the abandoned farmhouses in South Australia, this one, I gradually discovered, has a peculiar status. It appeared in 1987 on the cover of the award-winning Midnight Oil album, *Diesel and Dust*. Shot from a low angle, and up close, the house looms starkly out of its own shadow, while the hills behind form a distant ripple under a sky that shades from white to a deep enamel azure. In this photograph, there is no crop. Instead, the house is tightly encircled by furrows ploughed in the red earth, like a rock in a Zen garden. The stones of the house and its rusty roof and everything in the image except the sky and the yellow hills share the same red tone – and undoubtedly the hot north winds that Goyder feared would colour the air as well. As the photographer Ken Duncan observed, after its appearance in this form, that abandoned dwelling was 'quickly transformed into one of South Australia's most iconic houses!'[22] As a brief online search will demonstrate, it is literally a stock subject and has a place in Picture Australia, the national archive. In one shot it is surrounded by a ripe crop; in another it is ringed by stubble. In a modified form, the image of this house became the cover of a tourist guide to the Mid-North – 'inside country' – during the 1990s, where it was shown standing in a green-gold expanse of sprouting crop within the words: 'Classic country. Rediscover what's close to your heart ...'

During my investigation of the Line I came across the house itself, facing west at the turn-off to Spalding, south of Hallett, on the main road to Burra, and it *is* inside the Line, if only just. Goyder's Line runs roughly north–south immediately beyond the hills that stand behind it. As an icon, then, the house is ambiguous and exactly what it symbolises remains unclear. It is not, I think, too much to argue that this lack of clarity is not accidental, but representative of a culture still far from having assimilated a clear awareness and understanding of the Australian climate. The house is evidently the legacy of an at least modestly prosperous past, both in reality (as the fact that it stands in cropland attests) and in the promotional dream of 'classic country'; however, there can be no doubt that images of this house gain some of their resonance from the presence of the ruins not far away, on the other side of the Line, in what has been called not 'classic country', but 'heartbreak country'. Fortunately, the 'classic country' guide is no longer issued. Another ruin, weathered and crumbling, appeared on the cover of a later guide to the country around and beyond the Line, described as the 'edge of the outback', where red sand, saltbush and the 'forested ranges of the pastoral regions merge with the tamer greens and golds of the cropping lands'. That guide encouraged visitors to set off on the back roads, where 'ghost towns and the many stone ruins, which were once the treasured homes of the hard working pioneers' will be found.[23] Unfortunately, it did not explain why the ruins happen to be in that location in such numbers. The answer was on a page devoted to Orroroo ('just delightful'), although it was only obvious to someone already familiar with the history of the state. Orroroo, the tourist was informed, 'perches on Goyder's Line'.

Despite the number of people involved and the visible legacy of ruins, it is remarkable that the dramatic advance of the wheat farmers beyond the Line, and their subsequent annihilation, have not been integrated into the folk memory of Australia at large – although chance encounters outside South Australia with descendants of those involved made clear that this episode has not been forgotten in the families affected. Perhaps this is because the story is depressing from whichever way it is viewed, and from a narrowly urban perspective it probably looked like yet another story about farmers whose crops failed – the sort of thing that has happened to farmers from time immemorial. There was certainly nothing in it to appeal to those who insisted that the continent could be filled with teeming millions or those in search of formative rallying myths. It can hardly have helped that the events took place in South

Australia. While the state is geographically vast, at present it is home to only eight per cent of Australians – and a large proportion of these live in the coastal capital, Adelaide, as they have always done.

Perhaps there is another reason why the dramatic story of the failed attempt to farm the north has not become a well-known part of the story of European settlement, despite involving thousands of people. Michael Williams has alluded to the popular view that the rapid expansion beyond the Line was followed by a rapid retreat, the drought of the 1880s heralding the wholesale abandonment of the northern areas.[24] It is far more likely, he argues, that a dramatic course of events such as this would be widely remembered, but as Goyder's own words indicate, while many left, many more clung on desperately for the rest of their lives. The slow death was not remembered. According to Williams, depopulation of the area did not begin in earnest until the decade after federation.[25]

And now …

In the second half of the twentieth century, small farms were gradually consolidated into much larger holdings, and access to credit made those working these holdings better able to survive bad years than the farmers of Goyder's time. There have been changes, too, in agricultural techniques. Zero-till farming, which is now widespread, conserves soil moisture by barely disturbing the soil. Computer technology has enabled a range of new ways of managing the land and soil, along with access to sophisticated climate information and weather predictions. Although it is still not easy, with skill and attention farmers survive beyond the Line, but now they face new problems. Fossil salt rising to the surface is poisoning a significant proportion of agricultural land, while climate change threatens us all. One scientist has even suggested that if 'you like Clare Valley wines, you might be well advised to stock up while you can', because it has been estimated that the limit of agriculture could end up south of the Clare Valley.[26]

What then is the value of the Line today? Goyder's own words indicate that, as an agricultural limit, he placed the line of reliable rainfall not only in the context of the situation of the small farmers of the time, who needed an income from their crop every year, but in the context of what 'we know of growing cereals at present'.[27] In that respect his attitude was, in principle, flexible. Still, the Line has rightly been identified as an enduring symbol and should not be dismissed or forgotten. It was drawn to express a critically important climatic reality, and along with its position as an early step towards an understanding of the Australian climate, the cultural and historical value of the Line lies in pointing to that reality. In his Australia Day address in 2002, Tim Flannery observed that the period since 1788, or three human lifetimes:

> is simply not long enough for a people to become truly adapted to Australia's unique conditions, for the process of learning, of co-evolving with the land is slow and uncertain. Yet it has begun, and the transformation must be completed, for if we continue to live as strangers in this land – failing to understand it or live by its ecological dictums – we will forfeit our long-term future here by destroying the ability of Australia to support us.[28]

More pithily, a comment attributed to Burnum Burnum (the Aboriginal activist famous for having planted a flag on the cliffs of Dover to claim England for the Indigenous Australians) states that, while many people came to Australia, few actually arrived. George Goyder's significance is that, in relation to climate at least, he was one of those few. His experiences in the inland and Goyder's Line of Reliable Rainfall (as it should properly be known) represent important early steps in the process of adaptation to the realities of the continent. The story, or stories, of Goyder's passionate and hugely effective dedication to the public interest and the unparalleled astuteness of his perception of the natural environment make a refreshing addition to the frozen and often depressing canon of Australian colonial narrative, in which Burke and Wills are forever marching off towards the desert with their tons of equipment and cedar table, and Stuart and Sturt stagger endlessly and agonisingly across the desert. Goyder's story, which prefigures and engages contemporary concerns about adapting human societies to environmental realities, especially in relation to climate and water, properly belongs not in the byways of regional history but to the history of Australia.

Notes and abbreviations

From 1858 to 1895, only a single volume of parliamentary debates was published: *Debates in the Houses of Legislature*. This was suspended during 1863, 1864 and part of 1865. Transcriptions of the debates were provided by both the *Register* and *Advertiser*.

As most of Goyder's writings and recorded comments appear in the parliamentary papers, these have inevitably constituted a major source. Because so many have been used, an abbreviated form has been employed to cite them. For convenience, and to avoid confusion, the year of the parliamentary sitting has been placed ahead of the number of the paper. Citations appear in the form:

SAPP [year], [number], [page], [question or paragraph number where required].

The titles of the papers are given in the Source list of Parliamentary Papers at the end of the book.

Other abbreviations used in the notes are:

CCLI   Commissioner of Crown Lands and Immigration
GRG   Government Record Group (followed by numbers indicating the files series, year, and item number)
PRG   Private record group
SGO   Surveyor General's Office
SLSA   State Library South Australia
SRSA   State Records South Australia

South Australian newspapers

Although variations occur in the titles of the major South Australian newspapers, they have been cited in a uniform way. The *South Australian Advertiser* is referred to as the *Advertiser* (which became its title after 1889), and the *South Australian Register* as the *Register* (its official title from 1901). Similarly, the *Adelaide Observer*, which became the *Observer* in 1905, is identified as the *Observer* and the *South Australian Weekly Chronicle* (which underwent several changes in name) appears in notes as the *Chronicle*.

Source list of South Australian parliamentary papers

All of the parliamentary papers referred to in the text were published in *Proceedings of the Parliament of South Australia, with copies of documents ordered to be printed*, with the exception of *Hindmarsh Island and Lake Albert Peninsula*, which was published in 1856 as paper no. 14 of the earlier series, *Votes and proceedings of the Legislative Council*, and only later appeared as appendix A to SAPP 1890, no. 34A.

1857–58
72, *Northern exploration*; 153, *Exploration of the north-west*; 156, *Explorations by Mr. S. Hack*; 174, *Country adjacent to Lake Torrens*; 189, *North-western exploration*; 192, *Explorations by Messrs Miller and Dutton*; 193, *Northern exploration*

1858
25, *Northern exploration*

1859
21, *Select Committee to Inquire into and Report upon the Petition of B.H. Babbage*; 30, *Report on Poonindie Mission*; 53, *Instructions to Waste Lands Rangers–Burra Area.*; 119, *Geology of South Australia*; 162, *Petition against classification of runs*; 167, *Aboriginal appropriations and reserves*

1860
20, *Geological notes by A.R.C. Selwyn*; 38, *Railway route to Mount Remarkable*; 41, *Instructions to Mr. Goyder*; 116, *Report of the Select Committee of the House of Assembly ... on ... the Assessment of Stock Act*; 165, *Report of the Select Committee of the Legislative Council upon the Aborigines*

1860–61
177, *Mr. Goyder's report on northern triangulation*

1861
56, *Report on Wallaroo District*; 65, *J.M. Stuart's exploration, 1860*; 90, *Report respecting sales of land with deferred payments*; 101, *Report by the Estimator of Runs*; 108, *Assessment of Stock Act, 1861*; 109, *Report on Assessment of Stock Bill, 1861*

1862
51, *Report of the Select Committee of the House of Assembly on the present mineral laws*; 180, *Sale of pastoral leases by private contract*; 190, *Tipara mineral claims*; 220, *Tipara mineral claims*

1863
41, *Public works, South-East District*; 51, *Report of the Select Committee of the House of Assembly on the Tipara mineral claims*; 52, *Report of the Select Committee of the House of Assembly on drainage from Wallaroo mines*; 118, *Report on routes to River Murray*

1864

36, *Settlement of the Northern Territory;* **82,** *Instructions to Valuator of Runs;* **96,** *Report and journal of E.H. Hargreaves;* **102*,** *Revaluation of crown lands under lease;* **103,** *Revaluation of crown lands held under pastoral leases;* **103*,** *Memo. by Government on the revaluation of runs;* **104,** *Correspondence on revaluation of runs;* **105,** *Revaluation of crown lands under lease;* **105*,** *Revaluation of crown lands held under pastoral leases;* **105**,** *Tabulated statement of revaluation of runs;* **105A, 105B, 105C,** *Revaluation of crown lands under lease;* **106,** *Report on revaluation of runs;* **213,** *Progress reports of the Select Committee of the Legislative Council on Pastoral Lease Valuations*

1865

30, *Revaluation of crown lands under lease;* **72,** *Report of the Select Committee of the House of Assembly appointed to inquire into ... the System of selling the Crown Lands*

1865–66

30, *Northern Runs Commission* ; **57,** *Report of the Commissioners appointed ... to inquire into the State of the Northern Runs;* **62,** *Instructions to the Surveyor General;* **70,** *Petition for aid to Aborigines in the Far North;* **77,** *Revaluation of crown lands;* **78,** *Surveyor-General's report on demarcation of northern rainfall;* **82,** *Report of Surveyor-General on northern runs;* **105,** *Destruction of documents in the Survey Office;* **125A,** *Tabulated statement of runs revalued* ; **126,** *Detailed valuations of runs;* **133,** *Memo. of Surveyor-General on rainfall, &c.;* **137,** *Valuation of runs, South-East District;* **154,** *Map of the northern runs*

1866–67

17, *Report of Commission ... to inquire into the Management of the Northern Territory Expedition;* **65,** *Report of the Select Committee of the House of Assembly ... on the South-Eastern District improvements;* **128,** *Survey of land, Northern Territory, etc.;* **161,** *Reports of the Select Committee of the House of Assembly on Railway Extension;* **188,** *Tenders for Northern Territory surveys;* **190,** *Report of the Select Committee of the House of Assembly on Mineral Regulations*

1867

14, *Report of the Commission ... to inquire into the State of the Runs suffering from Drought;* **89,** *Reports by Valuators of Runs;* **141,** *Report of the Select Committee of the House of Assembly on Wallaroo and Clare Railway;* **142,** *Reports on South-Eastern district drainage works;* **145,** *Report on unfinished wells in North-East District*

1868–69

81, *Tenders for survey, Northern Territory;* **100,** *Tenders of survey, Northern Territory;* **101,** *Surveyor-General's estimate of Northern Territory survey;* **161,** *Agricultural areas under Waste Lands Amendment Bill;* **175,** *Northern Territory survey expedition*

1869–70

31, *Northern Territory survey progress reports;* **156,** *North Australian Telegraph;* **157,** *Survey of the Northern Territory;* **201,** *Bonus to Northern Territory survey party*

1870–71

23, *Report on Victoria land regulations* ; **78,** *Memo. on amendment of waste land laws;* **86,** *Report of the Select Committee of the House Assembly on the River Murray Trade;*

89, *Agricultural Area –Belalie*; **107,** *Drainage works, South-East District*; **144,** *Reports of suggested forest reserves*

1871
126, *Northern Territory goldfields*

1872
43, *Report of the Select Committee of the House of Assembly ... [on] Drainage Works in the South-East*; **202,** *Petition of Samuel White Sweet*

1873
94, *Report on Forest Tree Planting Encouragement Bill*; **105,** *Withdrawal of Surveyor-General's resignation*; **135,** *Report on forest reserves*; **153,** *Land east and north of the Burra*; **175,** *Examination of Murray mouth*

1874
34, *Petition for new agricultural areas*; **39,** *Report of Civil Service Commission*; **67,** *Civil service salaries*; **168,** *Memo. by Surveyor-General on lands in South-East*; **239,** *Survey of crown lands*

1875
22, *Report of the Commission ... on Railway Construction*; **85,** *Despatches on colonial timber*; **105,** *Withdrawal of Surveyor-General's resignation*

1875 Special Session
9, *Irrigation and drainage of Mayurra Agricultural District*; **26,** *Water drainage of eastern colonies* [map]; **32,** *Planting trees on reserves*

1876
19, *Journal of Mr. Lewis's Lake Eyre expedition, 1874–5, Macumba River*; **29,** *Correspondence with Bishop Bugnion*; **95,** *Reports on northern mineral land*; **108,** *Reports on Victor Harbor and Murray mouth*; **145,** *Correspondence concerning annual leases*; **160,** *Correspondence with Bishop Bugnion*; **203,** *Proposed new country and hundred*

1877
157, *Forest Board report 1876–7*; **159,** *Report on overflow of River Sturt*

1878
146, *The award of the umpire ... on the contracts for constructing Kingston and Rivoli piers*; **154,** *Forest Board report 1877–8*; **203,** *Reports on extending the South-East drainage works*

1879
31A, *Report on Mannahill Experimental Farm and surrounding country, with plan giving description of soil.*; **71,** *Report of the Select Committee of the House of Assembly on Credit Selection under Crown Lands Acts*; **83,** *Annual Progress report of the forest reserves and forest conservancy generally*

1880
139, *Annual report of the Forest Board with Conservator's progress reports and appendices*; **196,** *Report on the drained land in the South-East*

1881

43, *Annual progress report of the forests reserves*; **127,** *Report of the Select Commission of the House of Assembly [on the] Port Germein and Ororoo railway*; **144,** *Suggestions from heads of department re Civil Service Amendment Bill*

1882

25, *Final report [of the] Committee appointed to report on Public Works* ... ; **73,** *Petition for amendment of land laws*; **109,** *Annual progress report upon the State Forest Administration*

1883–84

52, *Water conservation and development*; **73,** *Annual report upon the State Forest Administration*

1885

124, *Report of the Select Committee of the Legislative Council ... on the Vermin Act Repeal Bill*

1887

34, *Report of the Commission on the Transcontinental Railway*

1888

28, *Report of the Commission on the Land Laws of South Australia*; **109,** *Woods and Forests Department ... annual report ... 1887–8*

1889

25, *Report of the Barossa Water Commission*; **113,** *Report of the Select Committee of the House of Assembly on Travelling Stock Reserves*

1890

30C, *Seventh Progress Report of the Public Service Commission*; **32,** *Report of the Mining Commission*; **34,** *Progress report of the Commission on the Utilisation of River Murray Waters*; **34A,** *Second progress report of the Commission on the Utilisation of River Murray Waters*; **60,** *Report on disposal of crown lands in South Australia* ; **64,** *Report, etc., on South-East drainage*; **94,** *Report of the Royal Commission on the Water Conservation Department*; **131,** *Information re construction of rip-rap dams* ; **144,** *River Murray[:] proposed Weir near sea mouth* [map]

1891

33, *Report of the Pastoral Lands Commission*; **41,** *Surveyor General's Report on land transactions*

1892

35, *Report of the Royal Commission on the South-Eastern Drainage System*; **152,** *Report on Pinaroo* [*sic*] *lands.*

1893

59, *Report of the Vermin-Proof Fencing Committee*; **83,** *Report re Pinnaroo lands*

1895

19, *Report of the Northern Territory Commission.*

1902

22, *Report of the Royal Commission into the Pinnaroo Railway*

Notes

Preface: Never to be forgotten

1 SAPP 1888, no. 28, p. 77, qu. 2201.

2 SAPP 1890, no. 60, p. 17.

3 South Australia, Parliament, *Debates*, 4 July 1882, c. 243.

4 ibid., 22 June 1882, c. 171. The phrase is drawn from a letter read in the Assembly, from Friedrich Kuttler, a farmer at Winninowie.

5 ibid., 7 Oct. 1897, p. 565.

6 G.W. Goyder, *Mr. Goyder's Report on Victoria Land Regulations*, Advertiser, Adelaide, [1870], p. 8, para. 55 (also as SAPP 1870–71, no. 23).

7 Although I was not aware of it at the time, Barry McGowan had published 'Goyder's Line: Australia's first environmental debate? An historical perspective', in the *Canberra Historical Society Journal*, no. 25, Mar. 1990, pp. 28–35.

8 Geoffrey Dutton, *A Taste of History: Geoffrey Dutton's South Australia*, Rigby, Adelaide, 1978, pp. 57, 95.

9 Geoffrey Dutton, *The Squatters: An illustrated history of Australia's pastoral pioneers*, Viking O'Neill, Ringwood, Vic., 1989, p. 14.

10 Donald W. Meinig, *On the Margins of the Good Earth: The South Australian wheat frontier 1869–1884*, facs. paperback edn, South Australian Government Printer, Netley, SA, 1988 (first published 1962), p. 217.

11 Michael Williams, 'George Woodroofe [*sic*] Goyder: A practical geographer', *Proceedings of the Royal Geographical Society (South Australian Branch)*, vol. 79, 1978, p. 2.

12 Michael Williams, *The Making of the South Australian Landscape: A study in the historical geography of Australia*, Academic Press, London, 1974, p. 66.

13 Williams, 'George Woodroofe [*sic*] Goyder', p. 20.

14 J.M. Powell, *Environmental Management in Australia, 1788–1914: Guardians, improvers, profit: An introductory survey*, Oxford University Press, Melbourne, 1976, p. 67.

15 J.M. Powell, *An Historical Geography of Modern Australia: The restive fringe*, Cambridge University Press, Cambridge, 1991, p. xv.

16 J.M. Powell, 'The genesis of environmentalism in Australia', in *Created Landscapes: Historians and the environment*, ed. Don Garden, History Institute Victoria Inc., Carlton, Vic., 1993, p. 12.

17 ibid.

18 'Death of Mr. G.W. Goyder, C.M.G.', *Observer*, 5 Nov. 1898, p. 32.

19 Donald W. Meinig, 'Goyder's Line of Rainfall: The role of a geographic concept in South Australian land policy and agricultural settlement', *Agricultural History*, vol. 35, October, 1961, p. 207.

20 T. Griffith Taylor, *The Australian Environment (Especially As Controlled By Rainfall): A regional study of the topography, drainage, vegetation and settlement, and*

of the character and origin of the rains, Advisory Council of Science and Industry, Melbourne, 1918, p. 98. An isopleth links points of the same numerical value, for any element or ratio of values.

21 A. Grenfell Price, *South Australians and Their Environment*, 2nd edn, Rigby, Adelaide, 1922, p. 33.

22 Meinig, *Margins*, pp. 45–6. Goyder is quoted by Meinig from SAPP 1867, no. 14, p. 113 and SAPP 1870–71, no. 23, p. 8.

23 Meinig, 'Goyder's Line of Rainfall', p. 210.

24 Williams, 'George Woodroofe [*sic*] Goyder', p. 10; Anne McArthur (ed.), *Through the Eyes of Goyder Master Planner: Transcripts of his 1864–5 detailed valuations of 79 pastoral runs in the South East of South Australia*, Kanawinka Writers and Historians Inc, Penola, SA, 2007, p. xx.

25 D.J. Mulvaney, *The Prehistory of Australia*, rev. edn, Penguin, Middlesex, 1975, p. 56.

26 SAPP 1888, no. 28, p. 77, qu. 2189.

27 Surveyor-General's Office, *Past and Present Land Systems*, Government Printer, Adelaide, 1881, p. 10.

28 Neville Nicholls, 'Climatic outlooks: from revolutionary science to orthodoxy', in Tim Sherratt, Tom Griffiths and Libby Robin (eds), *A Change in the Weather: Climate and culture in Australia*, National Museum of Australia Press, Canberra, 2005, p. 19. The writer quoted is William Stanley Jevons.

29 F.J.R. O'Brien, 'Goyder's Line', MA preliminary thesis, University of Adelaide, 1952, pp. 73–4, pp. 32–3.

30 ibid., p. 75. Goyder is quoted from SAPP 1888, no 28, p. 77, qu. 2186. O'Brien is correct in pointing out that in 1865 Goyder had not been asked to show where tolerably reliable rainfall was secured. That Goyder had claimed this in 1888 is better understood as evidence that this was the goal he had set himself and what he had intended to do.

31 John Andrews, 'Goyder's Line: A vanished frontier', *Australian Geographer*, vol. 3, no. 5, [1938?].

32 Charles Darwin, *The Autobiography of Charles Darwin 1809–1882*, ed. Nora Barlow, Collins, London, 1958, p. 70.

33 A.P. Sturman & N.J. Tapper, *The Weather and Climate of Australia and New Zealand*, Oxford University Press, Melbourne, 1996, p. 298.

34 Keith Colls & Richard Whitaker, *The Australian Weather Book*, 3rd edn, New Holland, Sydney, 2012, p. 25.

35 ibid. Michael Williams makes the same assertion in McArthur (ed.), p. xxi.

36 William Harcus, *South Australia: Its history, resources and productions*, Sampson, Low, Marston, Searle and Rivington, London, 1876, p. 413. Todd was the author of the chapter in which this appears.

37 Nicholls, p. 19, quoting Todd in *Australasian*, 1888, p. 1456.

38 'Death of Mr. G.W. Goyder', *Advertiser*, 4 Nov. 1898, p. 6, reprinted in *Chronicle*, 5 Nov. 1898, p. 16. The second quote is taken from a testimonial presented to Goyder on his retirement by staff of the Survey Department.

39   Marwyn S. Samuels, 'The biography of landscape: cause and culpability', in D.W. Meinig (ed.), *The Interpretation of Ordinary Landscapes: Geographical essays*, Oxford University Press, New York, 1979, p. 64.

40   ibid., p. 62.

41   SAPP 1890, no. 30C, p. vii.

Chapter one: Receiving the life of heaven

1   I.A. Robinson, *A History of the New Church in Australia 1832–1980*, Melbourne, [1980?], acknowledgments.

2   Bernard Burke, *A Genealogical and Heraldic History of the Colonial Gentry*, vol. 2, Harrison & Sons, London; E.A. Petherick, Melbourne, 1893, p. 690; Chris Goddard, 7 Mar. 2007, 'The Goyder family', viewed 20 Mar. 2007, http://www.webrarian. co.uk/goyder/index.html. Edward remains unnamed in his son's autobiography.

3   Burke, p. 690. It is also in Burke that the connection with 'Gwydir' is made.

4   D.G. Goyder, *My Battle for Life: The autobiography of a phrenologist*, Simpkin, Marshall, London, 1857, pp. 3–4.

5   Burke, p. 690; Goddard, 'The Goyder family', viewed 20 Mar. 2007.

6   Contemporary material about the school does not mention this Foundation but refers instead to 'Queen's scholars'.

7   D.G. Goyder, *My Battle*, p. 33.

8   ibid., p. 58.

9   ibid., p. 35.

10   ibid., pp. 80–1.

11   Jorge Luis Borges, 'Emanuel Swedenborg'. From an unidentified translation quoted in *Awaken from death*, Swedenborg Centre, Sydney, 1990, p. 50.

12   Emanuel Swedenborg, trans. John C. Ager, *Heaven and its Wonders and Hell: From things heard and seen*, standard edn, Swedenborg Foundation, West Chester, Pennsylvania, 1995 (first published 1758), p. 425, n. 528.

13   Peter Ackroyd, *Blake*, Minerva, London, 1996, pp. 100, 103.

14   D.G. Goyder, *My Battle*, pp. 109–10.

15   John Martin-Harvey, *The Autobiography of Sir John Martin-Harvey*, S. Low, Marston & Co., London [1933], pp. 12–13. Martin-Harvey also says that Sarah was 18 when she married, but a family tree shows her as born in 1794, and hence about 27.

16   D.G. Goyder, *My Battle*, p. 109.

17   Martin-Harvey, pp. 12–13.

18   D.G. Goyder, *My Battle*, p. 109.

19   ibid., p. 128; D.G. Goyder, *A Manual of the System of Instruction Pursued at the Infant School, Meadow Street, Bristol*, 4th edn, Longman, Hurst, Rees, Orme, Brown and Green, London, 1825.

20   Chris Goddard, 27 Nov. 2004, 'Sarah Anna Goyder', viewed 20 Mar. 2007, http://www.webrarian.co.uk/goyder/sarah_anna_goyder_1.html.

21   Ian Woodroffe Goyder, email, 8 Jan. 2007.

22   D.G. Goyder, *My Battle* – music: p. 13 and elsewhere; storytelling: pp. 43–4; gentleness: p. 424; hot-tempered and impulsive: p. 289.

23 'Goyderiana', *Quiz and The Lantern*, 10 Nov. 1898, p. 7.

24 R.A. Gilbert, 'Chaos Out of Order: The rise and fall of the Swedenborgian rite', *AQC* [*Ars Quatuor Coronatorum*], vol. 108, 1995, viewed 7 Oct. 2005, http//www.mastermason.com/luxocculta/swedenborg.htm.

25 D.G. Goyder, *My Battle*, pp. 400–1.

26 ibid., p. 250.

27 ibid., p. 263.

28 ibid., p. 114.

29 ibid., p. 118.

30 ibid., p. 30.

31 D.G. Goyder, *Manual*, p. 15.

32 D.G. Goyder, *My Battle*, p. 270. David defines his time in Glasgow as the seventh period in his life, from 1834 to 1848.

33 'Death of Mr. G.W. Goyder, C.M.G.', *Observer*, 5 Nov. 1898, p. 32.

34 D.G. Goyder, *My Battle*, p. 491.

35 Chris Goddard's website (http://www.webrarian.co.uk) also states that Goyder's elder sister Sarah was married by her father in the Parish Church of the Gorbals on 5 Dec. 1841.

36 'Death of Mr. G.W. Goyder, C.M.G.', *Observer*, 5 Nov. 1898, p. 32.

37 SAPP 1878, no.146, p. 40.

38 ibid.

39 F.M.L.Thompson, *Chartered Surveyors: The growth of a profession*, Routledge and K. Paul, London, 1968, p. 110, quoting *The Builder*, 22 Nov. 1845.

40 ibid., p. 111, quoting *The Builder*, 16 May 1846.

41 ibid., p. 113.

42 Martin-Harvey, p. 10.

43 D.G. Goyder, *My Battle*, p. 490.

44 ibid., pp. 380–1.

45 ibid., p. 364.

Chapter two: The climate of paradise

1 George Goyder's entry in *Burke's Colonial Gentry* (Bernard Burke, *A Genealogical and Heraldic History of the Colonial Gentry*, vol. 2, Harrison & Sons, London; E.A. Petherick, Melbourne, 1893, pp. 689–90) states that Sarah Anna had four children, all deceased – information that was probably provided by Goyder himself – and there is a gap between Sarah's first child and the next known child, also named Sarah Anna, born in 1850. The death notice for this Sarah Anna describes her as the 'third daughter', rather than the second child (*Register*, 18 Oct. 1871, p. 4).

2 Material relating to Sarah has been drawn from Chris Goddard, 27 Nov. 2004, Sarah Anna Goyder, viewed 20 Mar. 2007, http://www.webrarian.co.uk/goyder/index.html; from Margaret Goyder Kerr, *The Surveyors: The story of the founding of Darwin*, Rigby, Adelaide, 1971; from I.A. Robinson, *A History of the New Church in Australia 1832–1980*, Melbourne, [1980?]; and David George Goyder, *My Battle for Life: The autobiography of a phrenologist*, Simpkin, Marshall, London, 1857.

Dates given in Kerr and I.A. Robinson are unreliable. Both have Goyder arriving in Adelaide in 1845.

3   Goyder is not listed in the Index of Assisted Passengers, but his arrival is recorded in Registers of Assisted Immigrants from the U.K. (VPRS 7310), Book 4A, p. 42a and in Archives Authority of New South Wales, Persons arriving on bounty ships at Port Phillip 1848–9, AO reel no. 2144. He is included in Martin Adlington Syme, *Shipping Arrivals and Departures Victorian Ports*, vol. 2, 1846–1855, Roebuck, Melbourne, 1987. Goyder's arrival seems to have been first identified by Smith family historian Vladimir Derewianka. In *The Surveyors*, Margaret Goyder Kerr has her grandfather arriving in Sydney and even meeting J.B. Neales there. This idea seems to have originated in a construction in O'Brien's thesis based on an erroneous source (F.J.R. O'Brien, 'Goyder's Line', MA preliminary thesis, University of Adelaide, 1952, p. 9, note 33; *Pictorial Australian*, Feb. 1894, p. 19.).

4   'Shipping intelligence', *Geelong Advertiser*, 5 Apr. 1849, [p. 2], and evident in the *Osprey*'s entry in Registers of Assisted Immigrants from the UK.

5   Registers of Assisted Immigrants from the UK, Book 4A, p. 42a and 'Shipping intelligence', *Argus*, 10 Apr. 1849, [p. 2], 24 Apr. 1849, [p. 2].

6   Egon F. Kunz, *Blood and Gold: Hungarians in Australia*, Cheshire, Melbourne, 1969, p. 61.

7   Kerr, p. 2.

8   D.G. Goyder, *My Battle*, p. 394.

9   Kerr, p. 145.

10 'The overland route to Mount Alexander', *Register*, 15 Jan. 1852, p. 2.

11 Heather Curnow, *The Life and Art of William Strutt 1825–1915*, Alister Taylor, Martinborough, NZ, 1980, p. 134.

12 Heather Parker, *All in the Line of Duty: Danger and drudgery on the gold escort route Adelaide–Mount Alexander 1852–53*, Border Chronicle, n. p., 1971, map.

13 'Death of Mr. G.W. Goyder, C.M.G.', *Observer*, 5 Nov. 1898, p. 32.

14 'Diary of Edward George Day', *Proceedings of the Royal Geographical Society (South Australian Branch)*, vol. 29, 1927–28, p. 166.

15 W. Snell Chauncy, *A Guide to South Australia* [...], E. Rich, London, 1849, pp. 26–7.

16 ibid., pp. 29–30.

17 Quoted in Karen Moon, 'Perception and Appraisal of the South Australian Landscape 1836–1850', *Proceedings of the Royal Geographical Society (South Australian Branch)*, vol. 70, 1969, p. 47.

18 Moon, p. 49.

19 Geoffrey Gibbon, 'John Smith and Company: a Victorian family', 1975, pp. 1–2 (from a copy of the unpublished manuscript in the possession of Vaughn Smith).

20 Kerr, pp. 2–3.

21 John Smith, *A Treatise on the Artificial Growth of Cucumbers and Melons, Conjointly With That of Asparagus, Mushrooms, Rhubarb, &c.* [...], Edward Shalders, Ipswich, 1833. The dedication states that he had been employed by Alexander for nearly 20 years at that time.

22   Details about the Smiths are drawn from two unpublished family histories by
     Vladimir Derewianka: 'Index and directory: The Smith and Goyder families of
     South Australia (1850's): Their ancestors and descendants', 2nd draft, 1990, and,
     'The Smith and Wadham families of South Australia in the 1850's: A genealogical
     directory of their ancestors and descendants', first draft, 1992, both copies in the
     possession of Vaughn Smith. Both documents cite multiple primary sources in
     England and Australia.

23   *Paradise of Dissent* was the title of a book published by historian Douglas Pike, see
     note below.

24   D.G. Goyder, *My Battle*, p. 491.

25   Derewianka, 'Index and directory: The Smith and Goyder families', p. 28. This
     information appears to have been transmitted through the family.

26   John Smith is listed as a resident in Sands and MacDougall's directory for 1855, and
     has an early, undated entry in the assessments of the Walkerville Council. He is
     listed in the first dated assessments of 1858.

27   Edwin Mitchell Smith, 'Recollective diary', 1920, p. 1. In the Goyder family file,
     SLSA PRG 491/3.

28   *Register*, 1 Apr. 1907, p. 4.

29   I.A. Robinson, *A History of the New Church in Australia 1832–1980*, Melbourne,
     [1980?], pp. 20–1.

30   Derewianka, 'Index and directory: The Smith and Goyder families', pp. 35–6.

31   Goyder to colonial secretary Finniss, 4 Jan. 1853, SRSA GRG 24/6, and 'Death of
     Mr. G.W. Goyder, C.M.G.', *Observer*, 5 Nov. 1898, p. 32.

32   D.G. Goyder, *My Battle*, p. 491.

33   A. Grenfell Price, *The Foundation and Settlement of South Australia 1829–1845*, facs.
     edn, Libraries Board of South Australia, n.p., 1973 (first published 1924), p. 90,
     quoting an unidentified source.

34   William Epps, *Land Systems of Australasia*, Sonnenschein, London, 1894, p. 116.

35   Raymond Bunker, 'Town planning at the frontier of settlement', in Alan Hutchings
     & Raymond Bunker (eds), *With Conscious Purpose: A history of town planning in South
     Australia*, Wakefield Press in association with Royal Australian Planning Institute
     (South Australian Division), Adelaide, 1986, p. 23.

36   A Grenfell Price, *Founders and Pioneers of South Australia*, Mary Martin Books,
     Adelaide, 1978 (first published 1929), p. 139.

37   Light to Wakefield, 16 May 1838, quoted in M.P. Mayo, *The Life and Letters of Col.
     William Light*, F.W. Preece & Sons, Adelaide, 1937, p. 233.

38   ibid., p. 234.

39   William Harcus, *South Australia: Its history, resources and productions*, Sampson,
     Low, Marston, Searle and Rivington, London, 1876, p. 417. Todd was the author of
     the chapter in which the comment occurs, and quoted Kingston extensively.

40   Douglas Pike, *Paradise of Dissent: South Australia 1829–1857*, 2nd edn, Melbourne
     University Press, Carlton, Vic., 1967, p. 178, quoting a letter from Governor Grey.

41   Freeling to governor-general Sir W. Denison, 20 Oct. 1859, An account of the
     general system adopted in South Australia for the survey and sale of crown lands.
     SRSA GRG 35/20/1859/1729.

42   Quoted in B.C. Newland, 'Edward Charles Frome', *Proceedings of the Royal
     Geographical Society (South Australian Branch)*, vol. 63, 1961–62, pp. 52–69.

43   Edward Charles Frome, *Outline of the Method of Conducting a Trigonometrical
     Survey ...*, 2nd edn, rev. and enlarged, with an additional chapter upon colonial
     surveying, as adapted to the marking out of waste land, John Weale, London, 1850,
     p. 129.

44   Edward Charles Frome, *Instructions for the Interior Survey of South Australia*, Robert
     Thomas, Adelaide, 1840, p. 5.

45   Frome, *Outline of the Method*, pp. 125–6. This passage differs only slightly from
     paragraphs 8 and 9 of *Instructions for the Interior Survey of South Australia*, p. 7.

46   Goyder to colonial secretary Finniss, 4 Jan. 1853. SRSA GRG 24/6.

47   SAPP 1890, no. 30c, p. 7, qu. 864.

48   Freeling to CCLI Milne, 26 Sep. 1859, SRSA GRG 35/20/1859/1511 (and
     accompanying 35/2/1859/793); SAPP 1888, no 28, p. 77, qus 2182–3.

49   I.A. Robinson, p. 22.

50   'Building improvements. Private residence and general improvements', *Register*, 15
     Jan. 1867, p. 3. The assessments of the Walkerville Council for 1859 describe the
     house as possessing six rooms, and an earlier, undated entry mentions the vineyard.

51   Return of all officers employed in the Survey and Land Department ..., August
     1859. SRSA GRG 35/1/1859/679.

52   Michael Williams, *The Making of the South Australian Landscape: A study in the
     historical geography of Australia*, Academic Press, London, 1974, p. 76.

53   Quoted in Michael Williams, 'George Woodroofe [*sic*] Goyder: A practical
     geographer', *Proceedings of the Royal Geographical Society (South Australian Branch)*,
     vol. 79, 1978, pp. 3–4.

Chapter three: As far as the eye could reach

1   SAPP 1857–58, no. 72, p. 1.

2   ibid. p. 2.

3   G.W. Goyder, Field notebooks of G.W. Goyder as Assistant Surveyor General and
     Surveyor General 1856–92, vol. 1, p. [31] and 6 May 1857, p. [33]. SRSA GRG 35/256.
     In his field notebooks, Goyder used an ampersand to represent 'and' in sentences.
     It took the form of a small cross, like a plus sign, in which the horizontal arm is
     connected to the vertical bar as part of a loop. To preserve the informal character
     of the notebooks I have rendered this with a plus sign (+) rather than a printed
     ampersand (&).

4   ibid., 8 May 1857, p. [37].

5   SAPP 1857–58, no. 72, p. 2.

6   G.W. Goyder, Field notebooks, vol. 1, 15 May 1857, pp. [44–5].

7   SAPP 1857–58, no. 72, p. 1.

8   G.W. Goyder, Field notebooks, vol. 1, 21 May 1857, pp. [51–2].

9   ibid., 21 May 1857, p. [52].
10 The process of its creation is described, with a map, in Hans Mincham, *The Story of the Flinders Ranges*, 3rd edn, Rigby, Adelaide, 1983.
11 Edward John Eyre, *Journals of Expeditions of Discovery into Central Australia*, vol. 1, T. & W. Boone, London, 1845, pp. 111–12.
12 Quoted in Mincham, p. 29.
13 SAPP 1857–58, no. 72, p. 2.
14 Robert Forster, Rick Hosking & Amanda Nettelbeck, *Fatal Collisions: The South Australian frontier and the violence of memory*, Kent Town, SA, Wakefield Press, 2001, pp. 94–114. The chapter 'Fatal collisions in the Flinders Ranges' deals with this incident at length. Hayward is quoted there from his 'Reminiscences' (Mortlock Library of South Australiana, PRG 395, p. 92), and the traveller was John Bowyer Bull.
15 SAPP 1857–58, no. 72, p. 3.
16 Dorothy Tunbridge in assoc. with the Nepabunna Aboriginal School, *The Story of the Flinders Ranges Mammals*, Kangaroo Press, Kenthurst, NSW, 1991, p. 86 and note 2 on p. 93
17 'Survey of the Northern Country', *Advertiser*, 4 Apr. 1860, p. 3. This reproduced an official report from Goyder.
18 From a report to the governor by Goyder on the Aborigines of the Northern Territory, quoted in Margaret Goyder Kerr, *The Surveyors: The story of the founding of Darwin*, Rigby, Adelaide, 1971, p. 147. The original document could not be located, but it is quoted extensively in Kerr, pp. 147–8.
19 G.W. Goyder, Field notebooks, vol. 1, 30 May 1857, p. [66].
20 SAPP 1857–58, no. 72, p. 3.
21 ibid.
22 ibid.
23 ibid.
24 ibid., p. 4.
25 ibid.
26 G.W. Goyder, Field notebooks, vol. 1, 3 Jun.1857, p. [78].
27 SAPP 1857–58, no. 72, p. 4.
28 ibid.
29 ibid.
30 Quoted in Mincham, p. 33.
31 'The Discoveries in the North', *Register*, 30 Jun. 1857, p. 2.
32 SAPP 1857–58, no. 72, p. 1.
33 Bonney to chief secretary Finniss, 3 Jul. 1857, filed as an insertion in SRSA GRG 35/1/1857/247.
34 'The Discoveries in the North', *Register*, 10 July 1857, p. 2.
35 'Discoveries in South Australia', reprinted in *Register*, 6 Aug. 1857, p. 3.
36 'Preparations for Stocking Our Newly Discovered Country', *Register*, 3 Aug. 1857, p. 2.

37 'The Exploration of the Interior', *Register*, 30 Jul. 1857, p. 2.

38 'The Exploration of the Interior', *Register*, 24 Aug. 1857, p. 2.

39 'The Northern Explorations' [letter], *Register*, 16 Sep. 1857, p. 3.

40 'The Northern Explorations', *Register*, 26 Sep. 1857, p. 2.

41 SAPP 1857–58, no. 174, pp. 1–2. Freeling also produced a longer report, no. 193.

42 SAPP 1857–58, no. 153, p. 1.

43 ibid., p. 2.

44 ibid.

45 'The Northern Explorations', *Register*, 26 Sep. 1857, p. 2.

46 'Exploration of the Interior' [letter], *Register*, 12 Oct. 1857, p. 3.

47 Bessie Threadgill, *South Australian Land Exploration 1856 to 1880*, Board of the Governors of the Public Library, Museum, and Art Gallery of SA, [Adelaide], p. 9.

48 Frederick Sinnett, *An Account of the Colony of South Australia*, Government Printer, Adelaide, 1862, p. 26.

49 Alisa Bunbury, *Arid Arcadia: Art of the Flinders Ranges*, Art Gallery of South Australia, Adelaide, 2002, p. 72.

50 Mincham, p. 175; Bunbury, p. 66.

51 'Major-General Sir Arthur Freeling', *Public Service Review*, vol. 7, no. 3, 1900, p. 23.

52 SAPP 1860–61, no. 177, p. 3.

53 SAPP 1888, no. 28, p. 77, qu. 2192.

54 Tenison-Woods was also, for many years, the director of Catholic education in Australia, and has been himself the subject of biographies. He is perhaps most widely known as the 'Father Founder' of the Josephite order, of which Mary MacKillop, the first Australian to be canonised, was the first superior. In the light of that reputation it is worth pointing out that, in the authorised biography of MacKillop, the 'increasingly bizarre imprudence of his handling of spiritual problems' is contrasted with the 'sober judgement [Woods] exercised in his scientific publications' (Paul Gardiner, *Mary MacKillop: An extraordinary Australian*, Newton, NSW, E.J. Dwyer, 1993, p. 64.)

55 'Rivers of Interior Australia', *Register*, 28 Jan. 1864, p. 2.

56 J.E. Tenison-Woods, *A History of the Discovery and Exploration of Australia*, H.T. Dwight, Melbourne, 1864, p. 248.

57 ibid, p. 250.

58 Kerr, p. 34.

59 SAPP 1858, no. 25, p. 11.

60 'Adelaide Philosophical Society', *Register*, 1 Nov. 1875, p. 6.

61 SAPP 1883–84, no. 52, p. 2.

62 Andrew Garran, *Picturesque Atlas of Australasia*, facs. edn, vol. 2, Ure Smith, Sydney, 1974 (first published 1888), p. 424.

Chapter four: Systematic observation

1 G.W. Goyder, Field notebooks of G.W. Goyder as Assistant Surveyor General and Surveyor General 1856–1892, vol. 3, 12 Mar. 1859. SRSA GRG 35/256.

2 F.J.R. O'Brien, 'Goyder's Line', MA preliminary thesis, University of Adelaide, 1952, pp. 30, 38–9. The thesis contains a helpful map and a chronological list of Goyder's longer journeys made prior to 1865. However, the author apparently worked from the field notebooks only and missed the journey with Selwyn, which is out of sequence, and unidentified, at the back of vol. 3. It consists only of geological entries. The volumes were numbered differently when O'Brien was carrying out his research.

3 SAPP 1860, no. 20, p. 1.

4 ibid., p. 15.

5 From a presentation by Goyder reported in 'South Australian Association of Architects, Engineers & Surveyors', *Register*, 4 Aug. 1859, p. 2.

6 Robert Bruce, *Reminiscences of an Old Squatter*, W.K. Thomas, Adelaide, 1902. p. 18.

7 SAPP 1860, no. 20, p. 10.

8 SAPP 1857–58, no. 72, p. 2.

9 SAPP 1860, no. 20, p. 11.

10 SAPP 1860–61, no. 177, p. 2.

11 SAPP 1860, no. 20, p. 15.

12 SAPP 1859, no.119.

13 'South Australian Association of Architects, Engineers and Surveyors', *Register*, 8 July 1859, p. 2 and 4 Aug. 1859, p. 2.

14 SAPP 1860, no. 20, p. 15.

15 SAPP 1857–58, no. 153, p. 2.

16 Goyder to surveyor general Freeling, 24 Sep. 1859. SRSA GRG 35/1/1859/793.

17 Return of all officers employed in the Survey and Land Department …', August 1859. SRSA GRG 35/1/1859/679

18 Goyder to surveyor general Freeling, 24 Sep. 1859. SRSA GRG 35/1/1859/793.

19 ibid., appendix.

20 ibid.

21 SAPP 1860–61, no. 177, p. 2.

22 Goyder to surveyor general Freeling, 24 Sep. 1859. SRSA GRG 35/1/1859/793.

23 'Death of Mr. Wadham', *Register*, 9 Dec. 1895, p. 5.

24 SAPP 1863, no. 51, pp. 167–8, qu. 5741.

25 SAPP 1860, no. 41.

26 Freeling to governor general Sir W. Denison. SRSA GRG 35/20/1859 /1729.

27 'Survey of the Northern Country', *Advertiser*, 4 Apr. 1860, p. 3. The article reproduces a letter from Goyder to Freeling dated 16 Mar. 1860.

28 G.W. Goyder, Field notebooks, vol. 2, 9 Dec. 1859. SRSA GRG 35/256.

29 Michael Williams, 'George Woodroofe [*sic*] Goyder: A practical geographer', *Proceedings of the Royal Geographical Society (South Australian Branch)*, vol. 79, 1978, p. 13, quoting the letter. The same incident is described in Goyder's field notebook, vol. 2, 9 Dec. 1859. SRSA GRG 35/256.

30 G.W. Goyder, Field notebooks, vol. 3, 12 Apr. 1859. SRSA GRG 35/256. (The dates of the early volumes are not strictly consecutive.) Goyder's list appears to read 'mud

hut n. of kanyaka', although the 'n' could be 's' formed carelessly. It is unlikely that he would refer to any other mud hut in this way. Michael Williams, in *The Making of the South Australian Landscape: A study in the historical geography of Australia*, Academic Press, London, 1974, p. 343, states that the site had already been selected.

31 Lionel C.E. Gee, 'The South-East Fifty Years Ago. Vanished Aborigines', *Register*, 24 Mar. 1928, p. 5.

32 G.W. Goyder, Field notebooks, vol. 5, 17 July 1860. Also see Glen McLaren, *Beyond Leichhardt: Bushcraft and the exploration of Australia*, South Fremantle Arts Press, South Fremantle, 1996, pp. 150–2.

33 SAPP 1860–61, no. 177, p. 1.

34 G.W. Goyder, Field notebooks, vol. 2, 28–31 Dec. 1859.

35 ibid., vol. 2, 30 Jan. 1860.

36 Bessie Threadgill, *South Australian Land Exploration 1856 to 1880*, Board of the Governors of the Public Library, Museum, and Art Gallery of SA, [Adelaide], p. 29.

37 The two works (of which I am aware) are: Vincent Severnty, *The Desert Sea: The miracle of Lake Eyre in flood*, Macmillan, South Melbourne, 1985, and Vincent Kotwicki, *Floods of Lake Eyre*, Engineering and Water Supply Dept, Adelaide, 1986. The latter only mentions Goyder's seeing Lake Blanche in 1857 in the history of early exploration on p. 50. The appendix detailing records of flood and rainfall on p. 88 begins at 1864.

38 SAPP 1887, no. 34, p. 14, qu. 423.

39 SAPP 1860–61, no. 177, p. 3.

40 SAPP 1876, no. 19, p. 19.

41 SAPP 1860–61, no. 177, p. 3.

42 ibid.

43 ibid., p. 2.

44 G.W. Goyder, Field notebooks, vol. 4, 11–12 Mar. 1860.

45 'Survey of the Northern Country', *Advertiser*, 4 Apr. 1860, p. 3.

46 SAPP 1860–61, no. 177, p. 1.

47 ibid., p. 2.

48 ibid., p. 4.

49 *Advertiser*, 24 Jan. 1865, p. 2. In a note on springs, it was stated that the paper had been informed by the Land Office that Goyder had discovered these in 1859. However, the names 'Frances Springs' and 'Ellen Springs' no longer exist.

50 SAPP 1859, no. 21, p. 10, qu. 169.

51 SAPP 1860–61, no. 177, p. 4.

52 'Survey of the Northern Country', *Advertiser*, 4 Apr. 1860, p. 3.

53 SAPP 1864, no. 36, p. 14. Goyder described the method and map in similar terms again in Survey Department of South Australia, *Field Service Handbook for Government Surveyors*, Government Printer, Adelaide, 1880, pp. 4–5.

54 ibid.

55 SAPP 1860–61, no. 177, p. 6.

56 G.W. Goyder, Field notebooks, vol. 5, 14 Aug. 1860. SRSA GRG 35/256.

57 ibid., 15–16 Aug. 1860.

58 Margaret Goyder Kerr, *The Surveyors: The story of the founding of Darwin*, Rigby, Adelaide, 1971, p. 3. The *Observer* obituary places this incident in the explorations of 1857 (which took place in winter, not during a drought) in the vicinity of Lake Torrens (which Goyder was nowhere near in 1857 unless the name is used in its old sense). It is Kerr's account which indicates that this incident belongs to the northern triangulations.

59 'The Far North', *Register*, 20 Dec. 1860, p. 2.

Chapter five: Taking charge

1 Frederick Sinnett, *An Account of the Colony of South Australia*, Government Printer, Adelaide, 1862, pp. 55–6.

2 ibid., pp. 43–4.

3 SAPP 1858, no. 25, p. 47.

4 Quoted in William C. Foster, *Sir Thomas Livingston Mitchell and His World 1792–1855: Surveyor general of New South Wales 1828–1855*, Institution of Surveyors, Sydney, 1985, p. 164.

5 Neales, in House of Assembly, 28 Aug. 1863, reported in *Register*, 29 Aug. 1863, p. 3.

6 Freeling to CCLI Milne, 26 Sep. 1859. SRSA GRG 35/20/1859/1511.

7 Freeling to president of the Legislative Council, 25 Aug. 1859, Minutes of the proceedings of the Legislative Council, no. 29, 26 Aug. 1859.

8 'Death of Mr. G.W. Goyder, C.M.G.', *Observer*, 5 Nov. 1898, p. 32.

9 Goyder to CCLI Bagot, 25 Aug. 1860. SRSA GRG 35/1/1860/1011.

10 Michael Williams, 'George Woodroofe [*sic*] Goyder: A practical geographer', *Proceedings of the Royal Geographical Society (South Australian Branch)*, vol. 79, 1978, p. 6, quoting SRSA GRG 35/2/1859/938.

11 Freeling to CCLI Bagot, 7 Sep. 1859. SRSA GRG 35/20/1860/1627.

12 Goyder to surveyor general Freeling, 1 Oct. 1860. SRSA GRG 35/1/1860/1260.

13 ibid.

14 SAPP 1863, no. 52, p. 13, qu. 289.

15 'Official Appointments', *Register*, 18 Oct. 1860, p. 2.

16 'The New Surveyor-General', *Register*, 23 Nov. 1860, p. 2.

17 George Loyau, *Notable South Australians, or, Colonists – Past And Present*, facs. edn, Austaprint, Hampstead Gardens, SA, 1978 (first published 1885), p. 80–1.

18 SAPP 1863, no. 51, p. 170, qu. 5798.

19 ibid., p. 170, qu. 5806.

20 Goyder to surveyor general Freeling, 24 Sep. 1859. SRSA GRG 35/1/1859/793.

21 SAPP 1861, no. 90, p. 1.

22 Vladimir Derewianka, 'Index and directory: The Smith and Goyder families of South Australia (1850's): Their ancestors and descendants', 2nd draft, 1990; Death notice, *Register*, 26 Mar. 1861.

23 'Sketches of the Present State of South Australia. No. 11. Brighton and the coast', *Register*, 3 Mar. 1851, p. 2.

24 Derewianka, 'Index and directory: The Smith and Goyder families', p. 21.

25 Margaret Goyder Kerr, *The Surveyors: The story of the founding of Darwin*, Rigby, Adelaide, 1971, introduction.

26 *Register*, 1 Apr. 1907, p. 4. Walkerville Council records suggest things may not have been as simple as this and Sands and MacDougall's directory for 1855 lists Edwin, not John, as a resident, and a gardener. In the council assessments, Charles Ware appears as the occupier of a nursery on the parklands for the last time in 1863. While Goyder is shown to lease and then purchase property in 1863 and 1864, Edwin does not appear in these records until 1865.

27 Majorie Scales, *John Walker's Village: A history of Walkerville*, Rigby, Adelaide, 1974, p. 151. Smith's obituary in the *Register* (1 Apr. 1907, p. 4) states, very unclearly, that seven years after taking the Clifton Nursery on Robe Terrace, he 'bought the Clifton Nursery nearby in Walkerville' – perhaps meaning the nursery in Fuller Street. Robert F.G. Swinbourne, *Years of Endeavour: An historical record of the nurseries, nurserymen, seedsmen and horticultural retail outlets of South Australia*, South Australian Association of Nurserymen, [Adelaide?], 1982, p. 8, seems to say that the nursery on Robe Terrace had its name under Ware.

28 According to Mrs Meredith Farmer (a Spencer Smith), it was at 13 Northcote Terrace, and was close to the corner. Dan Farmer, email, 6 Dec. 2007.

29 Swinbourne, p. 8.

30 ibid.

31 Edwin is known to his descendants as Edwin Spencer Smith. (Dan Farmer, email, 6 Dec. 2007.)The middle name seems to have been unofficial, and used as a necessary point of differentiation.

32 Edwin Mitchell Smith, 'Recollective diary', 1920, p. 1. In the Goyder family file, SLSA PRG 491/3.

33 This last journey is recorded in the back of vol. 6 of the field notebooks (SRSA GRG 35/256). Others are noted in vols 7 and 8. For these small trips, I have relied on my own notes, made from the field notebooks, and the list included in F.J.R. O'Brien, 'Goyder's Line', MA preliminary thesis, University of Adelaide, 1952, appendix 1, p. 38.

34 SAPP 1876, no. 95, pp. 6–7. This includes Goyder's unpublished report of 1862.

35 Matthew Flinders, *A Voyage to Terra Australis* [...] *in the years 1801, 1802, and 1803*, G.W. Nichol, London, 1814, p. 116.

36 Glen McLaren, *Beyond Leichhardt: bushcraft and the exploration of Australia*, South Fremantle Arts Press, South Fremantle, 1996, chap. 5.

37 G.W. Goyder, Field notebooks, vol. 10, 5 July 1862. SRSA GRG 35/256.

38 ibid., 10 July.

39 ibid., 20 July.

40 ibid., 1 Aug.

41 ibid., 4 Aug.

42 ibid., 5 Aug. 1863; SAPP 1891, no. 33, p. 74, qu. 1925. Goyder also told the commission that he had written a report on this journey. It is not included among the parliamentary papers.

43 SAPP 1887, no. 34, p. 15, qus 432–3.

44 SAPP 1888, no. 28, p. 81, qu. 2288.

45 Ernest Favenc, *The History of Australian Exploration from 1788 to 1888*, facs. edn., Golden Press, Gladesville, NSW, 1983 (first published 1888), p. 236.

46 Goyder to CCLI Strangways, 10 Oct. 1862. SRSA GRG 35/1/1862/2081.

47 ibid.

48 'Personal Paragraphs. G.W. Goyder, C.M.G.', *Public Service Review*, July 1894, p. 93.

Chapter six: Bird's-eye view

1   G.W. Goyder, Field notebooks of G.W. Goyder as Assistant Surveyor General and Surveyor General 1856–1892, vol. 7, SRSA GRG 35/256; F.J.R. O'Brien, 'Goyder's Line', MA preliminary thesis, University of Adelaide, 1952, p. 38.

2   Edwin Mitchell Smith, 'Recollective diary', 1920, p. 3. In the Goyder family file, SLSA PRG 491/3.

3   [Leader], *Telegraph*, 30 May 1863.

4   J.J. Pascoe (ed.), *History of Adelaide and Vicinity*, Hussey & Gillingham, Adelaide, 1901, p. 315.

5   William Milne, 'Notes of a journey from Adelaide to the South-Eastern District of S.A. January, 1863', pp. 1, 6. SLSA PRG 746/2.

6   SAPP 1863, no. 41, pp. 3–4, p. 6.

7   SAPP 1892, no. 35, p. 77, qu. 2840.

8   SAPP 1865, no. 30, p. 12.

9   SAPP 1863, no. 41, p. 12.

10 Milne, p. 8; SAPP 1863, no. 41, p. 3.

11 Milne, p. 23.

12 ibid., pp. 24, 26.

13 ibid., p. 11.

14 ibid., p. 12.

15 'Death of Mr. G.W. Goyder', *Advertiser*, 4 Nov. 1898, p. 6, reprinted in *Chronicle*, 5 Nov. 1898, p. 16.

16 Lionel C.E. Gee, 'The South-East Fifty Years Ago. Vanished Aborigines', *Register*, 24 Mar. 1928, p. 5.

17 *Quiz and The Lantern*, 10 Nov. 1898, p. 6.

18 Milne, p. 26. In the typescript, the key word is missing, but the intent is clear enough. The sentence appears as: 'Goyder, not being [...], asked for one and got it'.

19 ibid., p. 36.

20 ibid., p. 2.

21 Kim Lockwood, *Big John: The extraordinary adventure of John McKinlay, 1819–1872*, State Library of Victoria, Melbourne, 1995, pp. 106–7.

22 Milne, p. 33.

23 'G.W. Goyder, Leader of the Expedition to North Australia', *Australian Journal*, vol. 5, July 1870, p. 659.

24 'George Woodroffe Goyder', *Advertiser*, 4 Nov. 1898, p. 4.

25 Milne, p. 44.

Chapter seven: Magnum opus: the people's grass

1   Eustace Reveley Mitford, 'Song of Praise', *Pasquin*, 24 Oct. 1868, in Mitford (ed.), *Pasquin: The pastoral, mineral and agricultural advocate vols I–III, January 1867 to November 1869*, Judd & Co., London, 1882, p. 608.

2   SAPP 1860, no. 116, p. 2, qu. 38.

3   SAPP 1861, no. 109, p. 1.

4   ibid., pp. 2–3.

5   SAPP 1864, no. 213, p. 41, qu. 1136.

6   SAPP 1861, no. 108, p. 1.

7   SAPP 1861, no. 101, p. 1.

8   ibid., p. 3.

9   K.R. Bowes, *Land Settlement in South Australia 1857–1890*, Libraries Board of South Australia, Adelaide, 1968, p. 124 (PhD thesis, ANU, 1963).

10   'Dissolution of Parliament – Public Meeting', *Register*, 11 Aug. 1864, p. 3.

11   'The Political Crisis – Meeting at Angaston', *Advertiser*, 7 Sep. 1864, p. 3.

12   SAPP 1864, no. 82.

13   ibid.

14   Frederick Sinnett, *An Account of the Colony of South Australia*, Government Printer, Adelaide, 1862, p. 45.

15   G.W. Goyder, Memorial on pastoral lands, Surveyor-General's Office 1868–1918, Confidential letter book, p. 144. SRSA GRG 35/19.

16   G.W. Goyder, Notebook compiled by G.W. Goyder containing detailed sketches and valuations of pastoral runs with ink sketches, [1864], pp. 319–20. SRSA GRG 35/653.

17   SAPP 1888, no. 28, p. 27, qu. 598.

18   SAPP 1864, no. 36, pp. 14–15.

19   SAPP 1864, no. 105, p. 2.

20   Margaret Goyder Kerr, *The Surveyors: The story of the founding of Darwin*, Rigby, Adelaide, 1971, p. 3.

21   SAPP 1864, no. 106.

22   Valuations of the South-East are reproduced, with transcriptions of the text, in Anne McArthur (ed), *Through the Eyes of Goyder Master Planner: Transcripts of his 1864–5 detailed valuations of 79 pastoral runs in the South East of South Australia*, Kanawinka Writers and Historians Inc, Penola, SA, 2007. One of the original volumes is held by State Records South Australia as SRSA GRG 35/19 (see above). Another is held by the Land Services Group, Department of Transport, Energy and Infrastructure.

23   G.W. Goyder, Notebook containing valuations of runs, p. 320.

24   SAPP 1864, no. 102*, p. 2.

25   G.W. Goyder, Notebook containing valuations of runs, p. 338.

26   SAPP 1864, no. 82.

27   [leader], *Advertiser*, 16 July 1864, p. 2.

28   Letter from John Jones, *Advertiser*, 25 July 1864, p. 3.

29   Glyde, CCLI, House of Assembly, 20 July 1864, reported in *Register*, 21 July 1864, p. 2.

30 [leader], *Advertiser*, 21 July 1864, p. 2.

31 Glyde, CCLI, House of Assembly, 20 July 1864, reported in *Register*, 21 Jul. 1864, p. 2.

32 'Defeat of the Ministry', *Register*, 21 July 1864, p. 2; [leader], *Register*, 28 July 1864.

33 Glyde, House of Assembly, 10 Aug. 1864, reported in *Register*, 12 Aug. 1864, p. 3.

34 House of Assembly, 5 Aug. 1864, reported in *Register*, 6 Aug. 1864, p. 2.

35 'Dissolution of Parliament: Public Meeting', *Register*, 11 Aug. 1864, p. 3.

36 'The Pastoral Tenants' [letter], *Advertiser*, 6 Sep. 1864, p. 3.

37 Edwin Hodder, *History of South Australia*, vol. 1, Sampson, Low, Marston, London, 1893, pp. 383–4.

38 'Revaluation of Runs', *Register*, 13 Aug. 1864, p. 2.

39 [leader], *Advertiser*, 13 Aug. 1864, p. 2.

40 'Visions of Office' [letter], from 'A. Shepherd', *Register*, 31 Jan. 1865, p. 3. The song 'Green grow the Hummocks O!' is included in the letter and is a joke at the expense of some of the paper's editors, who had shares in the Hummocks run, above the top of Gulf St Vincent.

41 'The Pastoral Leases', *Register*, 15 Aug. 1864, p. 2.

42 'The Pastoral Revaluations' [from the *Express*], *Advertiser*, 15 Aug. 1864, p. 2.

43 [leader], *Advertiser*, 31 Dec. 1864, p. 2.

44 Legislative Council, 16 Aug. 1864, reported in *Register*, 17 Aug. 1864, p. 2.

45 [leader], *Advertiser*, 18 Aug. 1864, p. 2.

46 Legislative Council, 6 Sep. 1864, reported in *Register*, 7 Sep. 1864, p. 2.

47 Legislative Council, 1 Sep. 1864, reported in *Advertiser*, 2 Sep. 1864, p. 2.

48 'Valuations of Runs' [letter], *Advertiser*, 25 Aug. 1864.

49 SAPP 1864, no. 106.

50 ibid.

51 SAPP 1864, no. 105, p. 2.

52 SAPP 1864, no. 104, p. 5.

53 ibid., p. 4.

54 Legislative Council, 6 Oct. 1864, reported in *Advertiser*, 7 Oct. 1864, p. 3.

55 SAPP 1864, no. 105, p. 2.

56 'Death of Mr. G.W. Goyder, C.M.G.', *Observer*, 5 Nov. 1898, p. 32. This unidentified autobiographical material was quoted identically in Goyder's obituaries in the *Observer*, *Advertiser* and *Chronicle*.

57 'The Pastoral Leases: Meeting at Port Gawler', *Advertiser*, 16 Sep. 1864, p. 3.

58 SAPP 1888, no. 28, p. 27, qus 602–6.

59 SAPP 1864, no. 105, p. 2.

60 [Oddfellows' anniversary dinner], *Advertiser*, 26 Sep. 1864, p. 6, c. 6.

61 House of Assembly, 27 Sep. 1864, reported in *Advertiser*, 28 Sep. 1864, p. 3.

62 [leader], *Advertiser*, 21 Sep. 1864, p. 2.

63 [second leader], *Australasian*, 15 Oct. 1864, p. 8.

64 'Declaration of Hundreds', *Register*, 9 Dec. 1864, p. 2.

65 Referred to in South Australia, Parliament, *Debates*, 17 Oct. 1865, cc. 85–86.

66 'A nice way of getting through a nasty job' [letter], *Register*, 30 Jan. 1865, p. 3.

67 'Doubtful Adherents', *Advertiser*, 7 Jan. 1865, p. 2. The article was reprinted from the *Express*, the *Advertiser*'s associated evening paper.

68 SAPP 1864, no. 106.

69 'Our Summary', *Register*, 26 Oct. 1864, p. 2, and 'Topics of the Day', *Advertiser*, 17 Nov. 1864, p. 2.

70 'Valuation of Runs', *Register*, 4 Jan. 1865, p. 2.

71 SAPP 1865, no. 30.

72 G.W. Goyder, Notebook containing valuations of runs, p. 242.

73 ibid., p. 3, Brown's run.

74 Lionel C.E. Gee, 'The South-East Fifty Years Ago. The early pioneers', *Register*, 3 Mar. 1928, p. 7.

75 SAPP 1865–66, no. 77.

76 'Death of Mr. G.W. Goyder, C.M.G.', *Observer*, 5 Nov. 1898, p. 32. This unidentified autobiographical material was quoted identically in his obituaries in the *Observer*, *Advertiser* and *Chronicle*. The letter is filed as SRSA GRG 35/1/1871/81.

Chapter eight: In search of the rainfall

1 'The diver and the duck', 'The cannabalistic father' and 'Valnaapa Wartalyunha and Yanggunha', in Dorothy Tunbridge, the Nepabunna Aboriginal School and the Adnyamathanha People, *Flinders Ranges Dreaming*, Australian Institute of Aboriginal Studies, Canberra, 1988, p. 21, pp. 19–20.

2 Robert Hoogenraad & George Jampijina Robertson, 'Seasonal calendars from Central Australia', in E.K. Webb (ed.), *Windows on Meteorology: Australian perspective*, CSIRO Publishing, Melbourne, 1997, pp. 34, 39.

3 Frederick Sinnett, *An Account of the Colony of South Australia*, Government Printer, Adelaide, 1862, p. 22.

4 ibid., p. 25.

5 'Pastoral Lamentations' (from the *Express*), *Advertiser*, 31 Aug. 1864, p. 2.

6 SAPP 1865–66, no. 57, p. 6.

7 'The Drought in the North', *Register*, 10 Oct. 1865, p. 2.

8 Dorothy Tunbridge in assoc. with the Flinders Ranges School, p. 21; Mincham, p. 115; Hans Mincham et. al., *The Flinders Ranges: A portrait*, St Peters, NSW, Little Hills Press, 1986, p. 22.

9 'Natives in the North' [letter], *Register*, 9 Dec. 1865, p. 2.

10 SAPP 1865–66, no. 70; Hans Mincham, *The Story of the Flinders Ranges*, 3rd edn, Rigby, Adelaide. 1983, p. 115.

11 South Australia, Parliament, *Debates*, 14 Dec. 1865, c. 559.

12 [leader], *Telegraph*, 15 Nov. 1862, p. 2.

13 'Drought in the North', *Register*, 23 June 1865, p. 2. The squatters were viewing their proposed line from the north, hence 'within'.

14 South Australia, Parliament, *Debates*, 29 Sep. 1858, c. 260.

15 SAPP 1865, no. 72, p. 70, qu. 2153.

16 ibid., qu. 2155.
17 ibid., qus. 2164, 2166.
18 ibid., p. 72, qu. 2200.
19 ibid., p. 75, qu. 2277.
20 ibid., p. 75, qus 2278–79 (mistakenly written as '2289'). The spellings 'Mount Browne', 'Mount Locke' and 'Onetree Hill' have been corrected in this passage.
21 SAPP 1864, no. 105, p. 5.
22 [leader], *Advertiser*, 5 Aug. 1865, p. 2.
23 'Another Crisis' (from the *Express*), *Advertiser*, 1 Nov. 1865; 'The Ministerial Difficulty', *Register*, 1 Nov. 1865, p. 2.
24 John Hart, Papers. SLSA PRG 218. In essence the papers are Hart's diaries, which are annual volumes with page-length entries for each day of the year. The quote is from that source. Waterhouse is quoted from a note to him read out by Hart (South Australia, Parliament, *Debates*, 20 Dec. 1865, c. 596, in relation to Glyde's comments of the same day, c. 594.)
25 Glyde's account of the events is given in South Australia, Parliament, *Debates*, 31 Oct. 1865, c. 198.
26 SAPP 1865–66, no. 77, pp. 2.
27 South Australia, Parliament, *Debates*, 12 Dec. 1865, c 522.
28 John Hart, Papers.
29 'Another Crisis!' (from the *Express*), *Advertiser*, 1 Nov. 1865, p. 2.
30 SAPP 1865–66, no. 62 (instructions); SAPP 1867, no. 14, p. 110, qu. 3843.
31 [second leader], *Advertiser*, 3 Nov. 1865.
32 'The Northern Runs', *Register*, 24 Nov. 1865, p. 2.
33 For example on 30 Nov. and 5 Dec. 1865.
34 G.W. Goyder, Field notebooks of G.W. Goyder as Assistant Surveyor General and Surveyor General 1856–1892, vol. 20. SRSA GRG 35/256.
35 SAPP 1865–66, no. 78.
36 In the back of field notebook vol. 20. 'Von Rieben's public house, at the bend' is mentioned in SAPP 1863, 118, p. 3.
37 SAPP 1865–66, no. 78.
38 'The Surveyor General', *Register*, 10 Nov. 1865, p. 2.
39 G.W. Goyder to Secretary, CCLI, 16 Nov. 1865. This letter is reproduced in O'Brien, pp. 26–7, and identified as SRSA GRG 35/1/1865/3040.
40 'The Northern Runs', *Register*, 30 Nov. 1865, p. 2.
41 SAPP 1865–66, no. 78.
42 'Death of Mr. G.W. Goyder, C.M.G.', *Observer*, 5 Nov. 1898, p. 32. This unidentified autobiographical material was quoted identically in Goyder's obituaries in the *Observer*, *Advertiser* and *Chronicle*. Although the source and date of Goyder's remarks are not given, the account refers to the 'rain-gauge records of the various post and telegraph offices', so the remarks must have been made after Todd become Post-Master General in 1870. The same sentence also implies a current state of drought, indicating that Goyder was speaking or writing either in the early 1880s, or

after his retirement in the drought of 1896. All the quoted material in this paragraph is from this source.

43 SAPP 1865–66, no. 78.

44 ibid.

45 The map appears to have accompanied GRG 35/1/1865/978, which was printed as SAPP 1865–66, no. 78, but without the map.

46 SAPP 1876, no.145, p. 1.

47 SAPP 1865–66, no. 78.

48 G.W. Goyder, Field notebooks, vol. 20, p. [44].

49 SAPP 1865–66, no. 82, p. 2.

50 SAPP 1902, no. 22, p. 59, qu. 2715.

51 ibid., p. 65, qu. 2964.

52 W.S. Kelly, *Centenary of Goyder's Line*, Libraries Board of South Australia, 1963 (reprinted from the *Chronicle* of 6, 13 and 20 June 1963). It is mentioned under the heading 'Goyder draws the line'.

53 Lionel C.E. Gee, 'The South-East Fifty Years Ago. Vanished Aborigines', *Register*, 24 Mar. 1928, p. 5.

54 *Plain of Contrast: A history of Willowie, Amyton, Booleroo Whim*, District Centenary Book Committee, n.p., 1975, p. 5.

55 ibid., p. 2.

56 SAPP 1864, no. 105, p. 5.

57 [Rodney Cockburn], *Pastoral Pioneers of South Australia*, vol. 1, Publishers Ltd, Adelaide, 1925, p. 129.

58 F.J.R. O'Brien, 'Goyder's Line', MA preliminary thesis, University of Adelaide, 1952, p. 83; Donald W. Meinig, *On the Margins of the Good Earth: The South Australian wheat frontier 1869–1884*, facs. paperback edn, South Australian Government Printer, Netley, SA, 1988 (first published 1962), p. 53, footnote 74.

59 See map 'The Province 1865' in T. Griffin & M. McCaskill (eds), *Atlas of South Australia*, correct. edn, S.A. Government Printing Division & Wakefield Press, [Adelaide], 1987, p. 17.

60 G.W. Goyder, Field notebooks, vol. 20, p. [43].

61 See map 'The Province 1865', cited above.

62 Surveyor-General's Office 1868–1918, Confidential letter book, 2 Aug 1883. SRSA GRG 35/19.

63 SAPP 1865–66, no. 82.

64 South Australia, Parliament, *Debates*, 14 Dec. 1864, c. 558.

65 ibid., 30 Nov. 1865, c. 417.

66 ibid., 14 Dec. 1865, c. 563.

67 'The Pastoral Resolutions', *Register*, 15 Dec. 1865, p. 2.

68 SAPP 1865–66, no. 133.

69 ibid.

70 SAPP 1865–66, no. 154.

71 SAPP 1867, no. 14, p. 110, qu. 3846.

72 ibid., p. 114, qu. 3925.

Chapter nine: Colonial morality: 1861–63

1 Geoffrey Blainey, *The Rush that Never Ended*, 3rd edn, Melbourne University Press, Carlton, Vic., 1978, p. 124.
2 'Wallaroo', *Register*, 27 Dec. 1860, [p. 2].
3 SAPP 1863, no. 51, p. 186, qu. 6259 (Goyder describes the state of the maps); p. 65, qus 2377–8 (the extent of the error); p. 165, qu. 5657 (error discovered).
4 SAPP 1863, no. 51, p. 157, qu. 5488.
5 ibid., p. 57, qus 2077, 2086; p. 97, qus 3461, 3466.
6 ibid., p. 61, qu. 2194.
7 ibid., p. 130, qus 4702–13; p. 132, qu. 4755; p. 161, qu. 5570.
8 'The Rumoured Error in the Tipara Survey', *Register*, 8 July 1861, p. 3. E.W. Andrews, who wrote the article, described his interview with Goyder to the Tipara Inquiry (SAPP 1863, no. 51, p. 210).
9 SAPP 1863, no. 51, p. 165, qu. 5654.
10 ibid., p. 165, qu. 5654; p. 62, qu. 2273; p. 21, qu. 739, p. 22, qu. 753.
11 ibid., p. 36, qu. 1290.
12 ibid., p. 37, qu. 1325; p. 21, qu. 742.
13 'E.R. Mitford', *Observer*, 6 Nov. 1869, p. 7.
14 SAPP 1863, no. 51, appendix 1, p. i.
15 Goyder wrote that he had been 'suffering severely from scurvy for several years afterwards', referring to the 14 months he spent in the north in 1859–60, in a later letter to the CCLI, Arthur Blyth, in which he requested nine month's leave to go back to England. SRSA GRG 35/1/1871/81.
16 SAPP 1863, no 51, p. 28, qu. 988; p. 77 qus. 2759–60.
17 E.R. Mitford to secretary CCLI, 26 Jan. [1862], in SAPP 1862, no. 190, p. 10. Mitford's letter is incorrectly dated 1861.
18 Eustace Reveley Mitford, 'Summary for Europe', *Pasquin*, 21 Dec. 1867, in Mitford (ed.), *Pasquin: The pastoral, mineral and agricultural advocate vols I–III, January 1867 to November 1869*, Judd & Co., London, 1882, p. 309.
19 Mitford, 'Hoang Ho', *Pasquin*, 18 July 1868, in Mitford (ed.), *Pasquin: The pastoral, mineral and agricultural advocate vols I–III, January 1867 to November 1869*, Judd & Co., London, 1882, p. 514.
20 Mitford, 'The Valuation of Gold Runs and Sheep Runs', *Pasquin*, 30 Mar. 1867, in Mitford (ed.), *Pasquin: The pastoral, mineral and agricultural advocate vols I–III, January 1867 to November 1869*, Judd & Co., London, 1882, p. 59; 'The Survey Department', *Pasquin*, 5 Sep. 1868, in Mitford (ed.), *Pasquin: The pastoral, mineral and agricultural advocate vols I–III, January 1867 to November 1869*, Judd & Co., London, 1882, p. 560.
21 Mitford, 'Apology for a Newspaper', *Pasquin*, 26 Jan. 1867, in Mitford (ed.), *Pasquin: The pastoral, mineral and agricultural advocate vols I–III, January 1867 to November 1869*, Judd & Co., London, 1882, p. 5.
22 SAPP 1863, no. 51, p. 185, qu. 6258.

23 ibid., p. 29, qus 1000–3; p. 77, qu. 2762.

24 [leader], *Telegraph*, 13 Oct. 1862.

25 South Australia, Parliament, *Debates*, 12 Sep. 1862, c. 843.

26 SAPP 1862, no. 220, p. 5.

27 SAPP 1863, no. 41, p. 11.

28 SAPP 1890, no. 60, p. 20.

29 SAPP 1863, no. 41, p. 12. (The account Goyder gave of this much later, in SAPP 1892, no. 35, p. 77, qu. 2840, differs in relation to the work in the south.)

30 SAPP 1863, no. 41, p. 12.

31 ibid.

32 Lionel C.E. Gee, 'The South-East Fifty Years Ago. Vanished Aborigines', *Register*, 24 Mar. 1928, p. 5.

33 Michael Williams, *The Making of the South Australian Landscape: A study in the historical geography of Australia*, Academic Press, London, 1974, p. 188.

34 SAPP 1890, no. 60, p. 20.

35 SAPP 1863, no. 51, p. 162, qu. 5580.

36 ibid., p. 73, qu. 2643; p. 185, qu. 6258.

37 ibid., pp. 167–8, qus 5735–42.

38 ibid., p. 168, qus 5741–2.

39 ibid., p. 169, qu. 5796.

40 ibid., p. 170, qu. 5807.

41 ibid., p. 170, qu. 5811.

42 ibid., p. 171, qu. 5831.

43 ibid., p. 185, qu. 6258.

44 ibid., p. 45, qu. 1645; also p. 3, qu. 92.

45 ibid., p. 175, qu. 5948.

46 ibid., p. 184, qus 6242–3.

47 ibid., p. 187, qu. 6283.

48 ibid., p. 187, qu. 6259. The complete statement covers pp. 185–7.

49 Edwin Hodder, *History of South Australia*, vol. 1, Sampson, Low, Marston, London, 1893, p. 347.

50 SAPP 1863, no. 51, p. I, para. 7.

51 ibid., p. II, para. 12.

52 ibid., p. 164, qu. 5640.

53 ibid., p. 60, qu. 2173.

54 ibid., p. 58, qu. 2098.

55 [leader], *Telegraph*, 4 Sep. 1863, p. 2.

56 'The Strangways-Phillipson Row', *Telegraph*, 6 Oct. 1863, p. 2.

57 House of Assembly, 29 Sep. 1863, reported in *Register*, 30 Sep. 1863, p. 3.

58 [leader], *Telegraph*, 29 Sep. 1863.

59 Jonathan Guinness & Catherine Guinness, *The House of Mitford*, Hutchinson, London, 1984, p. 28.

60 'The Surveyor-General and the Moonta Dispute', *Advertiser*, 27 Oct. 1863, p. 5.

61 Goyder to Barrow, 27 Oct. 1863, SRSA GRG 35/20 [Copies of letters sent] vol. 12/1863/169.

62 House of Assembly, 1 Oct 1863, reported in *Register*, 2 Oct. 1863, p. 3.

63 Reynolds to Goyder, 2 Oct. 1863, filed with SRSA GRG 35/20/63/223.

64 Goyder to Reynolds, 27 Oct. 1863, SRSA GRG 35/20/63/223.

65 House of Assembly, 1 Oct. 1863, reported in *Register*, 2 Oct. 1862, p. 3. Bagot's description matches Goyder's account, and both descriptions match the contents of Freeling's covering letter to Goyder's application. Goyder's account and the covering letter are quoted in chapter 5.

66 [Leader], *Age*, 8 Oct. 1863, p. 4.

67 SAPP 1863, no. 118, p. 1.

68 SAPP 1890, no. 34, p. 18, qu. 453.

69 SAPP 1856, no. 14, reprinted as SAPP 1890, no. 34A, p. 22 (App. A).

70 SAPP 1890, no. 34, p. 18, qu. 453.

71 SAPP 1863, no. 118.

72 'Adelaide to the Murray', *Register*, 21 Sep. 1863, p. 2.

73 In SAPP 1890, no. 34, p. 19, qu. 457 Goyder refers to the *Grappler* having done a great deal of good, but it is not stated when this work began.

74 'Jeames at the Levee', *Register*, 25 May 1865, p. 3.

75 'Attention! Dress!', *Telegraph*, 2 July 1863, [p. 3].

76 Mitford, 'The Gawler Prize History', *Pasquin*, 7 Dec. 1867, in Mitford (ed.), *Pasquin: The pastoral, mineral and agricultural advocate vols I–III, January 1867 to November 1869*, Judd & Co., London, 1882, p. 288.

77 Mitford, 'The Drought and the Valuations', *Pasquin*, 13 July 1867, in Mitford (ed.), *Pasquin: The pastoral, mineral and agricultural advocate vols I–III, January 1867 to November 1869*, Judd & Co., London, 1882, p. 147.

Chapter ten: Transition: 1866–68

1 SAPP 1865, no. 72, p. 74, qu. 2251.

2 Lionel C.E. Gee, 'The South-East Fifty Years Ago. Vanished Aborigines', *Register*, 24 Mar. 1928, p. 5.

3 South Australia, Parliament, *Debates*, 29 No. 1866, c. 1090. Milne quoted the letter at length in parliament.

4 ibid., 20 Nov. 1866, c.1001–2.

5 ibid., 29 Nov. 1866, c. 1090.

6 SAPP 1867, no. 14, p. 110, qu. 3841–2.

7 SAPP 1865–66, no. 137, p. 1.

8 SAPP 1866–67, no. 65, p. 9, qus 241–2.

9 ibid., p. 12, qu. 314.

10 J.M. Powell, 'Enterprise and dependency: water management in Australia' in Tom Griffiths & Libby Robin (eds), *Ecology and Empire: Environmental history and settler societies*, Melbourne University Press, Carlton, Vic, 1997, p. 106.

11 SAPP 1892, no. 35, p. 81, qu. 2965–6; p. 83, qu. 3015.

12   SAPP 1866–67, no. 65, p. 12, qu. 316; p. 81. qu. 2965

13   SAPP 1872, no. 43, p. 3, qu. 43.

14   ibid., pp. 3–4, qus 53–4; SAPP 1870–71, no. 107, p. 3, qu. 25.

15   SAPP 1867, no. 142.

16   Michael Williams, *The Making of the South Australian Landscape: A study in the historical geography of Australia*, Academic Press, London, 1974, p. 189.

17   SAPP 1872, no. 43. p. 50, qu. 1027.

18   Ebenezer Ward, *The South-Eastern District of South Australia: Its resources and requirements*, Ward, Adelaide, 1869, p. 9.

19   ibid., p. 49.

20   Williams, *Making*, p. 200.

21   SAPP 1879, no. 71, p. vi.

22   Williams, *Making*, p. 32.

23   K.R. Bowes, *Land Settlement in South Australia 1857–1890*, Libraries Board of South Australia, Adelaide, 1968, p. 170, 184 (PhD thesis, ANU, 1963). Bowes cites several letters in support of this account. Goyder later modified his opinion on tendering, favouring a modified auction system instead. The matter was raised by the select committee inquiring into the sale of crown lands (SAPP 1865, no. 72, pp. 69–77).

24   Goyder to CCLI Milne, 24 Aug. 1866. SRSA GRG 35/20/1866/463.

25   Attributed to Douglas Jerrold, from 'A Land of Plenty', *The wit and opinions of Douglas Jerrold* (1859), in Elizabeth Knowles (ed.), *Oxford Dictionary of Quotations*, Oxford University Press, 5th edn, Oxford, 1999, p. 407. (However, I was not able to locate the phrase in the copy I examined.)

26   Goyder to CCLI Strangways, Surveyor-General's Office1868–1918, Confidential letter book, p. 14. SRSA GRG 35/19. The letter is actually dated 19 Nov. 1869, a slip of the pen probably brought about by the year 1869 being mentioned in the first line of the text. The contents establish the correct year.

27   SAPP 1868–69, no. 161.

28   Bowes, p. 193.

29   SAPP 1864, no. 36, p. 14.

30   'The Ulysses of the Northern Territory', *Register*, 29 Feb. 1868, p. 2.

31   SAPP 1866–67, no. 128, p. 1.

32   SAPP 1866–67, no. 188, p. 3.

33   G.W. Goyder, memorandum, 31 Aug. 1868, Surveyor-General's Office 1868–1918, Confidential letter book, [p. 9]. SRSA GRG 35/19.

34   SAPP 1868–69, no. 101, p. 1.

35   'The Open Column. Noctes Arabianae', *Advertiser*, 2 Aug. 1864, p. 3.

36   'The Open Column. Discovery of a curious m.s.', *Advertiser*, 5 Sep. 1864, p. 3.

37   'A New Invention for Killing Squatters', *Register*, 15 Oct. 1864, p. 3.

38   'E.R. Mitford', *Observer*, 6 Nov. 1869, p. 7.

39   Goyder to Mitford, 26 Dec. 1866. SRSA GRG 35/20/13, p. 1314.

40   Eustace Reveley Mitford (ed.), 'Odds and Ends. Inspector of Mines', *Pasquin*, 1 Jun. 1867, in Mitford (ed.), *Pasquin: The pastoral, mineral and agricultural advocate vols I–III, January 1867 to November 1869*, Judd & Co., London, 1882, p. 111.

41   Mitford, 'The Survey Department', *Pasquin*, 5 Sep. 1868, in Mitford (ed.), *Pasquin: The pastoral, mineral and agricultural advocate vols I–III, January 1867 to November 1869*, Judd & Co., London, 1882, p. 561.

42   Mitford, 'Wakefield and Colonization', *Pasquin*, 16 Feb. 1867, in Mitford (ed.), *Pasquin: The pastoral, mineral and agricultural advocate vols I–III, January 1867 to November 1869*, Judd & Co., London, 1882, p. 21.

43   Mitford, 'Odds and Ends: Goyder's Valuations', *Pasquin*, 8 Feb. 1868, in Mitford (ed.), *Pasquin: The pastoral, mineral and agricultural advocate vols I–III, January 1867 to November 1869*, Judd & Co., London, 1882, p. 353.

44   Mitford, 'North Terrace', *Pasquin*, 22 Aug. 1868, in Mitford (ed.), *Pasquin: The pastoral, mineral and agricultural advocate vols I–III, January 1867 to November 1869*, Judd & Co., London, 1882, p. 552.

45   Mitford, 'Review', *Pasquin*, 10 Jul. 1869, in Mitford (ed.), *Pasquin: The pastoral, mineral and agricultural advocate vols I–III, January 1867 to November 1869*, Judd & Co., London, 1882, pp. 865–6.

46   'George W. Goyder, Esq., J.P.', *Illustrated Australian News*, 22 Feb. 1869, p. 52.

47   'G.W. Goyder, Leader of the Expedition to North Australia', *Australian Journal*, vol. 5, July 1870, p. 659.

48   'Death of Mr. E.R. Mitford', *Register*, 9 Nov. 1869, p. 4.

49   'E.R. Mitford', *Observer*, 6 Nov. 1869. p. 7.

50   Rob Linn, *A Diverse Land: A history of the Lower Murray, Lakes and Coorong*, Meningie Historical Society, [Meningie, SA], 1988, p. 100.

51   George Loyau, *The Representative Men of South Australia*, [facs. ed.], Austaprint, Hampstead Gardens, SA, 1978 (first published 1883), p. 177.

52   'Building Improvements. Private residence and general improvements', *Register*, 15 Jan. 1867, p. 3.

53   John Lewis, *The Walkerville Story: 150 years*, Corporation of the City of Walkerville, Adelaide, 1988, pp. 36, 40.

54   'Talk on the Flags', *Register*, 4 Nov. 1867, p. 3; 'Royal Visit. The Levee', *Register*, 2 Nov. 1867, p. 2.

55   'Shipping Intelligence. Cleared out', *Register*, 14 Dec. 1867, p. 2.

56   Margaret Goyder Kerr, *The Surveyors: The story of the founding of Darwin*, Rigby, Adelaide, 1971, p. 8. Kerr says that she 'never really recovered …', but the status of her comment is not clear.

57   'A Lady Accidentally Poisoned', *Times*, 11 Apr. 1870, p. 11.

58   Kerr, p. 33; E.J.R. Morgan, *The Adelaide Club, 1863–1963*, 2nd edn, Adelaide, [no date], p. 101.

Chapter eleven: Larrakia country: the founding of Darwin

1   This verse is quoted by W.S. Kelly, *Centenary of Goyder's Line*, Libraries Board of South Australia, 1963 (reprinted from the *Chronicle* of 6, 13 and 20 June 1963)

without giving the source or the context, but jokes about Goyder's bonus were commonplace and make the context unmistakeable.

2   'Port Darwin and its Antecedents', *Register*, 26 Apr. 1869, p. 2.
3   'The Northern Territory Expedition', *Register*, 12 Dec. 1868, p. 2.
4   Margaret Goyder Kerr, *The Surveyors: The story of the founding of Darwin*, Rigby, Adelaide, 1971, p. 38.
5   'The Northern Territory Expedition', *Register*, 12 Dec. 1868, p. 2.
6   'The Northern Territory Survey', *Register*, 5 Jan. 1869, p. 4.
7   Edwin Mitchell Smith, 'Recollective diary', 1920, p. 7. In the Goyder family file, SLSA PRG 491/3
8   'The Northern Territory Expedition', *Register*, 22 Dec. 1868, p. 2.
9   SAPP 1868–69, no. 175.
10 'The Northern Territory expedition', *Register*, 25 Dec. 1868, p. 3, includes a 'complete and correct list' which totalled 135, including six cooks. A list of 129 was included as an appendix to Kerr. This did not include the cooks.
11 'The Northern Territory Expedition', *Register*, 5 Jan. 1869, p. 4.
12 Smith, 'Recollective diary', p. 8. SLSA PRG 491/3.
13 'The Northern Territory Expedition', *Register*, 25 Dec. 1868, p. 3.
14 'The Northern Territory', *Register*, 26 Apr. 1869, p. 2. Pages 2–3 of this issue contain several letters from members of the expedition, and the official dispatches, reproduced in full.
15 SAPP 1869–70, no. 31, p. 1.
16 'The Northern Territory', *Register*, 26 Apr. 1869, p. 2.
17 From a report of the time, quoted in *The Encyclopaedia of Aboriginal Australia*, Aboriginal Studies Press for the Australian Institute of Aboriginal and Torres Strait Islander Studies, Canberra, 1994, p. 1196.
18 'The Northern Territory', *Register*, 26 Apr. 1869, p. 2.
19 G.W. Goyder, Diary of Surveyor General Goyder, 5th February, 1869 [–1 Mar. 1869], 6 Feb. 1869. This is from a copy of a typescript provided by the Northern Territory Library. The State Records of South Australia hold Goyder's field notebook, no. 12, Northern Territory 11 Feb. 1869 – 14 Oct. 1869. SRSA GRG 35/256.
20 Quoted in Kerr, p. 67.
21 G.W. Goyder, Diary of Surveyor General Goyder, 5th February, 1869 [–1 Mar. 1869], 6 Feb. 1869.
22 SAPP 1869–70, no. 157, p. 2.
23 Kerr, p. 60.
24 SAPP 1869–70, no. 31, p. 1.
25 ibid., p. 2.
26 ibid.
27 'The Arrival of the *Moonta*', *Register*, 24 Apr. 1869, p. 2.
28 *Register*, 26 Apr. 1869, p. 3. This was reported in a letter reprinted in the paper.
29 Smith, 'Recollective diary', p. 7.
30 SAPP 1895, no. 19, p. 22, qu. 476.

31   Kerr, p. 82.

32   'The Northern Territory', *Register*, 16 Nov. 1869, p. 2.

33   First annual report of the Colonization Commissioners of South Australia,(*British Parliamentary Papers*, 28 July 1836, vol. 491, pp. 8–9), quoted in T. Griffin & M. McCaskill (eds), *Atlas of South Australia*, correct. edn, S.A. Government Printing Division & Wakefield Press, [Adelaide], 1987, p. 30.

34   Appendix to second report of Commissioners of Colonization of South Australia, p. 16, clauses 34 and 35, quoted in SAPP 1860, no. 165, p. 5.

35   Second report from the Select Committee on South Australia (*British Parliamentary Papers*, 10 June 1841, vol. 394, p. 210), quoted in Griffin & McCaskill (eds), p. 30.

36   G.W. Goyder, Field notebooks, vol. 6, 8 Nov. 1860. SRSA GRG 35/256.

37   From a report to the governor by Goyder on the Aborigines of the Northern Territory, quoted in Kerr, p. 147. The original document could not be located, but it is quoted extensively in Kerr, pp. 147–8.

38   ibid., p. 148.

39   ibid.

40   ibid.

41   'The Aborigines' [letter], *Register*, 1 Oct. 1857, p. 3.

42   'Christians and Aborigines in the North' [letter], *Register*, 2 Aug. 1865, p. 2. This letter followed an earlier one, published under the same heading, on 12 July. Both letters document the mechanics and impact of dispossession.

43   SAPP 1868–69, no. 101, p. 2.

44   G.W. Goyder, Memo., Port Darwin, 20 Feb. 1869, Northern Territory Expedition out letters, vol. 1, Feb. – Dec. 1869, p. 2, no. 3/69. SRSA GRG 35/11.

45   Kerr, p. 91.

46   ibid., p. 96.

47   Von Mueller to Goyder, SRSA GRG 35/2/1868/1055, quoted in R.W. Home et al., *Regardfully Yours: Selected correspondence of Ferdinand Von Mueller*, vol. II: 1860–1875, New York, Peter Lang, 1998, p. 478.

48   Philip Jones, *Ochre and Rust*, Wakefield Press, Kent Town, SA, 2007, p. 155.

49   SAPP 1869–70, no. 31, p. 3.

50   Quoted in Kerr, p. 112; Jones, p. 171, from Hoare's diary.

51   Kerr, p. 138, quoting a letter from Bennett to a friend.

52   Among Goyder's field notebooks is an alphabetical listing of words of the 'Woolner Dialect Adelaide River' (SRSA GRG 35/256/11). The old cover title label, added some time after the volume's creation, attributes the book to Goyder, but it is obvious that the writing – very neat, with rounded, serifed letters – was certainly not his middle-aged scrawl, and Goyder had mentioned sending Bennett's notebook down. Philip Jones, p. 133, confirms that this is Bennett's notebook, which had been attributed to Goyder by N.B. Tindale.

53   SAPP 1869–70, no. 157, p. 1.

54   Kerr, chaps 20 and 21, contains several accounts of this incident quoted in full. They do not all agree. Kerr wryly observes of the expeditioners' letters concerning Bennett's death that 'nearly all claim that he died in their arms'.

55   SAPP 1869–70, no. 157, p. 2.

56   ibid.

57   ibid.

58   ibid.

59   Quoted in Kerr, p. 145.

60   SAPP 1869–70, no. 157, p. 2.

61   G.W. Goyder, Field notebooks, vol. 12, 12 June 1869.

62   ibid., 3 July.

63   ibid., 6 July.

64   ibid., 1 Aug.

65   Kerr, pp. 116–17.

66   'The Northern Territory', *Register*, 14 Mar. 1870, p. 5.

67   Goyder to A. Woods, 2 Apr. 1869, Northern Territory Expedition out correspondence, vol. 1, Feb. – Dec. 1869, p. 26, no. 22/69. SRSA GRG 35/11.

68   Lionel C.E. Gee, 'The South-East Fifty Years Ago. The early pioneers', *Register*, 3 Mar. 1928, p. 7.

69   'Dr Peel'[note], *Register*, 16 Nov. 1869, p. 2.

70   G.W. Goyder, Field notebooks, vol. 12, 3 Aug. 1869.

71   'The Northern Territory', *Register*, 16 Nov. 1869, p. 2.

72   Kerr, p. 101.

73   SAPP 1895, no. 19, p. 24, qu. 514; p. 26, qu. 571.

74   'The Northern Territory', *Register*, 16 Nov. 1869, p. 2.

75   Kerr, p. 168.

76   ibid., p. 170.

77   'The Northern Territory', *Register*, 16 Nov. 1869, p. 2.

78   SAPP 1869–70, no. 157, p. 1.

79   ibid., p. 2.

80   ibid., p. 1.

81   ibid., p. 2.

82   ibid.

83   ibid.

84   ibid.

85   ibid.

86   'Mr. Goyder's Return', *Register*, 16 Nov. 1869, p. 2.

87   ibid.

88   'Murder Palliated', *Pasquin*, 20 Nov. 1869.

89   'Mr. Goyder's Despatch', *Register*, 16 Nov. 1869, p. 3.

90   SAPP 1869–70, no. 157, p. 3.

91   From a report to the governor by Goyder on the Aborigines of the Northern Territory, quoted in Kerr, p. 148.

92   'Echoes from the Bush', *Register*, 24 Nov. 1869, p. 2.

Chapter twelve: Going home

1   'Plan of the Northern Territory Survey', *Register*, 23 Nov. 1869, p. 2.

2   SAPP 1872, no. 43, p. 5, qu. 79.

3   SAPP 1890, no. 60, p. 20.

4   'The Drainage Works' and 'The Unemployed Agitation Revived', *Register*, 15 Mar. 1870, p. 5.

5   SAPP 1872, no. 43, p. 7, qu. 125.

6   G.W. Goyder, Field notebooks, vol. 12, pp. [173–6]. SRSA GRG 35/256.

7   This account is based on 'A Lady Accidentally Poisoned', *Times*, London, 11 Apr. 1870, p. 11, 'An Overdose of Laudanum', *Western Daily Press*, Bristol, 11 Apr. 1870 and 'Death From Taking an Overdose of Medicine', *Daily Bristol Times and Mirror*, 11 Apr. 1870. The young David Goyder is identified as a doctor in Bradford in John Martin-Harvey, *The Autobiography of Sir John Martin-Harvey*, S. Low, Marston & Co., London, [1933], p. 24. For an understanding of how such an overdose could be intentional, but not deliberately suicidal, I am indebted to Alan Lovejoy, a pharmacist who has had both the anachronistic experience of preparing laudanum and considerable experience in relation to addiction.

8   Goyder to CCLI Cavenagh, 12 May 1870. SRSA GRG 35/1/1870/463.

9   'Arrival of the Mail', *Express and Telegraph*, 4 June 1871, 2nd edn, p. 2.

10   'Drainage Works', *Register*, 6 Aug. 1870, p. 5; SAPP 1870–71, no. 107.

11   'The Drainage Works', *Register*, 5 Sep. 1870, p. 5.

12   SAPP 1870–71, no. 107, p. 4, para. 35; p. 5, para. 38.

13   SAPP 1872, no. 43, p. 3, qu. 43; p. 5, qu. 82.

14   'Shipping Intelligence', *Register*, 19 Aug. 1870, p. 4.

15   Goyder to CCLI Townsend, 26 Jan. 1871. SRSA GRG 35/1/1871/81.

16   'Death of the Hon. Dr. Campbell', *Observer*, 5 Nov. 1898, p. 15.

17   Goyder to CCLI Townsend, 26 Jan. 1871 (enclosure). SRSA GRG 35/1/1871/81.

18   *Times*, 28 Apr. 1871, p. 9.

19   G.W. Goyder, 'Wreck of the Steamer Queen of the Thames', *Register*, 4 July 1871, p. 5.

20   ibid.

21   ibid.

22   *Times*, 28 Apr. 1871, p. 9.

23   G.W. Goyder, 'Wreck of the Steamer Queen of the Thames', *Register*, 4 July 1871, p. 5. These words were not included in the account published in the *Times* ('Wreck of The Screw Steamship Queen of the Thames', 22 May 1871, p. 6.)

24   'Journal of a Voyage from Melbourne to London. Commenced Feby. 18 1871.' Held in the La Trobe Library, Melbourne.

25   G.W. Goyder, 'Wreck of the Steamer Queen of the Thames', *Register*, 4 July 1871, p. 6.

26   ibid.

27   'The Queen of Thames', *Times*, 24 May 1871, p. 8. The article reproduces the report of the court.

28 'Wreck of the Queen of the Thames', *Register*, 15 June 1871, p. 6. The article contains Goyder's evidence.

29 'The Wreck of the Queen of the Thames', [letter from Westall and Roberts, solicitors], *Times*, 5 June 1871, p. 10.

30 Goyder to solicitors, Cullen & Murphy, 4 Nov. 1872, Surveyor-General's Office 1868–1918, Confidential letter book. SRSA GRG 35/19.

31 ibid.

32 'Shipping Intelligence' and 'Deaths', *Register*, 18 Oct. 1871, p. 4.

33 Douglas Pike (gen. ed.), *Australian Dictionary of Biography*, vol. 4, *1851–1890 D–J*, Melbourne University Press, Carlton, Vic., 1972, p. 257.

34 *Register*, 23 Nov. 1871, p. 2.

35 The first was Goyder's departure on 6 Jan. 1882 for England (and America) to purchase equipment to locate and exploit artesian water. The second set of dates, 19 May 1883 – 24 July 1883, record a visit to New Zealand not previously included in the record of Goyder's life.

36 These dates differ from the approximate dates assigned to these photographs by the State Library of South Australia. A photograph of Goyder while he was in the Northern Territory shows that his hair and beard were dark and in the style shown in the portrait in which he wears evening clothes. This indicates that the photograph in which his hair is shorter, lighter, and arranged in a curl would not have been taken after 1869.

Chapter thirteen: Nature's Line

1 SAPP 1891, no. 33, qu. 1997.

2 Eustace Reveley Mitford, 'Song of Praise', 24 Oct. 1868, in Mitford (ed.), *Pasquin: The pastoral, mineral and agricultural advocate vols I–III, January 1867 to November 1869*, Judd & Co., London, 1882, p. 608.

3 Goyder admitted this disappointment in the long unsourced passage by him, quoted in 'Death of Mr. G.W. Goyder, C.M.G.', *Observer*, 5 Nov. 1898, p. 32, and in his obituaries in the *Advertiser* and *Chronicle*.

4 SAPP 1867, no. 14, p. 113, qu. 3919.

5 ibid., pp. 113–14, qu. 3920.

6 SAPP 1865, no. 72, p. 75, qu. 2284.

7 SAPP 1867, no. 14, p. 111, qus 3874–5.

8 ibid., report, p. 3.

9 G.W. Goyder, *Mr. Goyder's Report on Victoria Land Regulations, Advertiser*, [Adelaide], p. 8, para. 56 (originally SAPP 1870–71, no. 23).

10 SAPP 1888, no. 28, p. 82, qu. 2308.

11 G.W. Goyder, *Report on Victoria Land Regulations*, p. 7. para. 53.

12 ibid., p. 8, para. 55.

13 SAPP 1870–71, no. 89.

14 Nancy Robinson, *Change on Change: A history of the northern highlands of South Australia*, Nadjuri Australia, Jamestown, SA, p. 104, quoting a letter in the *Observer*.

15 'Earth Hunger', *Register*, 1 Mar. 1872, p. 5.

16 Anthony Trollope ed. P.D. Edwards & K.B. Joyce, *Australia*, University of Queensland Press, St Lucia, Qld, 1967 (first published 1873), p. 656.

17 'Mr. Anthony Trollope', *Register*, 8 Apr. 1872, p. 4.

18 I.A. Robinson, *A History of the New Church in Australia 1832–1980*, Melbourne, [1980?], p. 184.

19 Trollope, pp. 650–1.

20 G.W. Goyder, *Report on Victoria Land Regulations*, p. 8. para. 54.

21 *Northern Argus*, 13 Feb. 1874, quoted in K.R. Bowes, *Land Settlement in South Australia 1857–1890*, Libraries Board of South Australia, Adelaide, 1968, p. 198 (PhD thesis, ANU, 1963).

22 Donald W. Meinig, *On the Margins of the Good Earth: The South Australian wheat frontier 1869–1884*, facs. paperback edn, South Australian Government Printer, Netley, SA, 1988 (first published 1962), p. 53, footnote 74, citing the *Farmers' Weekly Messenger* of 25 Sep. 1874.

23 SAPP 1874, no. 34.

24 *Northern Argus*, 22 May 1874, quoted in Meinig, *Margins*, pp. 53, 55.

25 South Australia, Parliament, *Debates*, 18 June 1874; Blyth, c. 579, 583; Ward, c. 582. (For Ward's involvement with rural newspapers see J.B. Hirst, *Adelaide and the Country 1870–1917*, Melbourne University Press, Carlton, Vic., 1973, p. 89.)

26 ibid., 14 July 1874, c. 905–6.

27 *The Areas' Express and Farmers' Journal*, 3 Nov. 1877, quoted in Meinig, *Margins*, p. 64.

28 Meinig, *Margins*, p. 53, quoting the issue of 22 May 1874.

29 Mitford, 'The Crisis', *Pasquin*, 26 Sep. 1868, in Mitford (ed.), *Pasquin: The pastoral, mineral and agricultural advocate vols I–III, January 1867 to November 1869*, Judd & Co., London, 1882, p. 581.

30 Mitford, 'Figures of Speech on North Terrace', *Pasquin*, 21 Sep. 1867, in Mitford (ed.), *Pasquin: The pastoral, mineral and agricultural advocate vols I–III, January 1867 to November 1869*, Judd & Co., London, 1882, p. 215.

31 Mitford, 'Our statesmen', *Pasquin*, 26 Sep. 1868, in Mitford (ed.), *Pasquin: The pastoral, mineral and agricultural advocate vols I–III, January 1867 to November 1869*, Judd & Co., London, 1882, p. 582.

32 [Letter], *Observer*, 4 Nov. 1865, p. 2.

33 South Australia, Parliament, *Debates*, 28 Jan. 1870, c. 1510.

34 'Original stories. The Inland Sea', *Register*, 29 July 1868, p. 3.

35 Mitford, *Pasquin*, 12 Sep. 1868, in Mitford (ed.), *Pasquin: The pastoral, mineral and agricultural advocate vols I–III, January 1867 to November 1869*, Judd & Co., London, 1882, pp. 572–3.

36 In 'Goyder's Line of Rainfall: The role of a geographic concept in South Australian land policy and agricultural settlement', *Agricultural History*, vol. 35, October, 1961, p. 214, Donald Meinig essentially agreed with the farmers, pointing out that Goyder's view was not that of 'science', and that, given the false predictions that had been made, it was entirely understandable that the people would demand the opportunity to conduct a mass empirical testing of the land and that the government

would agree. But, writing in the early 1960s, Meinig did not understand what Goyder's Line actually was, or how Goyder had come to the understanding which produced it.

37 Robert Frost, 'A Masque of Reason', in Robert Frost, *Collected poems, prose and plays*, Literary Classics of the United States Inc., New York, 1995, p. 380.
38 South Australia, Parliament, *Debates*, 18 June 1874, c. 584.
39 ibid., 14 July 1874, c. 906.
40 SAPP 1874, no. 239, p. 3.
41 South Australia, Parliament, *Debates*, 5 Nov.1874, c. 2235
42 SAPP 1888, no. 28, p. 77, qu. 2201.
43 'Extension of Agricultural Settlement Northward', *Advertiser*, 11 Nov. 1874, p. 2.
44 *Register*, 9 July 1874, p. 3. Nancy Robinson took the title for her history of the Mid-North, *Change on Change*, from the poem, which she attributed to 'Shepherd Borthwick' of Canowie.

Chapter fourteen: Following the plough

1 *Plain of Contrast: A history of Willowie, Amyton, Booleroo Whim*, District Centenary Book Committee, n.p., 1975, p. 74.
2 K.R. Bowes, *Land Settlement in South Australia 1857–1890*, Libraries Board of South Australia, Adelaide, 1968, p. 197 (PhD thesis, ANU, 1963).
3 ibid., pp. 201–2.
4 J.B. Hirst, *Adelaide and the Country 1870–1917*, Melbourne University Press, Carlton, Vic., 1973, pp. 20–1.
5 Bowes, pp. 52–3.
6 William Harcus, *South Australia: Its history, resources and productions*, Sampson, Low, Marston, Searle and Rivington, London, 1876, pp. 60–1. The same material appeared in the *Handbook for Emigrants Proceeding to South Australia* of 1873.
7 ibid., pp. 63–4.
8 ibid., pp. 66–7.
9 SAPP 1876, no. 145, p. 2.
10 ibid., p. 1.
11 ibid., p. 2.
12 SAPP 1879, no. 71, p. 22, qu. 549.
13 Donald W. Meinig, *On the Margins of the Good Earth: The South Australian wheat frontier 1869–1884*, facs. paperback edn, South Australian Government Printer, Netley, SA, 1988 (first published 1962), pp. 63–4.
14 Henry Nash Smith, 'Rain follows the Plow: The notion of increased rainfall for the Great Plains, 1844–1880', *Huntington Library Quarterly*, 10, 1947, pp. 186–9. Wilber is quoted from *The Great Valleys and Prairies of Nebraska and the Northwest*, 1881, p. 69. Nash claims that it was Wilber who distilled a folk belief into the notion that 'rain follows the plough', which he presented in 1881, but (for one example) the director of the Manna Hill Experimental Farm used the phrase in an annual report for an earlier year (SAPP 1879, no 31A).
15 Meinig, *Margins*, p. 59.

16 *Port Augusta Despatch*, 20 Apr. 1878, quoted in Meinig, *Margins*, p. 67.

17 SAPP 1879, no. 71, p. 21, qu. 529.

18 ibid., qu. 530.

19 ibid., qus 533–6.

20 SAPP 1881, no.127, p. 5, qu. 103.

21 ibid., qu. 104.

22 ibid., qu. 110.

23 ibid., qus 115, 117–18.

24 South Australia, Parliament, *Debates*, 11 Jul. 1882, c. 303.

25 ibid., 22 Jun 1882, c. 171.

26 ibid., c. 173.

27 SAPP 1882, no. 73.

28 Meinig, *Margins*, p. 236, quoting SRSA GRG 35/1/1882/1677. According to Bowes (note 10 on p. 374), the memo was undated, but was probably produced in February though not filed until September.

29 Bowes, p. 243.

30 Michael Williams, 'George Woodroofe [*sic*] Goyder: A practical geographer', *Proceedings of the Royal Geographical Society (South Australian Branch)*, vol. 79, 1978, p. 1.

31 SAPP 1888, no. 28, p. 80, qu. 2261; also SAPP 1890, no. 60, p. 2.

32 F.J.R. O'Brien, 'Goyder's Line', MA preliminary thesis, University of Adelaide, 1952, p. 69; SAPP 1902, no. 22, p. 65, qus 2970–3.

33 SAPP 1902, no. 22, p. 66, qu. 3006.

34 O'Brien, p. 72.

35 O'Brien, p. 72; SAPP 1892, no. 152; SAPP 1893, no. 83.

36 SAPP 1902, no. 22, p. 65, qu. 2970; p. 66, qu. 2973. According to O'Brien, Goyder modified the copy on which the Line was shown in red ink and which carried the inscription 'Certified correct' with his signature.

Chapter fifteen: Fresh water and peculiar country

1 SAPP 1891, no. 33, p. 79, qu. 2043.

2 SAPP 1889, no. 113, p. 1, qu. 4. (He began creating stock reserves about 1863, although they were not defined in law until 1878.)

3 SAPP 1857–8, no. 72, p. 4.

4 SAPP 1860, no. 20, pp. 9–10.

5 SAPP 1860, no. 41, and 'The Far North', *Register*, 20 Dec. 1860, p. 2.

6 SAPP 1860–61, no. 177, p. 4.

7 'Rivers of Interior Australia', *Register*, 28 Jan. 1864, p. 2.

8 SAPP 1867, no. 145.

9 SAPP 1867, no. 141, p. 5, qus 156–63.

10 SAPP 1861, no. 56, p. 3–4.

11 'Wallaroo', *Register*, 16 Jan. 1863, p. 3.

12 'Fresh Water at Wallaroo' [letter], *Register*, 23 Jan. 1863, p. 3.

13 'The Wallaroo District', *Register*, 24 Feb. 1863, p. 2.

14 'Australian Droughts', *Register*, 27 Apr. 1865, p. 2.

15 SAPP 1867, no. 141, p. 6, qu. 168.

16 SAPP 1890, no. 94, p. 2, qu. 27

17 SAPP 1867, no. 89, p. 2.

18 SAPP 1867, no. 14, p. 110, para. 3853. Cavenagh had expressed the same view, using the term 'peculiar country', the day before (p. 110, para. 3833).

19 SAPP 1867, no. 89, p. 3.

20 SAPP 1891, no. 33, p. 79, qu. 2043.

21 SAPP 1891, no. 33, p. 77, qu. 1992.

22 G.W. Goyder, Diary of Surveyor General Goyder, 5th Feb. 1869 [–1 Mar. 1869.], 8 Feb. 1869. This is from a copy of a typescript provided by the Northern Territory Library. The State Records Office of South Australia holds Goyder's field notebook, no. 12, Northern Territory 11 Feb. 1869 – 14 Oct. 1969. SRSA GRG 35/256.

23 SAPP 1875, Special Session, no. 26.

24 'Death of Mr. G.W. Goyder', *Advertiser*, 4 Nov. 1898, p. 6, reprinted in *Chronicle*, 5 Nov. 1898, p. 16.

25 SAPP 1890, no. 94, p. 1, qu. 6.

26 'Mr. Goyder in San Francisco', *Register*, 13 June 1882, p. 3.

27 ibid.

28 SAPP 1890, no. 94, p. 3, qu. 48.

29 ibid., qu. 39.

30 ibid., pp. 1–2, qu. 6; p. 2, qus 15–19.

31 ibid., p. 1–2, qu. 6; p. 3, qu. 40.

32 K.R. Bowes, *Land Settlement in South Australia 1857–1890*, Libraries Board of South Australia, Adelaide, 1968, p. 36 (PhD thesis, ANU, 1963).

33 South Australia, Parliament, *Debates*, 25 Oct. 1887, c. 1226.

34 SAPP 1883–84, no. 52.

35 SAPP 1889, no. 25, p. 50, qu. 1082.

36 SAPP 1883–84, no. 52, p. 2.

37 South Australia, Parliament, *Debates*, 24 Oct. 1883, c. 1406.

38 Marianne Hammerton, *Water South Australia: A history of the Engineering and Water Supply Department*, Wakefield Press, Netley, 1986, p. 56.

39 SAPP 1882, no. 25, Part II, pp. 42–3; SAPP 1889, no. 25, pp. 48–55.

40 SAPP 1889, no. 25, p. 55, qu. 1170.

Chapter sixteen: Tree theories

1 [leader], Advertiser, 21 Jan. 1882, p. 4.

2 Schomburgk to George Bentham, Aug. 1869, quoted in Pauline Payne, 'Dr Richard Schomburgk and the Adelaide Botanic Garden 1865–1891', PhD thesis, University of Adelaide, Department of History, 1992, p. 512.

3   W. Smillie, in *The Great South Land*, 1836, quoted in Michael Williams, *The Making of the South Australian Landscape: A study in the historical geography of Australia*, Academic Press, London, 1974, p. 129.

4   'A Visit to the Glen Para Saw-Mill', *Register*, 26 Apr. 1869, p. 3.

5   Williams, *Making*, pp. 133–4, quoting a letter to the *Register*, 3 June 1851.

6   Douglas Pike, *Paradise of Dissent: South Australia 1829–1857*, 2nd edn, Melbourne University Press, Carlton, Vic., 1967, p. 440.

7   George Loyau, *The Representative Men of South Australia*, [facs. edn], Austaprint, Hampstead Gardens, SA, 1978 (first published 1883), p. 134.

8   'Forest Culture', *Register*, 10 Jan. 1870, p. 3.

9   'Forest Trees', *Register*, 18 Jan. 1870, p. 6.

10   SAPP 1870–71, no. 78, p. 2.

11   'The Philosophical Society', *Register*, 10 Aug. 1870, p. 3.

12   ibid.

13   J.J. Pascoe (ed.), *History of Adelaide and Vicinity*, Hussey & Gillingham, Adelaide, 1901, p. 478–9.

14   'Forest Conservation and Forest Culture', *Register*, 22 Aug. 1870, p. 6.

15   Ian Auhl & Denis Marfleet, *Journey to Lake Frome 1843: Paintings and sketches by Edward Charles Frome and James Henderson*, Lynton Publications, Blackwood, SA, n.d., p. 86. Frome's watercolour *The Razor Back – from the top of Mt. Bryan* is reproduced on p. 89, and shows the hills dotted with gums. A photograph facing p. 11 shows the same view, with the hills bare.

16   SAPP 1859, no. 53.

17   SAPP 1873, no.153.

18   SAPP 1873, no. 94, p. 2.

19   ibid., p. 1.

20   ibid.

21   SAPP 1873, no. 94, p. 1.

22   SAPP 1873, no. 135, p. 1.

23   'Forest Culture', *Register*, 2 Jan. 1874, p. 10. This issue was the regular supplement for England. The article would have originally been published sometime in the previous month.

24   SAPP 1875, no. 85, pp. 1–2.

25   SAPP 1875 Special Session, no. 32.

26   SAPP 1875, no. 85, p. 2.

27   SAPP 1877, no. 157, p. 1, para. 2.

28   Nancy Robinson, *Change on Change: A history of the northern highlands of South Australia*, Nadjuri Australia, Jamestown, SA, p. 150.

29   SAPP 1877, no.157, p. 3, para. 24.

30   L.T. Carron, *A History of Forestry in Australia*, Australian National University Press, Rushcutters Bay, NSW, 1985, pp. 215.

31   SAPP 1877, no.157, p. 2, para. 18.

32   ibid., para. 21.

33 ibid., p. 3, para. 22. Para. 26 quotes Curnow at length. Curnow and his work are described in Nancy Robinson, pp. 149–50.

34 Pike (gen. ed.), *Australian Dictionary of Biography*, vol. 3, *1850–1890 A–C*, Melbourne University Press, Carlton, Vic., p. 261; SAPP 1879, no. 83, p. 1, para. 3.

35 SAPP 1878, no. 154, p. 1.

36 'Timber Licences' [letter], *Register*, 11 Feb. 1870, p. 7.

37 SAPP 1873, no. 135, p. 2, para. 12.

38 SAPP 1881, no. 43, p. 28.

39 The claim was made in a marginal note, according to J.M. Fielding, 'The Introduction of the Monterey Pine into Australia', *Australian Forestry*, vol. 21, no. 1, 1957, p. 15.

40 ibid., pp. 15–16.

41 E.B. Heyne, 'Forest Trees no. IV', *Register*, 31 Jan 1870, p. 6.

42 SAPP 1879, no. 83, p. 1, para. 9.

43 ibid., para. 10.

44 ibid., p. 20, para. 181.

45 ibid., p. 1, para. 9.

46 ibid., p. 69

47 SAPP 1880, no. 139, p. 35, para. 243.

48 ibid., para. 244.

49 ibid., para. 246.

50 'Tree-Planting in the Far North', *Register*, 20 Jan. 1882, supplement, originally published as SAPP 1882, no. 109, p. 20, para. 216.

51 SAPP 1882, no.109, p. 21, para. 224.

52 'Tree-Planting in the Far North', *Register*, 17 Jan. 1882, Supplement, p. 1.

53 [leader], *Advertiser*, 21 Jan. 1882, p. 4. The passage is quoted fully at the beginning of part four.

54 South Australia, Parliament, *Debates*, 20 July 1882, c. 427.

55 ibid., 15 Aug. 1882, c. 649–50.

56 SAPP 1883–84, no 73, p. 21, para. 187.

57 SAPP 1888, no. 109, p. 12.

58 SAPP 1890, no. 30C, p. 6, qu. 849.

Chapter seventeen: The universal genius

1 'The engineer', *The Wit and Opinions of Douglas Jerrold: Collected and arranged by his son Blanchard Jerrold*, London, W. Kent & Co., 1859, p. 139.

2 SAPP 1892, no. 35, p. 77, qu. 2840.

3 SAPP 1872, no. 43, p. 49, qu. 1006.

4 SAPP 1874, no. 168 and no. 1875, Special Session, no. 9.

5 SAPP 1878, no. 203, p. 2.

6 ibid.

7 ibid., p. 4.

8 SAPP 1892, no. 35, p. 83, qu. 3014; SAPP 1890, no. 60, p. 20.

9   SAPP 1883–84, no. 52, pp. 5–6.

10 SAPP 1880, no. 196, p. 1.

11 SAPP 1870–71, no. 86, p. i, para. 7.

12 SAPP 1876, no. 108, p. 7.

13 SAPP 1881, no. 127, p. 6, qu. 131. The same view is expressed in SAPP 1875, no. 22, pp. vii–viii, para. 6. The remark that the calculations had been made repeatedly is at SAPP 1870–71, no. 86, p. 38, qu. 990.

14 SAPP 1866–67, no.161, p. 1, qu. 7.

15 ibid., p. 1, qus 13–15; p. 27, qus 817–19.

16 ibid., p. 26, qu. 777; SAPP 1866–67, no. 190, p. 124.

17 Donald W. Meinig, *On the Margins of the Good Earth: The South Australian wheat frontier 1869–1884*, facs. paperback edn, South Australian Government Printer, Netley, SA, 1988 (first published 1962), p. 130.

18 SAPP 1866–67, no. 161, p. 27, qu. 801.

19 SAPP 1875, no. 22, p. xlvii, para. 542.

20 ibid., p. ix, para. 17.

21 Meinig, *Margins*, p. 139.

22 ibid., p. 140.

23 SAPP 1869–70, no.156.

24 'Overland Telegraph: Officers' dinner', *Register*, 25 Nov. 1872, p. 6.

25 'The Retiring Surveyor-General', *Public Service Review*, Feb. 1894, p. 51.

26 SAPP 1875, no. 22, report, p. 7, para. 8; SAPP 1887, no. 34, p. 16, qu. 466.

27 South Australia, Parliament, *Debates*, 8 Sep. 1858, c. 73.

28 SAPP 1864, no. 96, pp. 1, 3.

29 R.K. Johns (ed.), *History and Role of Government Geological Surveys in Australia*, SA Govt Printer, 1976, p. 64.

30 SAPP 1890, no. 32, p. 153, qus 3919–21.

31 Alexander Tolmer, *Reminiscences of an Adventurous and Chequered Career*, vol. 1, Sampson Low, Marston, Searle & Revington, London, 1882, p. 320.

32 Eustace Reveley Mitford, 'Northern Territory', *Pasquin*, 26 June 1869, in Mitford (ed.), *Pasquin: The pastoral, mineral and agricultural advocate vols I–III, January 1867 to November 1869*, Judd & Co., London, 1882, p. 850.

33 SAPP 1878, no. 146, p. 44 (Goyder); p. 67 (Symon). Symon's conclusion is a well-known quote from Molière, which translates as 'What the devil was he doing in that galley [ship]?'

Chapter eighteen: A house in the hills

1   K.R. Bowes, *Land Settlement in South Australia 1857–1890*, Libraries Board of South Australia, Adelaide, 1968, p. 111 (PhD thesis, ANU, 1963).

2   'Death of the Hon. Dr. Campbell', and 'The late Hon. Dr. Allan Campbell, M.L.C.', *Observer*, 5 Nov. 1898, pp. 15–16, and under the editorial, p. 24. Campbell's unofficial title is mentioned in the account of his funeral on p. 16.

3   'Religious: The late Hon. Dr. Campbell', *Register*, 7 Nov. 1898, p. 6.

4   Letter from Allan Campbell, 13 Mar. 1876, accompanying Goyder to CCLI
    Playford, 3 Mar. 1876, SRSA GRG 35/1/1876/336.

5   Cabinet reply (to Goyder's letter of 3 March), 27 Mar. 1876 with Goyder's response of
    28 Mar. 1876. SRSA GRG 35/1/1876/374.

6   His letter of intended resignation of 31 Mar. 1873 is reproduced in SAPP 1873, no.
    105.

7   Telegraph to CCLI Reynolds, 20 Mar. 1873. Surveyor-General's Office 1868–1918,
    Confidential letter book, p. 42. SRSA GRG 35/19.

8   'The Surveyor-General', *Register*, 5 Apr. 1873, p. 5.

9   'Civil Service Regulations', *Register*, 28 Aug. 1874, p. 4.

10 Goyder to CCLI Playford, 8 Mar., 19 Mar. 1878, Surveyor-General's Office 1868–
    1918, Confidential letter book, pp. 58–66. SRSA GRG 35/19.

11 Elfrieda Jensen & Rolf Jensen, *Colonial Architecture in South Australia: A definitive
    chronicle of development 1836–1890 and the social history of the times*, Rigby, Adelaide,
    1980, p. 695.

12 Jan Polkinghorne, *Mylor – Valley of Dreams*, the author, Aldgate, SA, 1991, p. 35.

13 Robert F.G. Swinbourne, *Years of Endeavour: An historical account of the nurseries,
    nurserymen, seedsmen and horticultural retail outlets of South Australia*, South
    Australian Association of Nurserymen, [Adelaide?], 1982, p. 10; H.T. Burgess, *The
    Cyclopedia of South Australia*, vol. 2, Cyclopaedia Co., Adelaide, 1909, p. 106.

14 *Our Inheritance in the Hills: Being a series of articles by our special correspondent*, 1889,
    W.K. Thomas, Adelaide, pp. 56–8.

15 'Death of Mr. G.W. Goyder', *Advertiser*, 4 Nov. 1898, p. 6, reprinted in *Chronicle*, 5
    Nov. 1898, p. 16.

16 *Our Inheritance in the Hills*, p. 56.

17 'Departure of Mr. Goyder', *Register*, 6 Jan. 1882, p. 4.

18 The claim seems to have been made first in W.S. Kelly, *Centenary of Goyder's Line*,
    Libraries Board of South Australia, 1963 (reprinted from the *Chronicle* of 6, 13 and
    20 June 1963) and has been repeated since then.

19 'The late Mr. G.W. Goyder', *Register*, 7 Nov. 1898, p. 6. He is reported to have made
    the journey 'daily during his fourteen or fifteen years residence in the hills', i.e.
    from 1883–84 to 1898. Goyder mentions being 'at train at 5 p.m.' in the few entries
    in Office diary of G.W. Goyder as Surveyor General 28.11–8.12.1893 (SRSA GRG
    35/255), and a brief note from 1891 concerns sending the trap to the station (SLSA
    PRG 491/4).

20 SLSA PRG 491/4. The last entry, which is clearly for G.W. Goyder in the
    Walkerville Council assessments is in 1880, but Goyders continue to be listed in
    Hawkers Road for years after. Marjorie Scales, *John Walker's Village: A history of
    Walkerville*, Rigby, Adelaide, 1974, p. 113, attests to the friendship between the
    Goyders and the Horns. (The line, which belongs in the second paragraph, has been
    accidentally transposed to the one below.)

21 ibid.

22 Information about the sale of the land on which The Myrtles stands is from the
    Walkerville Heritage Survey of 1987–88, Item Identification Sheet 185. But Scales,

p. 79, states that Goyder lived 'in the early ground-floor section of the present Myrtles', and when The Myrtles was auctioned in 2007 it was described in one newspaper article as having been 'built by Surveyor-General George Goyder'.

23   Scales, p. 47. Dan Farmer (a Smith descendant) drew attention to Ellen's work.

24   Goyder to Sec. CCLI, 19 Jan. 1874. SRSA GRG 35/1/1874/68.

25   Tom Dyster, 'Two walks in and around Mylor – Goyder's little village', a pamphlet produced by the Stirling District Council; Polkinghorne, p. 35.

26   Dyster.

27   Jensen & Jensen gleaned this 'slightly flippant digression' from the *Register* of 27 Nov. 1882.

28   'Death of Mr. G.W. Goyder, C.M.G.', *Observer*, 5 Nov. 1898, p. 32.

29   South Australia, Parliament, *Debates*, 5 Nov. 1887, c. 1470; SAPP 1888, 28, p. 82, qu. 2315; Goyder to Sec. CCLI, 1 Sep. 1888. SRSA GRG 35/20 [Copies of letters sent.] vol. 125, [Outgoing Letter Book no. 10 of 1888], folio 10572.

30   In his final resignation. SRSA GRG 35/1/1893/2715.

31   SAPP 1895, no. 19, p. 26, qu. 571.

32   'Goyderiana', *Quiz and The Lantern*, 10 Nov. 1898, p. 7.

33   ibid, p. 6.

34   SAPP 1866–67, no. 65, p. 10, qus 263, 273; p. 11, qu. 302.

35   Dirk van Dissel, 'The Adelaide gentry, 1850–1920' in *The Flinders History of South Australia: Social history*, ed. Eric Richards, Wakefield Press, Netley, SA, 1985, p. 363.

36   Goyder's arms are described in Bernard Burke, *A Genealogical and Heraldic History of the Colonial Gentry*, vol. 2, Harrison & Sons, London; E.A. Petherick, Melbourne, 1893, p. 690. Sir Henry Etherington's arms are described in George Crabb, *Universal historical dictionary: Or explanation of the names of persons and places in the departments of biblical, political and ecclesiastical history, mythology, heraldry, biography, bibliography, geography, and numismatics*, Vol. 1, Baldwin, Cradock and Joy, London, 1825.

37   Jan Todd, *Colonial Technology: Science and the transfer of innovation to Australia*, Cambridge University Press, Melbourne, 1995, especially pp. 169–73.

Chapter nineteen: Steward of all the Crown Lands

1   'The Northern Areas', *Advertiser*, 25 Nov. 1874, p. 2.

2   Michael Williams, *The Making of the South Australian Landscape: A study in the historical geography of Australia*, Academic Press, London, 1974, p. 65.

3   Margaret Goyder Kerr, *The Surveyors: The story of the founding of Darwin*, Rigby, Adelaide, 1971, p. 34.

4   The changes are described in detail in Williams, *Making*.

5   Michael Williams, *Making*, p. 77; 'George Woodroofe [*sic*] Goyder: A practical geographer', *Proceedings of the Royal Geographical Society (South Australian Branch)*, vol. 79, 1978, p. 4.

6   Williams, *Making*, p. 82; Anne McArthur (Ed.), *Through the Eyes of Goyder Master Planner: Transcripts of his 1864–5 detailed valuations of 79 pastoral runs in the South*

East of South Australia, Kanawinka Writers and Historians Inc, Penola, SA, 2007, p. xv.

7   Raymond Bunker, 'Town planning at the frontier of settlement', in Alan Hutchings & Raymond Bunker (eds), *With Conscious Purpose: A history of town planning in South Australia*, Wakefield Press in association with Royal Australian Planning Institute (South Australian Division), Adelaide, 1986, p. 26.

8   Williams, *Making*, p. 83.

9   *Proceedings of the Parliament of South Australia*, 1875, Assembly, Votes and Proceedings, 18 Aug.

10 Williams, *Making*, p. 85, although the date is incorrectly given as 1887.

11 Two hundred and seventy eight of 555 hundreds, according to Peter Kentish, the current Surveyor General, in McArthur (ed.), p. xv.

12 Williams, 'George Woodroofe [*sic*] Goyder', p. 11, quoting SRSA GRG 35/2/1859/1550.

13 Bunker, p. 32.

14 J.M. Powell, 'Enterprise and dependency: water management in Australia' in Tom Griffiths & Libby Robin (eds), *Ecology and Empire: Environmental history and settler societies*, Melbourne University Press, Carlton, Vic., 1997, p. 108.

15 Williams, 'George Woodroofe [*sic*] Goyder', p. 15.

16 SAPP 1860–61, no.177, p. 5.

17 Eustace Reveley Mitford, 'The Gawler Prize History', *Pasquin*, 7 Dec. 1867, in Mitford (ed.), *Pasquin: The pastoral, mineral and agricultural advocate vols I–III, January 1867 to November 1869*, Judd & Co., London, 1882, p. 288.

18 Williams, *Making*, p. 109, quoting SRSA GRG 35/2/1864/64.

19 SAPP 1863, no. 51; p. 186, qu. 6259.

20 Survey Department of South Australia, *Field Service Handbook for Government Surveyors*, Govt Printer, Adelaide, 1880, p. 4.

21 SAPP 1890, no. 30C, p. 6, qu. 854.

22 SAPP 1891, no. 41, p. 1.

23 SAPP 1890, no. 60, p. 15.

24 'The Licensing of Surveyors', *Advertiser*, 1 Dec. 1893, p. 4.

25 'Examining Board of Surveyors', *Observer*, 26 Mar. 1887, p. 38.

26 'Death of Mr. G.W. Goyder', *Advertiser*, 4 Nov. 1898, p. 6, reprinted in *Chronicle*, 5 Nov. 1898, p. 16.

27 F.M.L.Thompson, *Chartered Surveyors: The growth of a profession*, Routledge and K. Paul, London, 1968, chapter 1.

28 Bathurst to Brisbane, 1 Jan. 1825, *Historical Records of Australia*, series 1, vol. XI, pp. 437–8.

29 R. Wright, *The Bureaucrat's Domain: Space and the public interest in Victoria, 1836–84*, Oxford University Press, Melbourne, 1989, p. 4, 40.

30 SAPP 1890, no. 30C, p. vii.

31 'The Land Question in South Australia', *Register*, 2 July 1890, p. 4.

32 SAPP 1888, no. 28, p. 85, qu. 2396.

33 G.W. Goyder, *Mr. Goyder's report on Victoria land regulations*, p. 1, para. 3 (originally SAPP 1870–71, no. 23).

34 SAPP 1865, no. 72, p. 88, qu. 2556.

35 SAPP 1888, no. 28, p. 78, qu. 2224.

36 G.W. Goyder, *Report on Victoria land regulations,* p. 7, para. 54.

37 SAPP 1888, no. 28, p. 79, qu. 2243.

38 ibid., qu. 2231; p. 80, qu. 2274.

39 G.W. Goyder, *Report on Victoria land regulations*, p. 8, para. 54.

40 SAPP 1890, no. 60, p. 17.

41 SAPP 1891, no. 41, p. 6.

42 SAPP 1879, no. 71, p. 27, qu. 649. This was stated first at p. 21, qus. 533–4.

43 G.W. Goyder, *Report on Victoria land regulations*, p. 8, para. 54.

44 SAPP 1888, no. 28, p. 78, qu. 2221 (improving land); SAPP 1890, no. 60, p. 19 (small allotments).

45 SAPP 1865, no. 72, p. 86, qus 2517–21.

46 SAPP 1879, no. 71, p. 21, qus 535–6.

47 SAPP 1888, no. 28, p. 80, qu. 2271.

48 SAPP 1890, no. 30C, p. 74, qu. 2161; SAPP 1890, no. 60, p. 20. The relevant passages are quoted in the following chapter.

Chapter twenty: Final years

1 K.R. Bowes, *Land Settlement in South Australia 1857–1890*, Libraries Board of South Australia, Adelaide, 1968, p. 247–8 (PhD thesis, ANU, 1963).

2 ibid., pp. 248–9. 'CLO 1355 of 16 July 1884' (SRSA GRG 35/[1?]/1884/1355) is cited as evidence of Goyder's support.

3 SAPP 1888, no. 28, p. 78, qu. 2207; p. 80. qus 2262–6.

4 ibid., p. v.

5 SAPP 1891, no. 33, p. 76, qu. 1971.

6 ibid., p. 77, qu. 1980.

7 ibid., p. 79, qu. 2231.

8 SAPP 1891, no. 33, p. 78, qu. 2021.

9 G.W. Goyder, *Destruction of Rabbits: Instructions for using bisulphide of carbon and mode of preparing and using phosphorised grain*, Government Printer, Adelaide, 1885.

10 SAPP 1893, no. 59, p. 1, qus 18–19.

11 ibid., p. 2, qus 38–9.

12 Michael Williams, *The Making of the South Australian Landscape: A study in the historical geography of Australia*, Academic Press, London, p. 204.

13 SAPP 1890, no. 64.

14 Williams, *Making*, p. 204.

15 SAPP 1892, no. 35, p. 83, qu. 3014.

16 SAPP 1890, no. 34, p. 18, qu. 455.

17 ibid., qu. 456.

18   T. Griffin & M. McCaskill (eds), *Atlas of South Australia*, correct. edn, S.A. Government Printing Division & Wakefield Press, [Adelaide], 1987, p. 56.

19   SAPP 1890, no. 60, p. 17.

20   SAPP 1889, no. 25, p. 49, qu. 1070.

21   SAPP 1890, no. 60, p. 21. There is a heading missing from this page, and this material appears under 'forest reserves'.

22   Edwin Hodder, *History of South Australia*, vol. 1, Sampson, Low, Marston, London, 1893, p. 133.

23   Rob Linn, *A Diverse Land: A history of the Lower Murray, Lakes and Coorong*, Meningie Historical Society, [Meningie, SA], 1988, p. 143.

24   SAPP 1890, no. 30C, p. 6, qu. 856; p. 5, qu. 846.

25   ibid., p. 5, qu. 846.

26   ibid., p. 74, qu. 2162.

27   ibid., p. 31, qu. 1358.

28   ibid., p. 74, qu. 2161.

29   SAPP 1890, no. 60, p. 20.

30   SAPP 1891, no. 41, pp. 6–7.

31   SAPP 1865, no. 72, p. 70, qus. 2161, 2163.

Chapter twenty-one: A gentleman of the Civil Service

1   SAPP 1883–84, no. 52, p. 2.

2   SAPP 1890, no. 30C, p. 7, qu. 864.

3   ibid., p. vii, para. 17.

4   'Death of Mr. G.W. Goyder, C.M.G.', *Observer*, 5 Nov. 1898, p. 32.

5   SAPP 1890, 30C, p. vii, para. 17.

6   Michael Williams, 'George Woodroofe [*sic*] Goyder: A practical geographer', *Proceedings of the Royal Geographical Society (South Australian Branch)*, vol. 79, 1978, p. 6.

7   SAPP 1863, no. 51, p. 166, qu. 5681–2.

8   Williams, 'George Woodroofe [*sic*] Goyder', p. 2. Williams commented on this on the basis of his extensive work in the files, but it is quickly apparent to any researcher.

9   K.M. Reader, *The Civil Service Commission, 1855–1975*, HMSO, London, 1987, p.5, footnote 11; Ian Radbone & Jane Robbins, 'The history of the South Australian Public Service', in D. Jaensch (ed.) *The Flinders History of South Australia: Political history*, Wakefield Press, Netley, SA, 1986, pp. 448–451.

10   Lionel C.E. Gee, 'The South-East Fifty Years Ago. The early pioneers', *Register*, 3 Mar. 1928, p. 7; 'The South-East Fifty Years Ago. Vanished Aborigines', *Register*, 24 Mar. 1928, p. 5.

11   'Goyderiana', *Quiz and The Lantern*, 10 Nov. 1898, p. 7.

12   Margaret Goyder Kerr, *The Surveyors: The story of the founding of Darwin*, Rigby, Adelaide, 1971, p. 3.

13   SAPP 1892, no. 35, p. 81, qu. 2963.

14   G.W. Goyder, Field notebooks, vol. 4, 9 Mar. 1860. SRSA GRG 35/256.

15   'The Survey Office', *Register*, 7 Feb. 1870, p. 5.

16   SAPP 1890, no. 30C, pp. 8–10, qu. 866. Goyder's entire letter is reproduced in response to a question.

17   SAPP 1890, no. 60, p. 22.

18   ibid., p. 23.

19   SAPP 1881, no. 144.

20   Radbone & Robbins, p. 456.

21   'George Woodroffe Goyder', *Advertiser*, 4 Nov. 1898, p. 6, reprinted in *Chronicle*, 5 Nov. 1898, p. 16.

22   Lionel C.E. Gee, 'The South-East Fifty Years Ago. Vanished Aborigines', *Register*, 24 Mar. 1928, p. 5.

23   Goyder to Gee, 8 Sep. 1889, Surveyor-General's Office 1868–1918, Confidential letter book. SRSA GRG 35/19.

24   ibid., 19 Mar. 1890.

25   SAPP 1872, no. 202.

26   SAPP 1890, no. 30C, p. 23, qu. 1160.

27   ibid. p. 51, qus 1783–6.

28   SAPP 1863, no. 51, p. 114, qus 4191–2.

29   SAPP 1863, no. 51, p. 174, qu. 5918, and qus 5919–21.

30   Quoted in K.R. Bowes, *Land Settlement in South Australia 1857–1890*, Libraries Board of South Australia, Adelaide, 1968, p. 108 (PhD thesis, ANU, 1963). Bowes, identifies the source as a letter, 'SGO 1826 of 29 Apr. 1881' (SRSA GRG 35/[2?]/1881/1826). He added 'See also the comment on CLO 297 of 9 Feb. 1878'. (SRSA GRG 35/[1?]/1878/297.)

31   George Loyau, *The Representative Men of South Australia*, [facs. edn], Austaprint, Hampstead Gardens, SA, 1978 (first published 1883), p. 196.

32   SAPP 1888, no. 28, p. 281, qu. 9045.

33   J.J. Pascoe (ed.), *History of Adelaide and Vicinity*, Hussey & Gillingham, Adelaide, 1901, p. 323.

34   SAPP 1888, no. 28, p. 281, qus 9046, 9048.

35   According to Ian Goyder, the understanding in the Goyder family is that Queen Victoria could not have been expected to agree to knight a person whose marriage was incestuous according to the law of England.

36   'Death of Mr. G.W. Goyder', *Advertiser*, 4 Nov. 1898, p. 6.

37   ibid., p. 280. The journalist was Fred Johns.

38   The exchanges between Goyder, and Kingston and Playford, are recorded in Office diary of G.W. Goyder as Surveyor General 28.11–8.12.1893. SRSA GRG 35/19.

39   Office diary of G.W. Goyder.

40   Goyder to commissioner of crown lands Gillen, 13 Dec. 1893. SRSA GRG 35/1/1893/2715.

41   'Death of Mr. G.W. Goyder, C.M.G.', *Observer*, 5 Nov. 1898, p. 32.

42   South Australia, Parliament, *Debates*, 13 Dec. 1893, c. 3513–14.

43　See South Australia, Parliament, *Debates*, 22 Sep. 1887, c. 885.

44　South Australia, Parliament, *Debates*, 13 Dec. 1893, c. 3514.

45　'Retirement of Mr. G.W. Goyder', *Register*, 14 Dec. 1893, p. 4.

46　*Quiz and The Lantern*, 10 Nov. 1898, p. 6.

47　'Death of Mr. G.W. Goyder, C.M.G.', *Observer*, 5 Nov. 1898, p. 32.

48　*Public Service Review*, Jan. 1894, p. 45.

49　'The Retiring Surveyor General', *Public Service Review*, Feb. 1894, p. 52.

50　'Death of Mr. G.W. Goyder', *Advertiser*, 4 Nov. 1898, p. 6, reprinted in *Chronicle*, 5 Nov. 1898, p. 16.

51　'Personal Paragraphs', *Public Service Review*, July 1894, p. 93.

52　'Death of Mr. G.W. Goyder, C.M G.', *Observer*, 5 Nov. 1898, p. 32.

53　ibid.

54　'Personal Paragraphs', *Public Service Review*, July 1894, p. 93.

55　Melville Cornish, 'Fifty years – A Retrospect', *Proceedings of the Royal Geographical Society (South Australian Branch)*, vol. xxxvi, Adelaide, 1936, pp. 92, 95.

56　'Death of Mr. G.W. Goyder', *Advertiser*, 4 Nov. 1898, p. 6.

57　'The Overland Telegraph: Officers' dinner', *Register*, 25 Nov. 1872, p. 6.

58　'Death of Mr. G.W. Goyder', *Advertiser*, 4 Nov. 1898, p. 6.

59　*Quiz and The Lantern*, 10 Nov. 1898, p. 6.

60　This story, attributed to Ellen's grand-daughter Patricia Ross, is reported in the unpublished family history by Vladimir Derewianka, 'Index and directory: The Smith and Goyder families of South Australia (1850's): Their ancestors and descendants', 2nd draft, 1990, p. 38. This account states that Goyder was dead when found. My account has been influenced by the newspaper reports which states that he died quietly in the evening.

Chapter twenty-two: Remembering

1　'Death of Mr. G.W. Goyder, C.M.G.', *Observer*, 5 Nov. 1898, p. 32.

2　Michael Williams, *The Making of the South Australian Landscape: A study in the historical geography of Australia*, Academic Press, London, 1974, p. 2.

3　Michael Williams, 'George Woodroofe [*sic*] Goyder: A practical geographer', *Proceedings of the Royal Geographical Society (South Australian Branch)*, vol. 79, 1978, pp. 1–2.

4　Jill Ker Conway in a speech to the Sydney Institute in 1995, quoted by Chris Wallace in 'Clean, Orderly and Laminex Coloured', *Griffith Review*, Autumn 2008, pp. 138–9.

5　Edward Charles Frome, *Instructions for the Interior Survey of South Australia*, Robert Thomas, Adelaide, 1840, p. 8, para. 12.

6　'The Exploration of Lake Eyre', *Register*, 30 Sep. 1874, p. 4.

7　South Australia, Parliament, *Debates*, 9 Oct. 1872, c. 2210.

8　'Crabthorn's correspondence', *Register*, 30 Jul. 1872, p. 5.

9　Margaret Goyder Kerr, *The Surveyors: The story of the founding of Darwin*, Rigby, Adelaide, 1971, Introduction.

10 Hubert Wilkins, *Sir Hubert Wilkins: His world of adventure: an autobiography recounted by Lowell Thomas*, Arthur Baker, London, 1961–62, pp. 13–15.

11 SAPP 1902, no. 22, p. 64, qu. 2935; p. 65, qu. 2964.

12 Williams, *Making*, p. 54.

13 ibid., p. 55.

14 'Song of the Wheat', in *The collected verse of Banjo Paterson*, Viking-O'Neil (Penguin), South Yarra, Vic., 1992, pp. 99–101.

15 T. Griffith Taylor, *The Australian Environment (Especially As Controlled By Rainfall): A regional study of the topography, drainage, vegetation and settlement, and of the character and origin of the rains*, Government Printer, Melbourne, 1918, p. 98.

16 J.M. Powell, *Griffith Taylor and 'Australia Unlimited'*, the John Murtagh Macrossan Memorial Lecture, 1992, University of Queensland Press, St Lucia, Qld, 1993, pp. 9–11.

17 ibid., p. 25.

18 T. Griffin & M. McCaskill (eds), *Atlas of South Australia*, correct. edn, S.A. Government Printing Division & Wakefield Press, [Adelaide], 1987, p. 27. The description of the findings of the Soil Conservation Committee is an interpretation of the map reproduced here.

19 John Andrews, 'Goyder's Line: A vanished frontier', *Australian Geographer*, vol. 3, no. 5, [1938?], pp. 32, 35.

20 'Death of Mr. G.W. Goyder', *Advertiser*, 4 Nov. 1898, p. 6, reprinted in *Chronicle*, 5 Nov. 1898, p. 16.

21 The photograph is by Jim Frazier, in Mary E. White, *After the Greening: The browning of Australia*, Kangaroo Press, Kenthurst, NSW, 1994, p. 205.

22 Ken Duncan, *Life's an Adventure: The first twenty-five years*, Panograph Publishing, Wamberal, NSW, p. 81.

23 *Experience the Edge: Peterborough Orroroo Carrieton South Australia*, [no place], [no date], p. 2.

24 Michael Williams, 'The Northern Areas: The last fifty years' change', *Proceedings of the Royal Geographical Society of Australasia (South Australian Branch)*, vol. 75, 1974, p. 1. After deferring to Meinig's 'brilliant' and 'superlative' work, Williams nevertheless criticised him for presenting the popular view that retreat was sudden.

25 ibid., p. 4.

26 Ian Lowe, *Living in the Hothouse: How global warming affects Australia*, Scribe Publications, Melbourne, 2005, p. 60.

27 SAPP 1867, no. 14, p. 113, qu. 3919.

28 Tim Flannery, 'The Day, the Land, the People', Australia Day address, January 2002, in *An Explorer's Notebook: Essays on life, history and climate*, Text Publishing, Melbourne, 2007, p. 116.

List of illustrations

Frontispiece
George Woodroffe Goyder, as leader of the expedition to survey the Northern Territory
State Library of South Australia, Adelaide, B 16791/1.

page 3 Map: Goyder's Line and the limits of agriculture

page 17 The S.G. in pursuit of the Rainfall.
State Library of South Australia, Adelaide, B 14851.

page 53 A sketch from Goyder's 1857 field notebook showing the Mud Hut with Mount Brown and the Devil's Peak in the background.
State Records South Australia, Adelaide, GRG 35/256 vol. 1, [p. 107].

page 53 Point Bonney, Rawnsley Bluff and the Chace Range.
State Records South Australia, Adelaide, GRG 35/256 vol. 2, [p. 105].

page 56 The Horseshoe Lake.
State Library of South Australia, Adelaide, C 874.

page 64 The proprietors and editorial staff of the *Register* in 1857.
State Library of South Australia, Adelaide, B 5967.

page 67 Mr Goyder's Discoveries. Cartoon from *Melbourne Punch*, 8 October 1857.
State Library of Victoria, Melbourne.

page 85 Goyder's sketch of a fish on the endpaper of one of his field notebooks for 1860.
State Records South Australia, Adelaide.

page 87 Outlines of hills and a sketch of surrounding country from the top of Hamilton's Hill, 26 July 1860.
page 87 Goyder's calculations of the capacity and flow of the Blanche Cup, 26 July 1860.
State Records South Australia, Adelaide, GRG 35/256 vol. 5. 30 May 1860—31 August 1860 [Northern triangulations]

page 88 An illustration of the Blanche Cup from the *Australasian Sketcher* of 1883, accompanying material from Goyder's report of 1860. It shows the Overland Telegraph, completed in 1872, and the use of camels.
State Library of Victoria

page 90 Outlines of hills in the country south of Lake Eyre, from Goyder's field notebook, Monday 20 August 1860. The hills shown include Cadnia and Mount Nor'west.
State Records South Australia, Adelaide.

page 92 Plan of the country between Lake Torrens Lake Eyre and Termination Hill, the map accompanying Goyder's report on the northern triangulations (SAPP 1860-61, no. 177).
State Library of South Australia, Adelaide.

page 147 Hawker's grazing line (1858), Goyder's proto-line (June 1865), and the main route of Goyder's journey, November 1865.

page 168 The quadrangle within the Government Offices.
State Library of South Australia, Adelaide, B 22819.

page 168 King William Street, the Town Hall and the Government Offices from Victoria Square.
State Library of South Australia, Adelaide, B 691.

page 186 Goyder in his official uniform.
State Library of South Australia, Adelaide, B 6993.

page 202 The Ancient Mariner, Eustace Mitford.
State Library of South Australia, Adelaide, B 9256.

page 202 Design for a trophy to be erected at Tipara in honour of the 'Tax upon the grass' by grateful squatters, Eustace Mitford.
State Library of South Australia, Adelaide, B 9254.

page 203 Mutilated statue of the constitution, Eustace Mitford.
State Library of South Australia, Adelaide, supplement to *Pasquin,* 18 Jan. 1868.

page 214 Rough sketch of the ideal town.
South Australian Parliamentary Papers 1864, no. 36, p. 16.

page 234 Dreams of home.
State Records South Australia, Adelaide, GRG 35/256 vol. 12 [p. 173].

page 268 Map: The survey and occupation of South Australia: Declaration of hundreds to 1890.

page 331 A house in the hills.
State Library of South Australia, Adelaide, B 15158.

Plate section

1. Abandoned house.
Photo: Janis Sheldrick

2. *Land of the Salt Bush*, John White, 1898.
Art Gallery of South Australia, Adelaide.

3, 4, 5. David Goyder, Frances Goyder and the young George Goyder.
Photographs courtesy of Vaughn Smith.

6. George Goyder.
State Library of South Australia, Adelaide, B 11348/17.

7, 8. Sarah Anna MacLachlan *née* Goyder, Sarah Goyder *née* Etherington.
Photographs courtesy of Vaughn Smith.

9. *From the Razor Back Hill, looking south over Mt Bryan*, Edward Charles Frome, 1843.
Art Gallery of South Australia, Adelaide.

10. *First View of the Salt Desert—called Lake Torrens*, Edward Charles Frome, 1843.
Art Gallery of South Australia, Adelaide.

11. *The sandy ridges of Central Australia*, Samuel Thomas Gill, ca. 1846.
National Library of Australia, Canberra.

12. A theodolite, chain, and field notebook from the South Australian Survey Department.
Postcard from the (now defunct) Lands Office Museum.

13. *Sketch of the district south east of Lake Eyre.*
State Library of South Australia, Adelaide, C 102.

14. The nurseryman – Edwin Smith.
Photograph courtesy of Vaughn Smith.

15. The rosery at Clifton Nursery.
Sands & McDougall Directory, 1889. State Library of South Australia.

16. William Strawbridge, as surveyor general.
State Library of South Australia, Adelaide, B 26059.

17. Lionel Gee, who accompanied Goyder on trips.
State Library of South Australia, Adelaide, B 7103.

18. Goyder's Line, as first drawn.
Courtesy of the Surveyor General, South Australia.

19. *South Australia, Pastoral lease districts.*
State Library of South Australia, Adelaide.

20. *South Australia, Map of northern runs.*
State Library of South Australia, Adelaide.

21. Henry Bull Templar Strangways.
State Library of South Australia, Adelaide, B 11138.

22. Thomas Playford.
State Library of South Australia, Adelaide, B 5622/23.

23. Eustace Reveley Mitford.
State Library of South Australia, Adelaide, B 23411.

24. Members of the expedition to survey the Northern Territory.
State Library of South Australia, Adelaide, B 11348/5, B 16791/9, B 11348/23. Photograph
of George MacLachlan courtesy of Vaughn Smith.

25. Ellen Goyder; Ellen with the twins.
Photographs courtesy of Vaughn Smith.

26. G.W. Goyder.
State Library of South Australia, Adelaide, B 3198, B 14656.

27. Goyder in retirement.
State Library of South Australia, Adelaide, B 496.

28. Goyder's long-time friends: Charles Todd and Dr Allan Campbell.
State Library of South Australia, Adelaide, B 69996/70 and B 25678/3.

Index

A

Aboriginal groups
  Adnyamathanha (SA), 141
  Kaurna (SA), 39, 358
  Larrakia (NT), 209, 211, 230
  Ngarrindjeri (Narrinyeri) (SA), 355, 357
  Warnunger (NT), 230
  Woolna (NT), 212, 213, 220, 221, 230
Adelaide, 35, 39, 40, 43, 96
Adelaide Club, 208, 332
Advertiser (newspaper), 98, 126, 127, 128, 129, 130, 132, 136, 144, 146, 149, 182, 200, 236, 238, 252, 256, 258, 260, 265, 292, 307, 371, 374, 377
Age (newspaper), 183
Amytornis goyderi, 384
Anama (pastoral run), 124, 136
Andrews, John, 13, 390
Anlaby (pastoral run), 74, 124, 130
Anthony Forster, 173
Argus (newspaper), 64
Arkaba (pastoral run), 55, 265, 268
Aroona (pastoral run), 55, 57, 128
artesian water, 280, 281, 285, 286
Assessment on Stock Act (1858), 120, 121, 122, 140, 148
Auld, Patrick, 113
Australasian (newspaper), 136
Ayers, Sir Henry, 122, 127, 131

B

Babbage, Benjamin Herschel, 55, 59, 63, 66, 68, 71, 73, 78, 88, 280, 317
Bagot, J.T., 100, 101, 183, 250
Baker, John, 136, 137, 260, 370
Baker, Sir Richard Chaffey, 370
Balcarrie (pastoral run), 52, *See also* Mud Hut
Ball, A.G.
  The S.G. in pursuit of the rainfall, **17**, 47, 150
Barber, William, 289, 300, 313

Barrow, John Henry, 126, 127, 128, 129, 130, 131, 135, 136, 137, 143, 150, 182, 317
Bartlett, Henry, 286
Belair National Park, 308, 351
Bennett, J.W.O., 212, 221–22, 223, 227–29; **pl. 24**
Billiamook, 211–12
Blanche Cup, **87, 88**, 280
Blanchewater, 59, 69, 70
Blyth, Neville, 177, 182, 258
Blyth, Sir Arthur, 128, 144, 177, 258, 263, 264, 313
Bonney, Charles, 63, 68, 122, 129, 143, 190, 282
Booborowie (pastoral run), 75, 124
Booyoolee (pastoral run), 124, 133, 196, 226
Boucaut, Sir James Penn, 148, 156, 158, 159, 177–79, 181, 190, 367
Bowes, K.R., 380
Brady, Edwin James, 389
Brown, John Ednie, 12, 301–8
Bundaleer (pastoral run), 75, 124, 196, 219, 300, 301
Bungaree (pastoral run), 124, 136
Burra, 2, 36, 73, 74, 76, 124, 153, 191, 293, 384

C

Cadell, Francis, 65, 66, 198
Cameron, Alex, 115
Campbell, Allan, 237, 323–24, 368, 370, 372, 377
Canowie (pastoral run), 75, 124, 133, 253
Carr, John, 367
catastrophism (geological theory), 75
Cavenagh, Wentworth, 143
Chaffey brothers, 290
Chambers Creek, 83, 84
Chartered Institute of Surveyors, 340
Chauncy, William Snell, 36–37, 40, 80, 255, 366
Christie, W.H., 98, 101, 116, 281
Church of the New Jerusalem. *See* New Church
Clark, J.H., 231
Clifton (Bristol, UK), 235

Clifton Nursery, 104, 328, 329, 412 (note 27), **pl. 15**

coast disease, 113, 191

Coles, Sir Jenkin, 372

Colonization Commissioners, 43, 44, 45

Cooper, Arthur Bevan, 190

Coorong, 47, 175, 352, 355

D

Darwin, 383, 385

Darwin, Charles, 13

Davenport, Sir Samuel, 131, 132, 136, 137, 313, 314

Day, E.G., 36, 243, 366

Deane, George, 212, 216, 222, 226

Downer, Sir John, 373

drought (1859), 78, 81, 83, 121, 142, 153

drought (1864–67), 133, 142–44, 152–53, 160

Duffield, Walter, 146, 148, 177

Dutton, Frederick H., 49, 74, 124, 130, 183

Dutton, Geoffrey, 6

E

Echunga, 117

El Niño, 11, 14

El Niño-Southern Oscillation (ENSO), 11

Elder Stirling and Company, 166–67, 181

Elder, Sir Thomas, 143, 254

Emu Flats (pastoral run), 124–26, 130

Eyre Peninsula, 106, 119, 157, 195, 296, 339, 355, 387, 390

Eyre, Edward John, 55, 57, 65, 68, 69, 105, 384

F

Farina, 4, 273, 308, 315

Favenc, Ernest
    The History of Australian Exploration 1788–1888, 108

Fergusson, Sir James (governor), 230

Finniss, Boyle Travers, 44, 48, 89, 101, 197, 212, 217, 227, 300, 361–62

Fleurieu Peninsula, 73

Flinders Ranges, 10, 47, 51, 52–54, 55
    appearance and imagery, 52, 54, 57, 69, 381

Fort Point, 211, 214, 220, 222, 225, 284

Fowlers Bay, 105, 106, 108

Fred's Pass, 214, 215, 222, 225

Freeling, Arthur Henry, 47–48, 62–64, 66–69, 76, 79, 80, 97–102

Frome, Edward Charles, 45–47

First View of the Salt Desert – called Lake Torrens, 47; **pl. 10**

From the Razor Back Hill, looking south over Mt Bryan; **pl. 9**

G

Garlick, David, 326

Garran, Andrew, 71

Gawler, George (governor), 45

Gee, Lionel C.E., 111, 115, 139, 189, 224, 361, 365; **pl. 17**

Gillen, P.P., 370, 372, 376

Glasgow (Scotland), 29, 31, 33, 237, 323

Glyde, Lavington, 123, 126, 127, 128, 145, 148, 149, 159, 177, 242, 345

Gottlieb's Wells (pastoral run), 146, 156

Goyder Channel, 384

Goyder River (NT), 384

Goyder, David, 235

Goyder, David George (1796–1878), 19, 33, 34; **pl. 3**
    and New Church, 19, 22, 25, 37
    and Pestalozzi schools, 24–25, 27
    and phrenology, 25
    as pharmacist, 26, 29
    autobiography, 20
    children, 25, 26, 29
    early life, 20–22
    in Ipswich, Suffolk, 37, 38
    marriage, 24
    medical training and practice, 29
    writings, 26

Goyder, Edward (1740?-1800), 20

Goyder, Ellen Priscilla (neé Smith), 38, 40, 50, 79, 87, 91, 104, 206, 207, 233, 235, 237, 242, 243–44, 328, 368, 378, 383; **pl. 25**
    marriage and children, 243–44

Goyder, Frances Mary (neé Smith), 37, 38, 80, 87, 91, 119, 142, 178, 179, 207, 233–36, 383; **pl. 4**
    children, 41, 49, 50, 79, 118, 206
    marriage, 40

Goyder, Francis Etherington, 142, 365

Goyder, George Arthur, 49, 333, 365

Goyder, George Woodroffe; **pls 5, 6, 26**
    and Aboriginal people, 57, 58, 106, 217–21, 222–23, 227–30, 346, 353, 354–58, 382
    and Adelaide Club, 208, 332
    and Belair National Park, 308, 351
    and C.C. Kingston, 369–72

and Civil Service Association, 364
and Flinders Ranges, 52–54, 69, 72
and horses, 81–82, 86, 106, 125, 212
and introduction of forestry, 292, 294, 301
and marginal zone, 250, 275
and Moonta dispute, 119, 166–74, 176–84, 200, 204–5, 379
and New Church, 40, 49
and Overland Telegraph, 315–17
and pastoralism, 65, 279, 282–84, 350, 380
and permanent settlement, 15, 273, 343, 344–45, 382
and public service uniform, **186**, 188
and railways, 312–15
and River Murray, 185, 311, 352–53
and Smith family, 39, 40
and soil conservation, 256, 345, 389–90
and stranding of *Queen of the Thames*, 238–41
and the inland sea, 66, 84, 150, 200, 204, 259, 260, 261
and understanding Australian climate, 13–14, 386
and vegetation and climate, 9, 60, 154, 155, 271, 385
and water, 279–82, 284–91, 354, 385
and wild dogs and rabbits, 349–50
and William Light, 43
appearance, 13, 25, **331**
application for position of surveyor general, 98–101
appointment as surveyor general, 101
as an engineer, 309–19
as chief inspector of mines, 103, 166, 317–18
as deputy surveyor general, 48, 51
as engineer, 309–19, 37
as guide to A.R.C. Selwyn (1859), 73–78
as landscape author, 8, 15, 16, 382
as remembered in history, 5–8, 379–94
as valuator of runs, 103
at Adelaide Exchange, 40–41
at Margaret Street, North Adelaide, 40
at Medindie, 49, 104, 207, 328, 332
Australian Dictionary of Biography entry, 6, 79, 140
birth, 25
Brighton residence, 103
career as a whole, 14–16, 165
chairman of Forest Board, 299–308

chairman of Railways Commission (1875), 279, 314–15
character, 25–26, 32, 79, 116, 117, 189, 224, 323, 331–32, 361–63, 364–65, 373, 379–80
children, 41, 49, 50, 118, 119, 142, 244, 329, 384
death, 377, 378
describes 'proto' line of reliable rainfall, 146, **147**
discovery of water in the north (1857), **67**, 59–72
drainage of South-East, 175–76, 191–93, 232–33, 236–37, 309–11, 351–52
early education, 27–28
emigration, 31–32, 33–34
expedition to country behind Fowlers Bay (1862), 105–9
exploration north of Flinders Ranges (1857), 56–59
field notebooks, 51, **53**, 54, 61–62, 81, **85, 87, 88, 90**, 93, 107–8, 140, 150, 155, 218, **234**, 406 (note 3)
Hillside, 49, 206, 208, **234**
illness, 62, 91, 93, 95, 190, 213, 217, 226, 230, 237–38, 323–24, 327, 330, 368
in *Burke's Colonial Gentry*, 332
joins Office of the Colonial Engineer, 37
journey to investigate drought (1865), **147**, 148–51
later education and engineering training, 29–30
managerial style, 361–66
mapping the 1865 line, 157
marriage to Ellen Smith, 243–44
marriage to Frances Smith, 37, 40
model town plan, **214**, 215, 336
naming practice, 91, 243, 383
Northern Territory expedition, 198–99, 209–31, 385
opposition to tree planting in north, 292, 304–8
overland journey from Melbourne to Adelaide, 34–35
purchases drilling and excavating equipment overseas, 285–86
rapid promotion in public service, 48–50
Report on Disposal of Crown Lands (1890), 342, 353, 354, 364
resignations, 109, 324–26, 330, 372
response to unlimited expansion of farming, 270–72, 273–75, 347

retirement, 376; **pl. 27**
salary and income, 78, 79, 80, 97, 99, 101, 109, 110, 190, 199, 207, 208, 231, 324–26, 327
surveying Pichi Richi Pass road (1857), 51
surveying practice, 50, 79–80, 81, 88–91, 338–39; **pl. 13**
surveying training, 30–31
surveyor general and land steward, 334–46, 359–60
triangulation south of Lake Eyre (1859–60), **92, 93; pl. 13**
valuation books, 10, 125, 139
valuations of pastoral leases, 123–40, 148–49, 190, **202**
visit to South-East (1863), 111, 112–17, 175
visits and reports on Victoria (1870), 253
visits mines in Flinders Ranges (1862), 105
visits New Zealand, 324
Warrakilla. *See* Warrakilla
Water Conservation and Development (1883), 287–90, 328
Goyder, John Harvey, 244, **331**
Goyder, Margaret (neé Lloyd, 1750–1805), 20
Goyder, Margaret Diana Mary. *See* Harvey, Margaret (neé Goyder)
Goyder, Sarah (neé Etherington, 1794–1886), 24, 329; **pl. 8**
Goyder, Sarah Anna (1823–1909). *See* MacLachlan, Sarah Anna (neé Goyder)
Goyder, Thomas (1786–1849), 22
Goyders Lagoon, 83, 384
Goyder's Line, **3**, 2–6, 275–78; **pls 18, 19, 20**
  1882 version, 157, 277–78
  1893 restoration of 1865 line, 277
  abolition, 264
  Goyder's opposition to unlimited expansion, 270–72
  line of reliable rainfall, 11, 16, 146, **147**, 151, 189, 252, 390–91, 394
  line passes into law, 254
  meaning and interpretation, 8–13, 254–55
  opposition, 256–64
  origins and mapping of 1865 line, 142–61; **pl. 18**
  passage to enactment in law, 249–51
  vindication, 348
Goyder's Line of Reliable Rainfall, 394
Grainger, Henry, 373
Grappler (steamship), 185, 382, 421 (note 73)

Gulnare (ship), 216, 220, 226, 227
Guy, W., 221–22

H
Hallett, Alfred, 127, 313, 329
Hallett, George Frederick, 329, 365
Hallett, John, 127
Handbook for Government Surveyors, 97, 335, 338
Hanson, William, 113, 115, 208
Harcus, William, 269
Hargraves, Edward, 318
Hart, John, 135, 148, 149, 151, 152, 158, **202**, 296
Harvey, Margaret (neé Goyder), 29, 235–36, 242
Hawker, 267, 268
Hawker, George, 49, 136, 144, 275, 284
Hayward, J.H.('Fred'), 57
Henning, R.W.E., 275
Heyne, Ernst Bernhard, 294
Hoare, W.W., 220, 221
Hodder, Edwin, 129, 180, 354
Hodgkiss, John, 296, 297
Horn, Penelope Elizabeth (neé Belt), 328
Horn, William Austin, 167, 328
horseshoe lake, 54–56, 60, 68–69
Howe, J.H., 98, 373
Hughes, Herbert Bristow, 133
Hughes, John Bristow, 143, 219
Hughes, Sir Walter Watson, 166–69, 171, 172, 179, 180
Humboldt, Alexander von, 294, 304

I
Illustrated Australian News (newspaper), 205
Indian Ocean Dipole, 11
Ingleby, Rupert, 177
Institute of Surveyors (South Australia), 340
Ipswich, Suffolk, 37
Isaac, Mahala (neé Smith), 38, 40

J
Jones, J.W., 290

K
Kanyaka (pastoral run), 52, 151, 265
Kapunda, 73, 74, 102, 124, 312, 313
Kerr, Margaret Goyder, 6, 34, 37, 93, 104, 242, 362, 385
Kingston, 114, 116, 175
Kingston, Charles Cameron, 369–72, 373, 378

Kingston, Sir George Strickland, 44, 45, 134, 149, 177–80, 316, 369

Kooringa. *See* Burra

Krichauff, F.E.H.W., 274, 296, 297, 299, 347

L

Lacepede Bay, 114, 116, 175

Lake Blanche (salt lake), 60–61, 69, 83, 215

Lake Bonney, 192

Lake Eyre (salt lake), 83

Lake Eyre North (salt lake), 339, 384

Lake Eyre South (salt lake), 69, 78, 82, 83, 84, 87, 91, 108, 280, 283, 313, 338, 384

Lake Frome (salt lake), 47

Lake Torrens (1840–1862). *See* horseshoe lake

Land Laws Commission (1888), 11, 108, 311, 343, 347–48, 367

Light, William, 43–45, 381

M

Macdonald, George, 238–41

MacDonnell Creek, 57, 58, 59, 60

MacDonnell, Sir Richard Graves (governor), 63, 68, 69

MacLachlan, George, 210, 365; **pl. 24**

MacLachlan, Hugh Galbraith, 33, 41

MacLachlan, Sarah Anna, 236, 242

MacLachlan, Sarah Anna (neé Goyder), 25, 27, 32, 33, 41, 210, 235, 236, 242, 398–99 (note 1); **pl. 7**

Mais, Henry, 313

mallee, 9, 155–57, 194–95, 293, 299, 335, 342, 389

Manna Hill, 272, 315

Marree, 4, 308

Martin-Harvey, Sir John, 24, 31

McCulloch, Alexander, 155–56

McEwin, George, 295, 300

McFarlane, Alan, 177

Medindie, 39, 49, 105, 207, 242

Meinig, D.W., 7, 6–9, 11, 314, 380, 390, 429–30 (note 36)

Melbourne (Victoria), 34, 64, 65, 183, 236, 237, 251, 315, 329

Midge (steamship), 210, 216

Mid-North (region), 10, 74, 123, 129, 130, 134, 155, 276, 391

Mills, Samuel, 167, 169, 177, 180, 200

Milne, Sir William, 99, 112–17, 127, 175, 190, 192, 193, 331

Mincham, Hans, 69

Mira, 220–21

Mirage Creek, 61

Mitchell, Sir Thomas Livingstone, 8, 54, 97

Mitford, Eustace Reveley, 119, 170–71, 172–74, 176–77, 179, 180, 181, 182, 184, 187–88, **202**, **203**, 199–206, 259–60, 261, 272, 318, 337, 340; **pl. 23**

Moonta, 103, 105, 119, 165, 282, 293

Moonta (ship), 210, 211, 213, 215

Moonta (Tipara) Inquiry (1863), 177–81

Moonta mine, 165

Moonta mine dispute, 166–74, 176–77, 200

Morgan, William, 324

Morris, Henry, 120–22, 125, 132

Mount Babbage, 58

Mount Brown, 145

Mount Bryan, 144, 154, 155, 156, 261, 296, 297, 386

Mount Goyder (NT), 384

Mount Lofty, 80, 375

Mount Lofty Ranges, 75, 292, 335

Mount Remarkable, 62, 75, 80, 145, 151, 153, 157, 256, 264, 296, 375

Mount Serle, 51, 53, 54, 55, 57, 61, 74

Mount Serle – Mount McKinlay baseline, 53, 54

Mount Woodroffe, 384

Mud Hut, 52, **53**, 75, 81, 265, 336

Muecke, H.C.E., 328

Mueller, Sir Ferdinand von, 8, 220, 292, 295, 296, 302

Mulvaney, John, 10

Murray Lakes, 47

Murray Mallee (region), 195, 339, 387

Murray River. *See* River Murray

Mylor, 330

N

Narrow Neck, 193

Neales, John Bentham, 41, 97, 149, 158, 159, 177, 190, 204, 313

Needles (Aboriginal reserve), 355, 356

New Church, 22, 23

    Adelaide, 33, 35–36, 49

    Sydney, 33

New South Wales

Western Division, 388

North Flinders Ranges, 51

Northern Runs Commission (1865), 143, 148, 149, 159, 160
Northern Territory
    early attempts at settlement, 196–98
    survey, 224–26, 230
Nullarbor Plain, 108, 286

O
O'Brien, F.J.R., 12–13
Observer (newspaper), 8, 93, 258, 259, 375
Orroroo, 267, 275, 390, 392
Osprey (ship), 33, 34
Overland Telegraph, 14, **88**, 315–17, 368, 377
Owiendana (pastoral run), 56

P
Painter, J.M., 52, 53, 61
Palmerston, 214, 215
Parry, Samuel, 78, 96–97
Pasquin (newspaper), 119, 200–201, 200–201, **202**, **203**, 204, 206, 229, 249, 260, 261, 391
Past and Present Land Systems of South Australia, 342
Pastoral Association, 127, 131
Paterson, A.B. 'Banjo'
    Song of the Wheat, 387–88
Peel, Dr Robert, 216, 221, 226
Pewsey Vale, 74
Phillipson, Montague, 170, 172, 179, 182
Pichi Richi Pass, 51
Picturesque Atlas of Australasia, 71
Pinnaroo, 277, 386, 387
Pitman, Jacob, 49
Playford resumptions, 284, 367, 368
Playford, Sir Thomas, 284, 310, 366, 367, 368, 367–71, 374, 378; **pl. 22**
Point McLeay (Aboriginal reserve), 354, 355, 356, 357
Point Pearce (Aboriginal reserve), 355
Poonindie Native Institution, 355, 357, 358
Port Augusta, 51, 64, 75, 81, 142, 151, 280
Port Darwin, 211
Port Lincoln, 139, 145, 151, 196, 296, 358
Powell, J.M., 6–8, 15, 380, 390
Price, A. Grenfell, 9
Public Service Commission (1888–91), 16, 308, 342, 355
Public Service Review, 69, 375, 376

Q
Queen of the Thames (steamship), 238–41, 379

R
Ragless brothers, 52
rain follows the plough (belief), 272, 275
Raukkan, 354, *See also* Point McLeay (Aboriginal reserve)
Rawnsley Bluff, 52, **53**
Reade, Charles, 336
Register (newspaper), 55, 62–63, **64**, 65, 66, 68, 71, 101, 127, 129, 130, 136, 137, 142, 143, 149, 150, 159, 169, 173, 185, 187, 197, 200, 205, 206, 207, 215, 219, 229, 231, 232, 236, 238, 258, 261, 265, 282, 285, 294, 299, 301, 324, 325, 327, 361, 369, 374, 375, 384
Reynolds, Thomas, 128, 183
River Murray, 184–85, 311–12, 352–53, 355
Rocky and Reedy springs, 59, 280
Rowe, John, 56
Rowe, William, 56, 106, 150
Royal Engineers, 31, 43, 45, 47, 98, 101
Royal Geographical Society, 63
Royal Geographical Society of South Australia, 5, 7, 377
ruins, 1, 4, 7, 392, 391–93; **pl. 1**
Ryan, Patrick, 166, 167, 169, 177

S
Salt Creek, 175, 176
saltbush, 2, 4, 9, 86, 154, 155, 156, 257, 270–72, 273, 275, 386; **pl. 2**
Samuels, Marwyn, 15
Schomburgk, M. Richard, 292, 294–96, 300
Schultze, Friedrich, 211, 220
Scott, Abraham, 133
Scrub Lands Act (1867), 195, 196, 347
Selwyn, A.R.C., 73–78, 80, 89, 280, 317–18
Sinnett, Frederick, 69, 95–96, 112, 123, 129, 141, 142, 144, 173, 182
Smith family
    emigration, 38
Smith, Arthur Henry, 38, 39, 210, 222, 286, 365; **pl. 24**
Smith, Edwin, 38, 39, 40, 104, 207, 298, 326, 329; **pl. 14**
    and Charles Ware's nursery, 49, 412 (note 26)
Smith, Edwin Mitchell, 38, 39, 105, 111, 154, 210, 216, 365, 370, 376, 386; **pl. 24**

Smith, Ellen Priscilla. *See* Goyder, Ellen Priscilla (neé Smith)

Smith, Frances Mary. *See* Goyder, Frances Mary (neé Smith)

Smith, John (1793–1861), 37, 38, 39, 103–4

Smith, Mahala. *See* Isaac, Mahala (neé Smith)

Smith, Marion, 233, 237

Smith, Susannah (neé Underwood, 1790–1884), 38, 39, 49, 104, 329

Smyth, Robert Brough, 362

South Australia

    Aboriginal people, 218, 219

    Crown Lands Office, 173

    Forest Board, 299–308

    forestry, 293–94

    forests, 292–93

    government survey, 41–47, 48, 51, 61, 78, 81, **268**, 339, 334–40

    Land Office, 47, 48, 165, 167, **168**, 171, 173, 190

    marginal country, 2, 6, 250, 275, 390

    parliament, 138

    pastoral lease valuations, 120–40, 148–49

    pastoralism, **3**, 119–20, 123–24, 350, 382

    public service, 360–61

    Public Works Department, 176, 192, 309, 310, 351, 376

    role of surveyor general, 41–42

    Survey and Land Department, 47

    Survey Department, 173, 340, 360

    wheat farming, **3**, 193–94, 253, **268**, 267–78, 393

South Australian (ship), 235, 242

South Australian Society of Architects, Engineers and Surveyors, 76

South-East (region), 111–12, 175, 192, 296, 380, 382, *See also* Goyder, George Woodroffe – drainage of South-East

springs, 86–88, 280, 410 (note 49)

squatting. *See* pastoralism

St Mary's Pool, 59, 61

Stirling, Edward, 167, 180

stock routes, 279

Stow, Randolph Isham, 126, 127

Strangways Act (1869), 196, 232, 253, 310, 347

Strangways, Henry Bull Templar, 109–10, 128, 144, 159, 171–74, 179, 182–83, 190, 195, 196, **202**, 312, 366, 367; **pl. 21**

Strawbridge, William, 115, 154, 264, 340, 360, 364, 366, 376, 377, 386, 387; **pl. 16**

Stuart Creek (formerly Chambers Creek), 83, 84

Stuart, John McDouall, 78, 196, 280, 315, 316, 383

Sturt, Charles, 39, 45, 55, 59, 64, 65, 68, 84, 383, 394

Swedenborg, Emanuel, 23, 95, 309

Swedenborgians. *See* New Church

Symon, Sir Josiah Henry, 319, 370

systematic colonisation, 41–42

T

Taplin, George, 354

Taylor, John, 167, 169, 180

Taylor, T. Griffith, 9, 388–89

Telegraph (newspaper), 112, 144, 173, 174, 182, 187

Tenison-Woods, Julian, 70, 71, 261, 281, 318, 408 (note 54)

The Myrtles, 328

Thompson, F.M.L., 340

Times (newspaper), 236, 239, 240, 241

Tipara. *See* Moonta

Todd, Sir Charles, 14, 45, 115, 149, 315, 316, 340, 368, 375

Trebilcock, James, 56, 67

tree planting to increase rainfall, 273, 295, 296, 303–4, 305, 306, 308

Trollope, Anthony, 254–55, 256

tuberculosis, 24, 38, 40, 44, 103, 236, 241, 242

U

Umballa, 211–12

Umberatana (pastoral run), 56

V

Valentine, Charles, 143, 190, 282

W

Wadham, William, 80, 178–79, 207, 328

Wakefield, Edward, 41, 42, 44

Wallace, Alfred Russel, 255

Wallace's Line, 256

Wallaroo, 100, 103, 105, 165, 169, 293, 317, 389

Warburton, P.E., 78, 88, 280

Ward, Ebenezer, 193

Warrakilla, **234**, 288, 326–28, 330, **331**

Waterhouse, George Marsden, 144, 148, 159, 183

Way, Sir Samuel James, 323, 370
Weathered Hill, 59, 60, 61
Wilber, Charles Dana, 272, 430 (note 14)
Wildman, E.T., 159
Wilkins, Sir George Hubert, 386
Williams, John, 146
Williams, Michael, 6–8, 10, 13, 380, 381, 382, 387, 393
Willochra Plain, 51, 52, 75, 81, 265
Willowie Plain, 153, 155, 267

Wilpena (pastoral run), 55, 128, 265
Wilpena Pound, 52, **53**, 75, 76, 105, 280, 306, 308
Wirrabara, 294, 296, 300, 301, 380
World's End (pastoral run), 2, 73, 265

Y

Yorke Peninsula, 132, 138, 145, 157, 165, 166, 167, 194, 195, 196, 265, 281, 293, 296, 355

www.ingramcontent.com/pod-product-compliance
Lightning Source LLC
Chambersburg PA
CBHW051946270326
41929CB00015B/2552